Locomotion of Tissue Cells

The Ciba Foundation for the promotion of international cooperation in medical and chemical research is a scientific and educational charity established by CIBA Limited – now CIBA-GEIGY Limited – Basle. The Foundation operates independently in London under English trust law.

Ciba Foundation Symposia are published in collaboration with Associated Scientific Publishers (Elsevier Scientific Publishing Company, Excerpta Medica, North-Holland Publishing Company) in Amsterdam.

Associated Scientific Publishers, P.O. Box 211, Amsterdam

Locomotion of Tissue Cells

Ciba Foundation Symposium 14 (new series)

Withdrawn from UF. Surveyed to Internet Archive

1973

Elsevier · Excerpta Medica · North-Holland

Associated Scientific Publishers · Amsterdam · London · New York

ISBN Excerpta Medica 90 219 4015 9
ISBN American Elsevier 0-444-15010-2

Library of Congress Catalog Card Number 73-80386

Published in 1973 by Associated Scientific Publishers, P.O. Box 211, Amsterdam, and 52 Vanderbilt Avenue, New York, N.Y. 10017.
Suggested series entry for library catalogues: Ciba Foundation Symposia.

Ciba Foundation Symposium 14 (new series)

Printed in The Netherlands by Van Gorcum, Assen

Contents

Participants

Symposium on Locomotion of Tissue Cells held at the Ciba Foundation, London, 30th August-1st September 1972

Chairman: M. ABERCROMBIE Strangeways Research Laboratory, Wort's Causeway, Cambridge CB1 4RN

G. ALBRECHT-BÜHLER Friedrich Miescher-Institut, Postfach 273, CH-4002 Basel, Switzerland

A. C. ALLISON Clinical Research Centre, Watford Road, Harrow, Middlesex HA1 3UJ

E. J. AMBROSE Department of Cell Biology, Chester Beatty Research Institute, Fulham Road, London SW3 6JB

D. BRAY MRC Laboratory of Molecular Biology, University Postgraduate Medical School, Hills Road, Cambridge CB2 2QH

A. S. G. CURTIS Department of Cell Biology, The University, Glasgow W2

G. A. DUNN Strangeways Research Laboratory, Wort's Causeway, Cambridge CB1 4RN

M. H. GAIL Biometry Branch, National Cancer Institute, National Institutes of Health, Landow Building, Room C-519, Bethesda, Md. 20014, USA

*I. M. GELFAND Laboratory of Mathematical Methods in Biology, Corpus 'A', Moscow State University, Moscow V-234, USSR

D. GINGELL Department of Biology as Applied to Medicine, Middlesex Hospital Medical School, London W1P 7PN

R. D. GOLDMAN Department of Biology, Case Western Reserve University, Cleveland, Ohio 44106, USA

T. GUSTAFSON Wenner-Grens Institute, University of Stockholm, Norrtullsgatan 16, Stockholm, VA, Sweden

A. K. HARRIS Department of Zoology, University of North Carolina, Chapel Hill, North Carolina 27514, USA

* Contributed *in absentia*.

JOAN HEAYSMAN Department of Zoology, University College London, Gower Street, London WC1E 6BT

H. E. HUXLEY MRC Laboratory of Molecular Biology, University Postgraduate Medical School, Hills Road, Cambridge CB2 2QH

LILIANA LUBIŃSKA Department of Neurophysiology, Nencki Institute of Experimental Biology, 3 Pasteur Street, Warsaw 22, Poland

C. A. MIDDLETON Department of Zoology, University College London, Gower Street, London WC1E 6BT

S. DE PETRIS Basel Institute of Immunology, Grenzacherstrasse 487, Basel CH 4058, Switzerland

K. R. PORTER Department of Molecular, Cellular and Developmental Biology, University of Colorado, College of Arts and Sciences, Boulder, Colorado 80302, USA

M. S. STEINBERG Department of Biology, Princeton University, Princeton, New Jersey 08540, USA

J. P. TRINKAUS Department of Biology, Yale University, Kline Biology Tower, New Haven, Connecticut 06520, USA

*JU. M. VASILIEV Institute of Experimental and Clinical Oncology, 6 Kashirskoje Schosse, Moscow 115478, USSR

N. K. WESSELLS Department of Biological Sciences, Stanford University, Stanford, California 94305, USA

K. E. WOHLFARTH-BOTTERMANN Institut für Cytologie und Mikromorphologie, Universität Bonn, 53 Bonn 1, Gartenstrasse 61a, West Germany

L. WOLPERT Department of Biology as Applied to Medicine, Middlesex Hospital Medical School, London W1P 7PN

Editors: RUTH PORTER and DAVID W. FITZSIMONS

* Contributed *in absentia*

Chairman's introduction

M. ABERCROMBIE

Strangeways Research Laboratory, Cambridge

Research on the machinery responsible for the locomotion of cells on a solid substratum has for many years been focused mainly on the large rhizopod protozoa. Students of metazoan cells, particularly pathologists interested in the chemotaxis of leucocytes, tissue culturists interested in the expansion of cell colonies *in vitro*, and embryologists interested in morphogenesis have had virtually no information with which to argue about rival hypotheses of locomotion. They have had to stand on the side-lines while the protozoologists debated.

The situation is now beginning to change. Knowledge of how widespread actin and myosin are in metazoan cells has combined with the recent great advances in understanding of muscle contraction to put theorizing at a molecular level on solid ground. At the same time the electron microscope has been producing a steadily improving picture of the architecture of microfilament and microtubule systems within the cell, which refined observation and experiment on living cells are beginning to convert into functional terms.

It seemed therefore an opportune moment to bring together, in what is apparently the first international meeting on the subject, some of those engaged in research on the movement of tissue cells. Our concentration on tissue cells and relative neglect of leucocytes was an attempt to restrain the spread of discussion. Everybody believes in the importance of leucocyte movement. But the importance of tissue cell locomotion is still greatly underestimated. Pioneers like Vogt, Holtfreter and Paul Weiss gave the notion deep roots in embryology. But the propensity of animal cells to form, and in the case of malignancy to destroy, supracellular organization by their autonomous power of crawling has not penetrated far into biological and especially into biochemical thinking. This is our justification for concentrating on the movement of tissue cells, including the closely analogous extension of nerve fibres. For the same reason the response of the locomotory machinery to environmental cues, which governs the morphogenetic outcome of cell movement, is a natural extension of the discussion.

Cell surface movements related to cell locomotion

ALBERT K. HARRIS

Strangeways Research Laboratory, Cambridge

Abstract The cells of most tissues are capable of autonomous locomotion, especially when placed in tissue culture. Cultured cells such as fibroblasts flatten and move on solid substrata by the tractional forces exerted by the thin outer cell margins (lamellae). The effects of these forces can be seen by culturing cells on flexible substrata, such as a plasma clot or silicone fluid, which stretch or flow in response to the forces imposed. By this means it is found that some cells exert much stronger forces than others. The cell surface undergoes several types of motion in connection with this traction, including ruffling (the repetitive upfolding of the cell margin), blebbing (the repeated herniation of the surface membrane) and particle transport (the continuous rearward or centripetal flow of objects which become attached to the cell surface). Particle transport seems to reflect the continuous rearward flow of the surface membrane on both dorsal and ventral surfaces. This would require the rapid reassembly of membrane at the leading margin and disassembly of membrane on more central areas of the cell surface. It is suggested that this membrane flow propels the cell forward in a manner analogous to a tractor tread, except that the membrane 'tread' flows rearward on both outer surfaces, while moving forward within the cell. The forces pulling the membrane inward are possibly the same as those which cause ruffling and blebbing and might be produced by actin-like cytoplasmic filaments.

The circumstances in which cells of higher animals locomote are many. During early embryonic development, there are the mass movements of cells during gastrulation, the migration of the neural crest cells and the outgrowth of nerve processes. In the adult, however, cell locomotion is confined primarily to such cells as leucocytes and macrophages. The cells of adult solid tissues do, however, retain dormant locomotory abilities which presumably assist in wound repair; an excessive degree of cell motility is probably one aspect of neoplasia. Cell locomotion is most commonly studied in tissue culture, where cells can be

---

* *Present address:* Department of Zoology, University of North Carolina, Chapel Hill, USA

observed free of the obstruction and interference of the surrounding body, and also where cell motility seems to be stimulated by the culture environment.

The mechanism by which cells propel themselves is not yet known with certainty. Nevertheless several important clues are provided by certain movements of the cell surface and of materials adhering to it. These surface movements and their physical interpretation are the subject of this paper.

I shall be primarily concerned with fibroblastic cells, those relatively undifferentiated stellate cells, such as those which dominate the outgrowth zone around explanted tissues (e.g. heart) and those which are carried as lines, (e.g. 3T3 and BHK). Such cells flatten and spread on glass, plastic or other solid substrata by the locomotory activity of their margins, which stretch the cell body out between several points of adhesion around the cell margin. A spread fibroblast has a thickened central area containing the nucleus and other cytoplasmic organelles. Surrounding this is a thinner marginal area, typically consisting of two or more elongate flattened projections, which I shall call lamellae (Fig. 1). These lamellae somehow exert traction against the substratum,

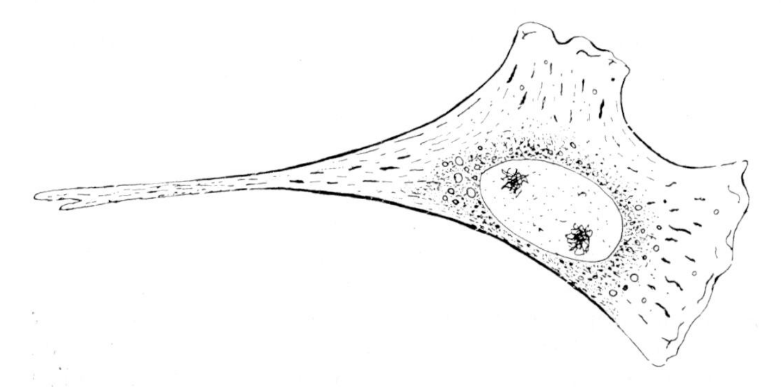

FIG. 1. (*a*) A typical chick heart fibroblast with two large flattened lamellae extending toward the upper and lower right. Such a cell would be moving gradually toward the right at about 0.5 μm/min. The long process trailing behind is a previously extended lamella which has become inactive, narrowed greatly, and will soon detach and be retracted.

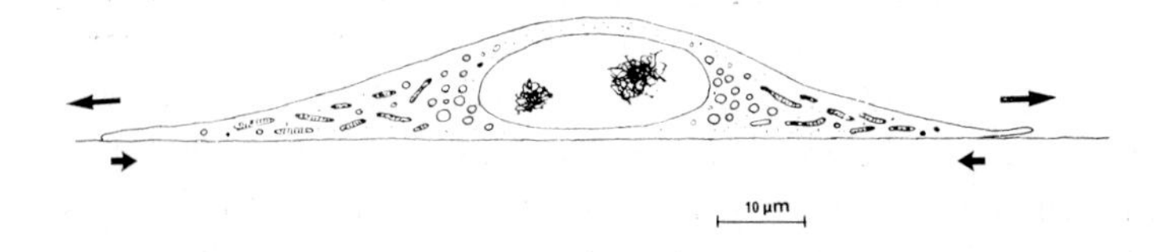

(*b*) Side view of a chick heart fibroblast. The thin lamellae at either end exert traction against the substratum, extending themselves outward and stretching the thicker cell body. The arrows above the cell indicate the direction of marginal extension, while the arrows below the line indicate the direction of the tractional force.

and each one tends to pull the cell body outward in its own direction. If the force of opposing lamellae is of equal strength, the cell simply remains stretched between them, but if the lamellae on one side dominate, the cell is pulled in that direction. As the adhesions at the rear margin are detached, the movement is often characterized by sudden jumps. If all its adhesions are broken, the cell simply rounds up. How does the cell margin exert this tractional force and thus pull itself forward?

THE FORCES EXERTED ON THE SUBSTRATUM

A good analogy to the situation in a spread fibroblast would be a tug of war with an elastic rope; the contractility of the cell, tending to pull the margins inward, is balanced by the tractional forces of locomotion by which the margin pulls itself outward. The effects of these forces can be seen by culturing cells on flexible substrata, such as plasma clots and silicone fluids, which are deformed by the tension imposed upon them.

By this method it was found (as expected) that the cell margin pulls the substratum rearward (inward) as the cell moves forward (outward). This tractional force tends to align nearby plasma clot fibres along the axis of cell movement, and this alignment can be seen in polarized light. Presumably this clot alignment is equivalent to the radial alignment observed by Weiss (1955) surrounding explanted tissues. Although Weiss attributed this orientation to metabolic dehydration, there are at least two reasons for attributing this alignment to the direct physical 'pull' of the cells instead. First, around bipolar cells the clot becomes oriented longitudinally along the axis of the cell, rather than radially as it should if dehydration were responsible. Secondly, the stretched clot springs outward elastically away from the cell whenever the cell detaches, enters mitosis or dies.

When fibroblasts are overlain with plasma clot, the dorsal cell surface exerts a rearward tractional force in the same way as the ventral surface does. Likewise, cells which are suspended within a clot produce this same type of stretching and orientation. This is to be expected, since *in vivo* cells are confronted on all sides by potential substrata, and not just on one side as they are when cultured on a surface.

Although traction is exerted by all motile cells, surprisingly large differences are observed in the strength of the forces exerted, some cell types being stretched by much weaker forces than others. For example, plasma clots which are only negligibly disturbed by moving mouse peritoneal macrophages are highly distorted and even torn by spreading 3T3 fibroblasts, the inference being that fibroblasts are much more strongly contractile and thus must exert greater tractional forces than the macrophages.

A more quantitative estimate of the relative forces exerted is based on the inability of cells to spread on silicone fluid if the contractile tension they exert causes the fluid to flow inward more rapidly than their margins can extend outward. The more contractile a cell, the more viscous the substratum must be to support its spreading. Fortunately the differences observed dwarf the rather crude simplifying assumption on which this analysis must be based. For example, rabbit peritoneal leucocytes can spread on fluid with a viscosity of 0.03 St(stokes), mouse peritoneal macrophages require a viscosity of 0.1–0.5 St, Hep-2 cells (a human carcinoma line) 125–500 St, and chick heart fibroblasts about 2 500 St. 3T3 fibroblasts can spread only partially on a fluid with a viscosity of 10 000 St. Clearly the differences in tension are great, though it remains to be seen whether this difference can really be of a factor of 10^5.

Elastic substrata also offer an opportunity to ascertain whether cytochalasin B interferes with cell contractility, as has been suggested (Wessells *et al.* 1971). When fibroblasts which have spread on clots are perfused with medium containing cytochalasin B (1 μg/ml), an immediate relaxation of the cells is observed, the exposed cells being elongated somewhat by the elastic clot. Full contractility is rapidly restored within a few minutes after cells are returned to drug-free medium.

RUFFLING

Probably the most obvious feature of fibroblast locomotion, especially when viewed by time lapse cinemicrography, is the cell surface movement known as ruffling (Abercrombie 1961). Ruffling consists of the repeated formation and movement of long narrow thickened areas, usually at the cell margin. These thickened areas, ‘ruffles’, appear as dark wavy lines under phase contrast microscopy. They form at the margin at intervals of 30–60 s, and move inward across the cell surface for a few μm before disappearing. A most striking fact about ruffling is that it occurs principally along those parts of the cell margin which are advancing forward, giving the subjective impression that cells are somehow pulled along by the ruffling movements. Similarly, when cells move on flexible substrata, they can be seen to pull the substratum inward primarily along ruffling areas of the cell margin, so it is clear that ruffling movements are somehow linked to the traction of the cell.

By observing fibroblasts in side view, both Ingram (1969) and I (Harris 1969) found that ruffles form by the upfolding of the cell margin (Figs. 2 and 3).

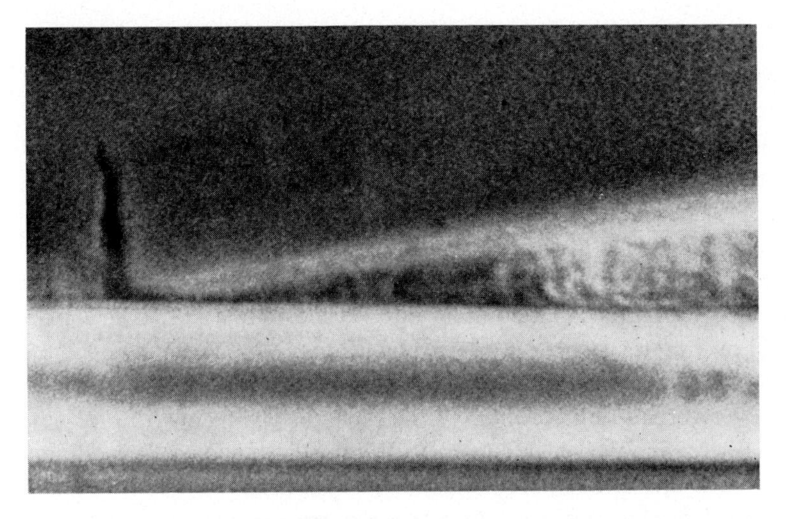

FIG. 2. A side view of a ruffling BHK cell showing an upward folded ruffle. The cell is moving to the left on a single fibre of glass wool about 10 μm in diameter.

The upfolded region corresponds to the thickened area, and much of its apparent rearward 'propagation' consists of gradual folding. The three-dimensional structure of ruffles can also be observed by viewing cells from above with an objective of high numerical aperture (low depth of focus) and repeatedly focusing up and down through the ruffle so that its vertical dimension is revealed by sequential optical sections.

The speed of the upfolding of the ruffle is somewhat irregular and frequently jerky. The impression given is of two opposed and slightly imbalanced forces. The average speed of folding is 2–3 deg/s. Along the length of a given ruffle,

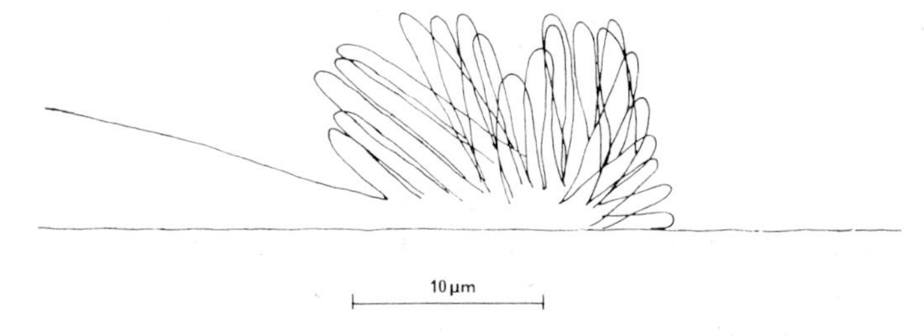

FIG. 3. Sequential positions of a backward folding ruffle of a 3T3 cell seen in side view. The cell body is to the left and the cell is moving toward the right. The time interval between each ruffle position is 4 s.

different portions often fold backward at slightly different speeds, assuming an irregular wavy or scalloped form, for which the term ruffle is most apt and descriptive. Ruffles may even reverse the direction of their folding, sometimes flopping back down onto the substratum.

The length of a ruffle ranges over 2–10 μm and sometimes up to 20 μm. The ruffle tip, appearing as a phase-dense line when seen in vertical profile, moves inward for a distance about twice the length of the ruffle, but in most cases this is due to folding about a nearly stationary axis. This axis, that is the basal section of the upfolded region, usually moves inward no more than about 1 μm. In some cases, however, the whole ruffle may propagate inward over the upper cell surface, following an irregular path of 20–30 μm or more in length, especially in cells with very broad thin lamellae. Ruffles may also elongate upward into the medium, increasing their length from 5–10 μm to as much as 30 μm. This elongation is accompanied by narrowing, and gives the impression of the ruffle being squeezed laterally. Since ruffles which are 'propagated' inward always lean in the direction of their movement, it is possible that this propagation is a special case of elongation. Microspikes (Taylor & Robbins 1963) can also participate in the marginal upfolding; usually they extend outward from the cell margin along the substratum and then fold upward as part of a ruffle, extending from it like fingers from a hand.

Ruffles sometimes form by direct upward extension of the dorsal cell surface rather than by upfolding. Such non-marginal ruffles often form 5–10 μm in from areas of marginal ruffling, but occasionally elsewhere. For example, when cells whose ruffling has been completely inhibited by cytochalasin B are returned to medium free of this drug, they often undergo a period of vigorous extension of the ruffles, not only at their margins but all over their dorsal surface.

Pinocytosis is often observed with ruffling (Abercrombie 1961). Large vesicles of medium appear to become engulfed among the ruffle upfoldings and are taken into the cell. These vesicles then move inward, eventually collecting around the nucleus, although they often burst and disappear after a few minutes.

Unfortunately this observation that ruffles form by marginal upfolding does little to explain the relation of ruffling to locomotion. In itself ruffle upfolding entails a slight retreat of the margin rather than an advance. Meticulous statistical analyses of cell margin movements by Abercrombie *et al.* (1970*a*, *b*) showed that the cell margin constantly wavers, sometimes extending and sometimes retracting. They found that net movement is forwards because the outward protrusions are of longer average duration than the retractions. They also showed that ruffle upfolding is predominant during the phases of withdrawal.

Ingram (1969) suggested that backfolding was due to a contraction on the dorsal surface of the cell, or more probably to a slightly stronger contraction on the dorsal surface than on the ventral surface. The basic idea was that the ventral contraction pulled the cell forward, while the movement of the dorsal surface was wasted slippage resulting from the cell's abnormal situation of being attached to a substratum on one side only. Alternatively, Abercrombie *et al.* (1970*c*) proposed that the ruffle was pushed upward and backward by an active expansion of the cell membrane just beneath the margin. They also proposed that this membrane pressure might serve to project the cell margin outward. (The evidence for membrane expansion will be discussed later.)

Of these two hypotheses, Ingram's would appear to correspond more closely to the known contractility of the cell lamella. It is also doubtful if the membrane could exert or transmit a push, since it seems to be quite flexible. Nevertheless, it is clear that the cell margin is pushed outward by some force, which raises the question of how the cell's overall contractility can be reconciled with its ability to project extensions outward. One possible relation between contractility and extension is suggested by another type of surface movement, known as blebbing.

BLEBBING

Blebs are almost hemispherical herniations of the cell surface, which bulge out rapidly, filling with fluid from the cytoplasm. Blebs usually expand rapidly within 5–10 s, reaching diameters of 2–10 μm. They contract more slowly, over 20–60 s, their surfaces gradually wrinkling as they retract. Blebs may form anywhere on the cell surface, often on the surface of other blebs, developing chains 20–30 μm in length. The most usual site of blebbing, however, is the cell margin, where blebbing activity frequently alternates with ruffling (Fig. 4). As cells respread after being detached, their margins first bleb, until ruffling gradually takes over as the cell becomes more flattened. Intermediate forms between blebs and ruffles are often observed in such circumstances. Although ruffles may swell into blebs, blebs never flatten to form ruffles directly. Marginal blebs, like ruffles, move rearward across the cell surface as they are retracted.

The significance of blebbing is apt to be ignored because it is characteristic of unhealthy cells or poor culture conditions. However, blebbing is also seen, at least occasionally, in all cells, for example after detachment, during respreading and after mitosis, etc. Even in unhealthy cells, the occurrence of blebbing must reveal important physical properties of the cell and its surface.

The explanation of blebbing appears to be that hydrostatic pressure within the cell is momentarily released by expansion of weak points in the cell surface. If the culture medium is made strongly hypertonic with sorbitol, this pressure

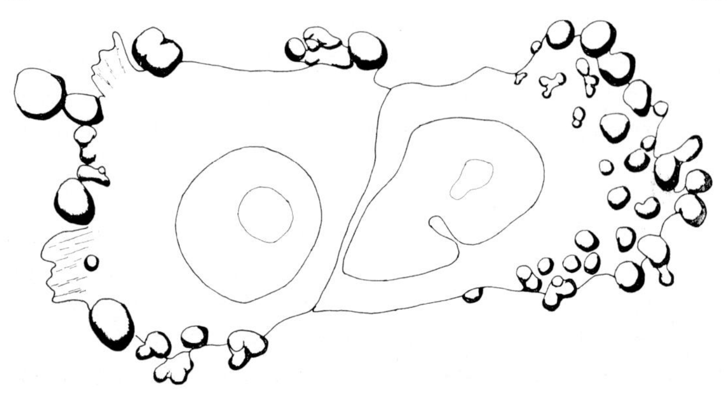

FIG. 4. Two adjacent Hep-2 cells respreading on glass after mechanical detachment. Many blebs are forming along the cell margins, except where the cells are in contact. These blebs are gradually moving inward over the dorsal cell surface as they are resorbed, especially on the right. Ruffles are beginning to replace blebs along the left hand margin.

differential across the membrane is reversed, with the result that existing blebs collapse within a few seconds and further blebbing ceases until cells are returned to normal medium. 0.4M-Sorbitol is sufficient to stop blebbing completely; less inhibition is produced by less hypertonic media.

Since blebs erupt preferentially along the cell margin, this part of the surface must be the weakest part of the cell and hence most susceptible to expansion. This suggests that normal outward extension of the margin is also produced by a localized release of hydrostatic pressure. Indeed, if a pressure differential does exist across the cell membrane, it cannot fail to contribute energy to any extension of the cell surface.

Such a model has the advantage of explaining outward projections from the cell surface in terms of contractions at other points on the surface. For example, the cell contractility which distorts flexible substrata or which retracts detached lamellae should also serve to generate and maintain pressure within the cell. Of course, some rigid, outward-pushing structures within the cytoplasm could also be responsible for extending the cell margin, but such structures need not be necessary except possibly to reinforce hydrostatic projections during or after their formation.

MOVEMENT OF THE CELL SURFACE MEMBRANE

The occurrence of surface movements such as ruffling and blebbing raises the question of how the membrane itself moves during locomotion. By placing various small particles on the cell surface and observing their motion by light

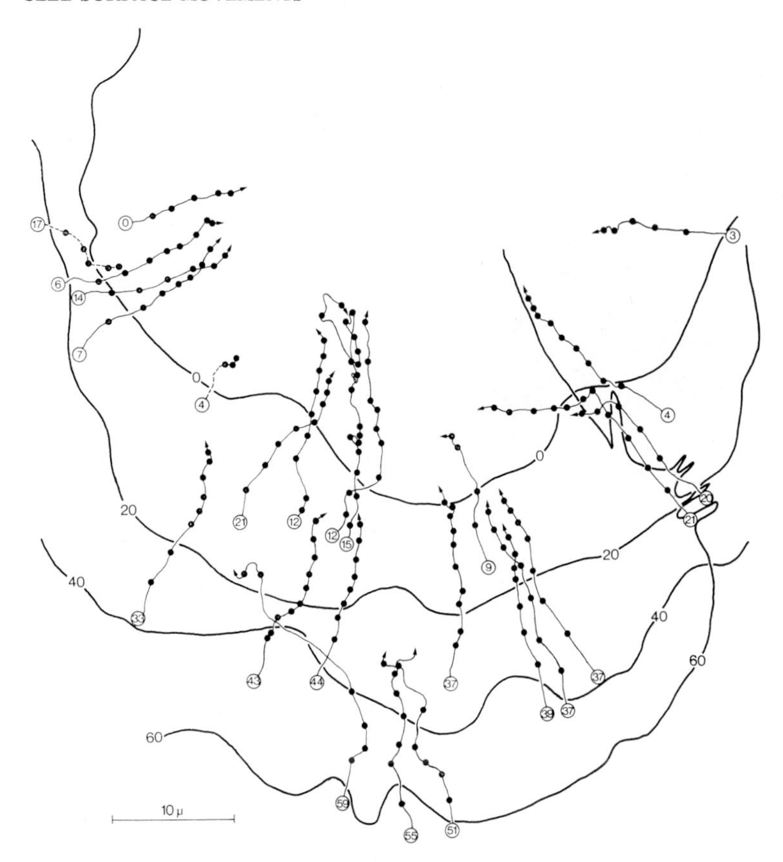

FIG. 5. Paths of anion exchange resin particles transported rearward over the surface of a moving chick heart fibroblast. Thicker wavy lines show the position of the advancing cell margin at 20 min intervals over one hour. The larger uncircled numbers indicate the number of minutes elapsed when the margin reached each position, while the small numbers within circles indicate the elapsed time at which an individual particle was picked up on the cell surface. The small black circles show the position of these particles at one minute intervals and the connecting lines describe their paths. The two broken lines toward the upper left indicate the paths of two particles which moved on the ventral surface of the cell.

microscopy, a highly consistent pattern of surface particle transport has been observed (Abercrombie *et al.* 1970*c*; Ingram 1969; Bray 1970; Harris & Dunn 1972). Particles lying on the substratum which are encountered by a moving cell are more often than not picked up by the advancing cell margin and transported rearward over the dorsal surface of the cell (Fig. 5), eventually accumulating either in the vicinity of the nucleus or at the trailing margin of the cell, and sometimes elsewhere (Figs. 6 and 7). Some particles remain stuck to the substratum as the cell moves over them.

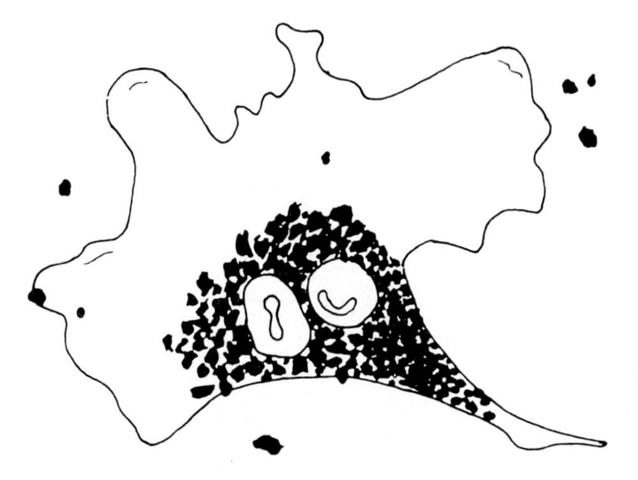

FIG. 6. Tracing from a time lapse film of a 3T3 fibroblast which is moving toward the upper left corner of the picture and which has picked up a large number of glass particles on its surface. These particles have been accumulated in a large pile on the rear part of the dorsal surface of the cell, but curiously do not overlie the two nuclei.

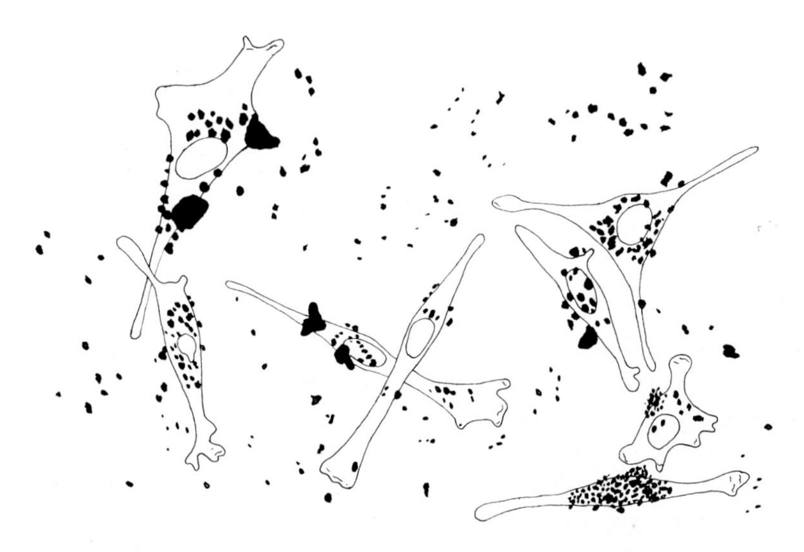

FIG. 7. A group of S-180 cells which have been picking up glass particles from the substratum and accumulating them in patches, primarily on their dorsal surfaces.

Since pinocytosis and phagocytosis are common at the cell margin, and since phagocytized objects also accumulate around the nucleus, it is essential to distinguish this rearward transport of particles on the cell surface from that within the cell. Fortunately we have several grounds for being certain that the transported particles actually remain on the cell's exterior. For one thing, such particles not infrequently come to lie in profile, especially in elongate cells, and also they can be viewed from the side, like ruffles. Clumps of accumulated particles sometimes obligingly fall off cells or become attached to neighbouring cells. Such clumps can also be displaced by vigorous pipetting of the medium or by micromanipulation. Abercrombie *et al.* (1971) have published electron micrographs of sections clearly showing transported particles outside the plasmalemma.

Particles which are picked up by the fibroblast margin are moved rearward initially at speeds of about 3–4 μm/min relative to the substratum (Fig. 5). Their movement gradually slows down until they come to rest after moving about 15–20 μm over the cell surface (Harris & Dunn 1972). Occasional rapid acceleration and changes of direction are not infrequently observed, however. The particles used include charcoal, molybdenum sulphide, ground glass, ground anion and cation exchange resins, colloidal gold and nickel. The movement of all these particles is indistinguishable, except that some detach from the cell surface more readily, and also is independent of particle size over a wide range (at least 0.2–2.0 μm diameter). Particle transport is not influenced by the gravitational orientation of the cell; in inverted cultures the same phenomenon is observed.

This ability to transport attached particles appears to be virtually universal among actively moving tissue cells, at least in culture. Particle transport has now been observed on a variety of fibroblastic cells (A. Harris, unpublished data), on outgrowing nerve axons (Bray 1970), on epithelial cells (A. Di-Pasquale, unpublished data) and even on macrophages and leucocytes (A. Harris, unpublished data). Particle transport is a concomitant of cell locomotion, and particles are always transported in the direction opposite to that of locomotion. The details of this transport vary somewhat with cell morphology and mode of locomotion, however. For example, on outgrowing nerve axons, particles are only transported rearward a few μm rather than all the way to the cell body. On leucocytes, the attached particles move rearward more slowly than the cell as a whole moves forward so that particles collect at the rear uropod primarily because the cell has moved forward beneath them. In contrast fibroblasts extend their margins at maximum rates of 0.5–1.0 μm/min, which is much slower than the rate of particle transport. Cells need not undergo net locomotion in order to transport particles. For example, bipolar cells, with ruffling lamellae at opposite ends, frequently pick up particles at either end and transport them toward the centre. If particles settle from suspension and attach to the cell surface, they will be moved in the same way as particles picked up at the margin. Surprisingly, particles can be transported on the cell's ventral surface, between the cell and substratum (Harris & Dunn 1972), although the particles move inward only short distances (2–10 μm) before they apparently become wedged between the cell and the substratum.

MECHANISM OF PARTICLE TRANSPORT

What forces could produce such particle transport? Several possible explanations can be discarded at once. It cannot be due to an electrophoretic phenomenon, since particles of opposite charge are transported in the same way (e.g. particles of either anion or cation exchange resins). Nor can it be due to local currents in the culture medium, such as might be generated by ruffling, since particles do not begin to move until they are contacted by the advancing cell margin. During their inward movement they are firmly stuck to the cell surface and do not slide or fall off if the culture is tipped or inverted.

Ambrose's theory (1961) that cells are propelled by undulatory waves of contraction cannot be used to explain particle transport: not only do particles continue to move past the zone of marginal ruffling, but peristaltic undulations would be incapable of moving attached particles unless these particles were able to slide over the cell surface in response to gravity. For example, particles might slide down the advancing side of each ruffle as a surfboard slides down the face of a wave. If this were so, then tipping a culture would cause all particles on the cell surface to slide down to the lower end of each cell. This does not happen and cells can transport particles regardless of their orientation with respect to gravity. Furthermore, particles do not move in any sort of oscillatory pattern as might be expected if contractile waves contributed to their motion.

How, then, are the particles transported? It appears that the remaining possible mechanisms can be divided into three categories: (i) the surface membrane remains stationary and the particles are transferred from point to point over the surface; (ii) parts of the surface membrane which become attached to external objects are induced to flow inward, carrying the particles with them while the remainder of the membrane circulates forward around these points; or (iii) the whole cell surface membrane flows continuously in a rearward (centripetal) direction, with membrane material being disassembled in the central areas and reassembled at the advancing margin.

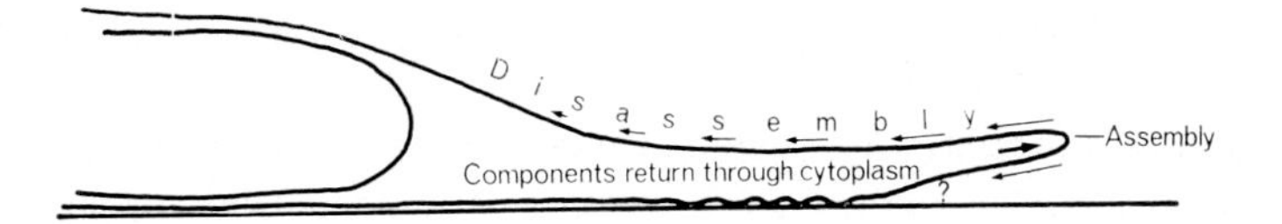

FIG. 8. Hypothetical 'tractor tread' interpretation of particle transport. The cell surface membrane is constantly being assembled at the tip of the advancing margin (right). From there this membrane flows slowly inward over both dorsal and ventral surfaces, probably being pulled by some component of the cytoplasm. At least on the dorsal surface the membrane is gradually disassembled as it flows inward so that transported particles slow down and eventually stop before reaching the nucleus (left). These disassembled membrane components pass into the cytoplasm and return to the advancing margin, perhaps by diffusion. Although it has not been determined where membrane is disassembled on the ventral surface, it is suggested that membrane movement in this area serves to exert traction.

Currently available evidence seems to support most strongly the third of these possibilities, namely continuous membrane assembly and flow. This was proposed recently by Abercrombie *et al.* (1970*c*), and it is analogous to a theory Shaffer (1962) set forth to account for a somewhat similar pattern of particle movement found on slime mould amoebae (*Polysphondylium*). The most surprising, or even revolutionary, aspect of such a theory is the high rate of surface membrane turnover (assembly and disassembly) which would be required. We (Harris & Dunn 1972) calculated the necessary rate of membrane assembly of some chick heart fibroblasts and concluded that more than a tenth of the total surface membrane would have to be reassembled each minute. Of course, other cells might move more slowly and the expected rate of turnover might be less for them.

Incorporation studies, such as those of Warren & Glick (1968), show that the overall rate of biochemical turnover is very much slower than this, materials being found to remain part of the membrane for periods of days or more. It is obvious therefore that any high rate of membrane assembly would mean repeated re-use of the same membrane components. These would be 'disassembled' in the central area of the cell surface and taken into the cytoplasm either as vesicles or as soluble materials, perhaps sub-units, before transfer to the cell margin for reassembly, From there they would again flow centripetally over the surface, this process being repeated indefinitely (Fig. 8).

More direct evidence in favour of membrane flow is that of Abercrombie *et al.* (1972) who used concanavalin A bound to peroxidase as a surface marker. When chick heart fibroblasts were fixed immediately after exposure to this material, their whole dorsal surfaces were coated. If these cells were then placed in medium free of concanavalin A for ten minutes before fixation, the dorsal surface near their margin became clear of the coating, indicating that this part of the membrane had been newly formed in the intervening period.

If the cell membrane is continuously being reassembled in this way, then it is to be expected that any newly synthesized membrane component will appear first at the cell margin and gradually spread inward from there. There are several instances in which exactly this pattern of surface behaviour has been observed. For example, Marcus (1962), studying a haemagglutinin in the cell membrane of myxovirus-infected HeLa cells, found that the haemagglutinin first enters the membrane at the cell margin, and from there moves progressively inward, gradually occupying the entire cell surface. Similarly, studies of gene reactivation in fused heterocaryons by H. Harris and his co-workers (1969) showed that newly synthesized surface antigens also appear first at the cell margin, particularly 'at the tips of elongated cytoplasmic processes'.

Any material which is attached to the surface of motile or potentially motile cells should be gradually accumulated at foci just as attached particles are (unless these materials are ingested, of course). This may be the explanation of the capping phenomenon in which antibodies and lectins bound to the surfaces of lymphocytes become localized into small caps. Conversely, membrane components such as surface antigens would presumably be disassembled and reassembled together with the rest of the membrane. This would be expected to result in fairly rapid intermingling of surface antigens in fused cells, as was observed by Frye & Edidin (1970). If experiments such as theirs were performed on spread cells (instead of suspended ones), the theory of membrane flow would predict that surface antigens would be mixed first at the cell periphery and subsequently in the central areas.

One phenomenon which I find particularly hard to account for in any other way than by membrane flow, is the continuous rearward movement of blebs from the cell margin. As blebs collapse and the membrane of their surface shrivels, they are moved across the cell surface one after another, often in close succession. It is not too difficult to visualize how the relatively simple shape of a ruffle could move over the surface, without the membrane itself moving. This could take place in the same way a wrinkle in a carpet can be pushed about without net displacement movement of the carpet material. However, it is nearly impossible to believe that such complex shapes as shrivelled blebs could be moved, without their component material (i.e. the membrane) moving as well, especially since particles attached to the outside of the membrane are also moved rearward.

However, it may be premature to conclude that particle movement is definitely due to membrane flow, and some consideration should be given to the other possibilities. The idea that particles move by adhering successively to

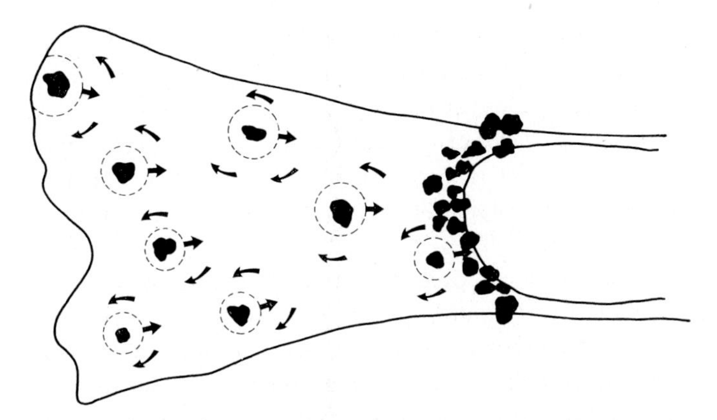

FIG. 9. Alternative hypothesis for particle transport. The attachment of particles would induce the underlying patch of membrane to be pulled inward toward the nucleus. Continual assembly of membrane would not be required at the leading margin since membrane could flow forward on the surface, passing between the rearward moving particles.

different points is particularly difficult to disprove, partly because the movement is very difficult to imagine in physical terms. Possibly a 'wave of adhesion' sweeps inward across the surface with different parts of the membrane becoming temporarily adhesive, one after another. If particles were drawn rearward by such a mechanism, they probably should undergo rolling motions as different parts of the particle surface adhered to the membrane. No such rolling is observed, even though it could be detected by light microscopy if it were present.

Alternatively, if particles were carried rearward by small patches of membrane (Fig. 9), it would be necessary to suppose that these patches were induced to move by the presence of the particles, otherwise particles would move outward as well as inward. The tendency of cells to round up completely when covered with a large number of particles can be considered as evidence for such induced flow. If the whole surface is covered with particles, then the whole surface should be induced to move toward the cell centre, and retraction should be the result. But it is found that nearby particles tend to move in almost parallel paths, changing direction and speed simultaneously, and this would not be expected if particles were carried by independent patches of membrane.

If the membrane does flow inward, what force propels it and how is this related to locomotion? Abercrombie *et al.* (1970*c*) suggested that the membrane itself might be pushed backward on the dorsal surface by the pressure of its own assembly and pulled rearward on the ventral surface by strands of filaments which they observe in the cytoplasm. For several reasons, I would think it more likely that the membrane is pulled rearward on both surfaces. First, the membrane is flexible and so cannot transmit much pressure. Secondly, I have found that traction can be exerted against overlying materials by the cell's dorsal surface and that particles can be transported by its ventral surface. This suggests that particle transport is a special case of traction in which the rearward force is applied to a movable object (the particle) rather than to an immovable one (the substratum).

The movements of the surface membrane seem to be analogous to those of a tractor tread propelling the cell forward. There is of course the important difference that the membrane flows rearward on both surfaces rather than rearward on the lower surface and forward on the upper as in the case of a tractor. For surface flow to exert such traction the membrane would have to be pulled rearward with considerable force. The actin-like filaments which often underlie the cell membrane seem to be the most likely agents for exerting this force, especially since they are typically oriented along the axis of cell locomotion and particle transport (Ishikawa *et al.* 1969; Shaffer 1962; Wessells *et al.* 1971). These filaments might pull on the membrane directly or they might interact with the myosin-like proteins which have been reported to be associated with the cell membrane (Jones *et al.* 1970). In the latter case, membrane flow could be produced by a sliding filament mechanism analogous to that in the contraction of striated muscle.

THE FORCES WHICH PRODUCE TISSUE CELL LOCOMOTION: A HYPOTHESIS

Speaking more generally, all the cell surface movements discussed here can be explained as the result of a centripetal (rearward or inward) pull applied to the membrane by some component of the cytoplasm. This inward force would pull steadily on the substratum and cause lamellae to retract immediately if they became detached so that completely detached cells round up into spheres.

In addition, ruffling seems to be produced by an inward pull becoming temporarily greater on the dorsal surface than on the ventral. Particle transport could result from this pull drawing the membrane gradually inward, and traction could be produced by this same force acting on the ventral surface membrane. By pulling the membrane inward this force would also compress the cell contents somewhat, causing them to squeeze out at weak points on the cell surface. This pressure could also serve to project the advancing cell margin forward, if this margin were for some reason especially weak (Fig. 10). If an equilibrium existed between disassembled membrane components in the cyto-

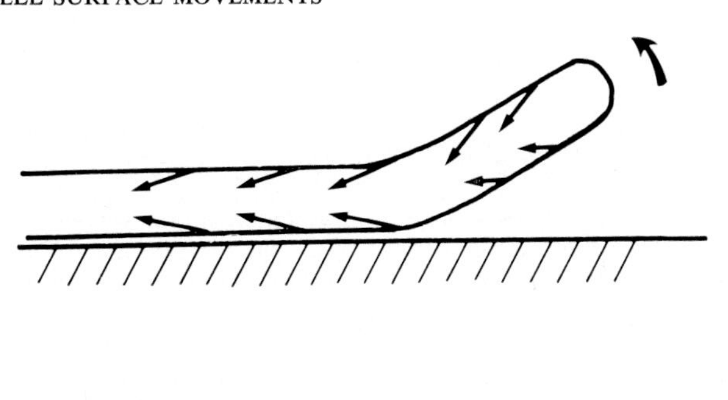

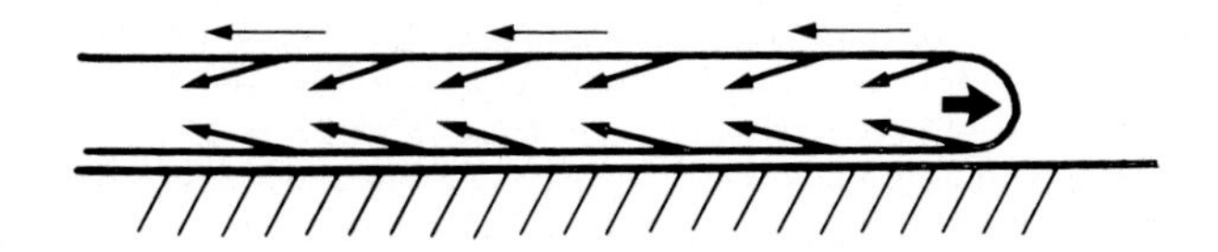

FIG. 10. Diagram of the hypothetical contractile forces producing cell surface movements and locomotion. These outlines indicate two side views of a fibroblast advancing toward the right. Contractile elements of the cytoplasm pull the membrane inward on both dorsal and ventral surfaces. If the forces on the dorsal surface temporarily become stronger, ruffle backfolding occurs (top). These forces squeeze the cytoplasm so that it sometimes bulges out in blebs, or causes the cell margin to protrude outward. Membrane assembly at the leading margin allows this protrusion and permits the surface to flow gradually rearward, on the dorsal side producing particle transport and on the ventral side resulting in locomotion. A concentration of stress at the leading margin might serve to localize membrane assembly there.

plasm and the cell membrane itself, then tension imposed upon this membrane at particular points would tend to favour assembly there. As a result, assembly would be localized at points of greatest tensional stress (such as the margin, perhaps). A pull directed normal to the membrane would tend to produce an inpocketing and in this way might contribute to pinocytosis.

Obviously it will require much additional work to determine whether all these varieties of surface movement are really consequences of a single inward pulling force. Further work will also be required to identify the cell components responsible for producing such forces.

ACKNOWLEDGEMENTS

The author thanks the Damon Runyon Memorial Fund for Cancer Research, Inc., for a post-doctoral fellowship. Original research reported here was supported by grants from the Medical Research Council to Michael Abercrombie of the Strangeways Research Laboratory, and by grants from the National Science Foundation to J. P. Trinkaus of Yale University.

References

ABERCROMBIE, M. (1961) The bases of the locomotory behaviour of fibroblasts. *Exp. Cell Res.* **8** (Suppl.) 188-198

ABERCROMBIE, M., HEAYSMAN, J. E. M. & PEGRUM, S. M. (1970*a*) The locomotion of fibroblasts in culture. I. Movements of the leading edge. *Exp. Cell Res.* **59**, 393-398

ABERCROMBIE, M., HEAYSMAN, J. E. M. & PEGRUM, S. M. (1970*b*) The locomotion of fibroblasts in culture. II. 'Ruffling'. *Exp. Cell Res.* **60**, 437-444

ABERCROMBIE, M., HEAYSMAN, J. E. M. & PEGRUM, S. M. (1970*c*) The locomotion of fibroblasts in culture. III. Movements of particles on the dorsal surface of the leading lamella. *Exp. Cell Res.* **62**, 389-398

ABERCROMBIE, M., HEAYSMAN, J. E. M. & PEGRUM, S. M. (1971) The locomotion of fibroblasts in culture. IV. Electron microscopy of the leading lamella. *Exp. Cell Res.* **67**, 359-367

ABERCROMBIE, M., HEAYSMAN, J. E. M. & PEGRUM, S. M. (1972) *Exp. Cell Res.*, **73**, 536-539

AMBROSE, E. J. (1961) The movements of fibrocytes. *Exp. Cell Res.* **8** (Suppl.), 54-73

BRAY, D. (1970) Surface movements during the growth of single explanted neurons. *Proc. Natl. Acad. Sci. U.S.A.* **65**, 905-910

FRYE, L. D. & EDIDIN, M. (1970) The rapid intermixing of cell surface antigens after formation of mouse-human heterokaryons. *J. Cell Sci.* **7**, 319-335

HARRIS, A. (1969) Initiation and propagation of the ruffle in fibroblast locomotion. *J. Cell Biol.* **43**, 165a-166a

HARRIS, A. & DUNN, G. (1972) Centripetal transport of attached particles on both surfaces of moving fibroblasts. *Exp. Cell Res.*, **73**, 519-523

HARRIS, H., SIDEBOTTOM, E., GRAVE, D. M. & BRAMWELL, M. E. (1969) The expression of genetic information: a study with hybrid animal cells. *J. Cell Sci.* **4**, 499-525

INGRAM, V. M. (1969) A side view of moving fibroblasts. *Nature* (*Lond.*) **222**, 641-644

ISHIKAWA, H., BISCHOFF, R. & HOLTZER, H. (1969) Formation of arrowhead complexes with heavy meromyosin in a variety of cell types. *J. Cell Biol.* **43**, 312-328

JONES, B. M., KEMP, R. B. & GRÖSCHEL-STEWART, U. (1970) Inhibition of cell aggregation by antibodies directed against actomyosin. *Nature* (*Lond.*) **226**, 261-262

MARCUS, P. I. (1962) Dynamics of surface modification in myxovirus-infected cells. *Cold Spring Harbor Symp. Quant. Biol.* **27**, 351-365

SHAFFER, B. M. (1962) The Acrasina. *Adv. Morphog.* **2**, 109-182

TAYLOR, A. C. & ROBBINS, E. (1963) Observations on microextensions from the surface of isolated vertebrate cells. *Dev. Biol.* **7**, 660-673

WARREN, L. & GLICK, M. C. (1968) Membranes of animal cells. II. The metabolism and turnover of the surface membrane. *J. Cell Biol.* **37**, 729-746

WEISS, P. (1955) in *Analysis of Development* (Willier, B. H., Weiss, P. A. & Hamburger, V., eds.), pp. 346-401, Princeton University Press, Princeton, New Jersey

WESSELLS, N. K., SPOONER, B. S., ASH, J. F., BRADLEY, M. O., LUDUEÑA, M. A., TAYLOR, E. L., WRENN, J. T. & YAMADA, K. M. (1971) Microfilaments in cellular and developmental processes. *Science* (*Wash. D.C.*) **171**, 135-143

Discussion

Steinberg: Do you think there is a difference between a bleb and a ruffle? It occurred to me that a ruffle might be a longitudinal bleb; that is to say, a bleb drawn out in one dimension.

Harris: Yes, one often sees projections of the cell surface intermediate in shape between blebs and ruffles, and ruffling activity may change gradually into blebbing activity. Ruffles may be changed directly into blebs by introducing hypotonic medium or cytochalasin, for example, but I have never seen a bleb converted directly into a ruffle. Blebs can probably be considered as essentially 'swollen' ruffles in which the internal cytoskeleton is weak or absent, so that the membrane simply bulges into a spherical shape.

Trinkaus: It is very arresting that blebs and ruffles appear at the leading edge of the cell. Do you know anything about the fine structure of blebs as opposed to ruffles? Spooner *et al.* (1971) have shown that ruffles possess a cytoskeleton, a filamentous mosaic or network, which might give them structure and prevent them from rounding up. In contrast, blebs on *Fundulus* deep cells appear to be relatively empty, or at least devoid of major organelles (Lentz & Trinkaus 1967). Are the blebs on your cells similar to those on *Fundulus* cells?

Harris: Fibroblast blebs sometimes contain mitochondria, other granules and occasionally vacuoles, and even bits of the endoplasmic reticulum may enter very large blebs. Price (1967) has studied blebs on tissue culture cells by electron micrography and shown some differences in ultrastructure.

Porter: In my experience blebs are loaded with ribosomes; whereas mitochondria, endoplasmic reticulum and other particulates are excluded from them, and one does not find many ribosomes in ruffles.

Trinkaus: What about fibrillar components?

Porter: They are not apparent in blebs and certainly not to the extent that one finds them in ruffles.

Trinkaus: That would fit the hypothesis that a protrusion which does not have any skeletal structure would adopt a hemispherical form; that is, a bleb, rather than an elongate ruffle.

Porter: I am surprised by the relation Dr Harris proposes between blebs and ruffles. Until now, it never occurred to me that they were related in the least. The blebbing you showed certainly resembles some of the ruffling in its behaviour. Do the blebs, like ruffles, form rapidly and then migrate back from the advancing edge?

Harris: Yes, and we have noted that particles attached to the cell membrane are transported away from sites of blebbing as if new membrane were being assembled there, so that in this respect also blebbing seems analogous to ruffling.

Porter: Do you relate blebs to the tonicity of the medium?

Harris: Blebbing can be prevented by raising tonicity with sorbitol.

Abercrombie: What is the rate of turnover?

Harris: From the observed rates of particle transport, moving cells appear to be continually reassembling their surface membrane at rates of 3–6 μm^2/min over each μm of their leading margin. In small, rapidly moving cells, this would mean that as much as a tenth of the total surface would turn over every minute (Harris & Dunn 1972). In large or slowly moving cells the turnover rate would be lower.

Wolpert: I have argued for a long time that the cell surface turnover is not an essential feature of cell movement (Wolpert 1971; Wolpert & Gingell 1968). Consider an alternative explanation (cf. p. 14 and Fig. 9). A filamentous system underneath the surface could pull at the point of attachment of a particle, drawing that part of the surface back, in an analogous way to the contraction of a pseudopod on contact with a substratum. The interpretation of the behaviour of particles would be more satisfactory if you could label the surface. Using large fresh water amoebae, which perhaps are not good analogues, we have shown that new surface is not formed when a pseudopod is put out. When we labelled the surface of the cells of the cellular slime moulds, which at least resemble a tissue cell surface more, with fluorescent antibody (Garrod & Wolpert 1968), we found that the surface label persisted as the cell put out a pseudopod. To prove that there is turnover of the surface you must show that the label goes inside the cell, because clearly the cell surface is fluid and much surface flow is possible. I suggest that your observations are artifacts due to the particles, which induce localized contractions.

Porter: Is it possible that at its advancing margin the cell is producing an exudate which stabilizes immediately? The cell might then move along under it, so that what you are watching is the fixed exudate with the cell moving by.

Trinkaus: But then the label should not stop at the nucleus. Moreover, particles on the surface of stationary ruffling cells also move backwards.

Abercrombie: It should also be said that these particles move backwards in relation to the substrate, not just in relation to the cell (Abercrombie *et al.* 1970).

Trinkaus: DiPasquale (1972), working on epithelial cells in our laboratory, has found that cytochalasin B, which inhibits both locomotion and ruffling, also inhibits backward movement of adhering particles. In addition, where cells are in contact with other cells in a sheet, with their ruffling inhibited by contact, adhering particles cease to move backward.

Allison: A relevant point is the phenomenon of 'capping' by membrane constituents, which Dr de Petris will be talking about. H-2 antigens in mouse fibroblasts show the same type of movement which you have described: 'capping' by bivalent antibody results in accumulation of the antigens on the upper surface of the cells just in front of the nucleus. In other words proteins such as H-2 antigens move *within* membranes although this does not necessarily reflect concurrent movement of membrane lipids. Membrane proteins or glycoproteins could represent sites of adhesion to particles and substratum and these could likewise move along the plane of the membrane through lipids.

You suggested that blebs and ruffles are projected by the momentary release of hydrostatic pressure within the cell. Would just squeezing or pleating be enough to cause forward movement? If the upper and lower parts of the plasma membrane at the front of a cell were brought together, the leading edge would flatten, extend forward and establish a new point of adhesion; further contraction of the upper membrane would then cause ruffling. Wouldn't this explain the motility, without invoking a pressure phenomenon?

Harris: But attached particles move continually inward from the margin on both top and bottom of the cell.

Allison: Each time the ruffle folds back, new membrane constituents are required.

Harris: Where is the new membrane coming from, then?

Allison: Unattached membrane protein constituents could move forward by diffusion and perhaps together with some new membrane protein and lipid be assembled into the membrane required at the leading edge for cell motility. Microfilaments attached to the membrane proteins could pull them backwards through the lipid bilayer.

Harris: But since attached material moves inward on the bottom as well as the top of the cell, the new membrane would have to move forward through the cytoplasm.

Allison: No. I am suggesting that much of the movement takes place within the membrane itself rather than through the cytoplasm.

Harris: But then why would you get the steady backward movement of particles, ruffles and blebs, one after another, on the upper surface?

Allison: Because the lower surface is held by adhesion to the substratum and only the proteins in the upper part are free to move backward, carrying the particles, as you observe. The flow of membrane constituents, probably glycoproteins, from the front of the cell towards the nucleus would be intermittent. Adherent particles would cross-link membrane proteins in a manner similar to antibody. If you suppose that the whole membrane moves backwards, do you think that all constituents of the membrane, the proteins, the phospholipids and so on, disappear into the cytoplasm?

Harris: Yes, I suspect that these components are dissolved in the cytoplasm along the central and rearward surfaces and diffuse from there back to the leading edge where they are reassembled. The material could go in as vacuoles or vesicles, but since we do not see such vesicles moving forward through the cytoplasm, I suspect the membrane is dissolved. I see no other way to explain the simultaneous rearward movement of particles attached to the membrane *and* of wrinkles in the membrane itself.

Allison: This, it seems to me, is the main difficulty of your theory.

Goldman: Dr Harris, the paths followed by the particles added to the surface of cells (Fig. 5) are similar to the paths taken by moving mitochondria and other particles within the cytoplasm (Goldman & Follett 1969). Microtubules, 10 nm filaments and bundles of 4–6 nm microfilaments found in the fibroblastic processes of BHK-21 cells are oriented parallel to the paths taken by the moving organelles and we have suggested that they provide the motive force for such movements (Goldman & Knipe 1973). The submembranous bundles of microfilaments possibly control movements such as those exhibited by the surface adherent particles.

Harris: The movement of attached particles along the axes of the cytoplasmic fibres is to be expected if the membrane is propelled inward by these fibres.

Goldman: Another possible mode of surface movement back from the edge of a cell towards the nucleus is that suggested by Marcus (1962; and personal communication) for the migration of haemagglutinin over the cell surface. The particles adhering to the cell surface might be passed back towards the nucleus in a similar fashion.

Harris: Marcus' observations could be explained as the result of the new virus-produced haemagglutinin being assembled into the membrane at the cell margin and moving inward from there together with the rest of the membrane components.

Goldman: Or the particle is passed from one adhesive point to the next.

Harris: It is hard to be certain that adhering materials are not passed from point to point. Such an explanation could not account for the rearward movement of surface antigens, however, or for the movement of blebs and ruffles.

Curtis: Though I would like to agree with Professor Wolpert, some recent work of ours (unpublished) convinces me that your explanation, Dr Harris, is at least potentially possible. We found that 30% of the phosphatidyl lipids are turned over, with respect to their fatty acids, in the cell surface within 20 min in cells in cultures which behave in a very similar way to yours. Evidence exists (Pasternak & Friedrichs 1970; Fischer *et al.* 1967; Curtis, unpublished data) that there is rapid breakdown and repair synthesis of the membrane from the lysophosphatidyl compounds. One can only speculate on Dr Harris' hypothesis that the lysophosphatidyl compounds drift back internally to the front of the cell. We have measured turnover in two ways: by looking first at the rate of incorporation of labelled fatty acids into plasmalemmal phospholipids, and secondly at the rate of loss of incorporated label from plasmalemmal fractions.

Wolpert: Your results differ markedly from those of Warren (1969). He found that surface membrane turnover in growing mammalian cells was rather slow.

Curtis: We are looking in different ways and at different components. Pasternak & Friedrichs (1970) have already commented on reasons for the apparent discrepancy.

Wolpert: Are you looking at all the membrane components or just one?

Curtis: We are looking at one group: the phosphatidyl and lysophosphatidyl compounds.

Harris: In my model the molecules would be rapidly reassociated but not entirely resynthesized.

Curtis: Suppose you label a cell surface very heavily with oleolyl groups in the R^1 and R^2 positions in the phosphatidyl lipids and then transfer the cell to a medium lacking oleic acid. You would then see a rapid loss of plasmalemmal oleic acid. If you now put in a different labelled fatty acid (say myristic acid) together with CoA and ATP, the cell would take these rapidly into the plasmalemmal fraction.

Wohlfarth-Bottermann: Do the marker particles gather in the vicinity of the nucleus? In small amoebae (e.g. *Hyalodiscus simplex*) the situation seems to be similar: both detrital and added marker particles gather on the rear of the cell above the endoplasmic hump, in which the nucleus is located.

In *Hyalodiscus simplex* the cell membrane rolls, in contrast to the motion of *Amoeba proteus*. During locomotion, the cell membrane of *Hyalodiscus* moves all around the cell like a rubber sheet in a balloon half filled with water and which is rolling forward over the ground (Hülsmann & Haberey 1973).

References

ABERCROMBIE, M., HEAYSMAN, J. E. M. & PEGRUM, S. M. (1970) The locomotion of fibroblasts in culture. III. Movements of particles on the dorsal surface of the leading lamella. *Exp. Cell Res.* **62**, 389-398

DIPASQUALE, A. (1972) *An Analysis of Contact Relations and Locomotion of Epithelial Cells*, Ph. D. Dissertation, Yale University, New Haven, Connecticut, USA

FISCHER, H., FERBER, E., HAUPT, I., KOHLSCHÜTTER, A., MODOLELL, M., MUNDER, P. G. & SONAK, R. (1967) Lysophosphatides and cell membranes. *Protides Biol. Fluids Proc. Colloq. Bruges* **15**, 175-184

GARROD, D. R. & WOLPERT, L. (1968) Behaviour of the cell surface during movement of the preaggregation cells of the cellular slime mould *Dictyostelium discoideum* studied with fluorescent antibody. *J. Cell Sci.* **3**, 365-372

GOLDMAN, R. D. & FOLLETT, E. A. C. (1969) The structure of the major cell processes of isolated BHK-21 fibroblasts. *Exp. Cell Res.* **57**, 263-276

GOLDMAN, R. D. & KNIPE, D. M. (1973) The functions of cytoplasmic fibers in non-muscle cell motility. *Cold Spring Harbor Symp. Quant. Biol.* **37**, 523

HARRIS, A. K. & DUNN, G. (1972) Centripetal transport of attached particles on both surfaces of moving fibroblasts. *Exp. Cell Res.* **73**, 519-523

HÜLSMANN, N. & HABEREY, M. (1973) Phenomena of amoeboid movement. Behavior of the cell surface of *Hyalodiscus simplex* Wohlfarth-Bottermann. *Acta Protozool.* **12**, 71-82

LENTZ, T. L. & TRINKAUS, J. P. (1967) A fine-structural study of cytodifferentiation during cleavage, blastula, and gastrula stages of *Fundulus heteroclitus*. *J. Cell Biol.* **32**, 121-138

MARCUS, P. I. (1962) Dynamics of surface modification in myxovirus-infected cells. *Cold Spring Harbor Symp. Quant. Biol.* **27**, 351-365

PASTERNAK, C. A. & FRIEDRICHS, B. (1970) Turnover of mammalian phospholipids. *Biochem. J.* **119**, 481-488

PRICE, Z. H. (1967) The micromorphology of zeiotic blebs in cultured human epithelial (HEp) cells. *Exp. Cell Res.* **48**, 82-92

SPOONER, B. S., YAMADA, K. & WESSELLS, N. K. (1971) Microfilaments and cell locomotion. *J. Cell Biol.* **49**, 595-613

WARREN, L. (1969) The biological significance of turnover of the surface membrane of animal cells. *Curr. Top. Dev. Biol.* **4**, 197-222

WOLPERT, L. (1971) Cell movement and cell contact. *Sci. Basis Med. Annu. Rev.* 81-98

WOLPERT, L. & GINGELL, D. (1968) Cell surface membrane and amoeboid movement, in *Aspects of Cell Motility* (*XXII Symp. Soc. Exp. Biol.*), pp. 169-198

Fluidity of the plasma membrane and its implications for cell movement

S. de PETRIS* and M. C. RAFF

MRC Neuroimmunology Project, Department of Zoology, University College London

Abstract The study of the distribution of immunoglobulin on the surface of mouse spleen cells, with fluorescein- or ferritin-labelled bivalent anti-mouse Ig antibody, or their univalent Fab fragments, has shown that surface Ig molecules are mobile in the plane of the membrane. Surface Ig, labelled with univalent antibody, appears to be diffusely distributed over the entire surface at all temperatures, whereas it clusters in discrete patches when labelled and cross-linked with bivalent antibody in the cold or in conditions of metabolic inactivity. At or above 20 °C these patches are actively transported to one pole of the cell (the tail or 'centrosomal' region), and segregate from the unlabelled membrane, forming a 'cap' of labelled molecules, and are eventually pinocytized. Other surface components of lymphoid cells (e.g. concanavalin A receptors, H-2 or HL-A antigens, θ antigen) behave similarly. The evidence suggests that the labelled patches are linked to a cytoplasmic contractile system and actively moved backwards, while the fluid unlabelled components are free to flow in the opposite direction. It is suggested that countercurrent flow and localized transitions from a 'linked' ('gel'-like) state to a fluid state of the plasma membrane, and *vice versa*, characterize membrane behaviour also during normal cell movement.

Knowledge of the physical characteristics of the plasma membrane is of obvious importance for an understanding of the function of the membrane in cell movement and cell-to-cell contact. Several investigations have supported the idea that most of membrane lipids are arranged in a molecular bilayer, according to the classical Davson–Danielli model (Wilkins *et al.* 1971; Singer 1971). At physiological temperatures the hydrophobic chains of the phospholipid molecules are in a disordered fluid state, and sometimes these molecules are capable of translational diffusion in the plane of the membrane (cf. Oldfield & Chapman 1972). Detailed data on the structural properties of the other major membrane component, the proteins, are scant, but some physical measurements together

---

* *Present address:* Basel Institute of Immunology, Switzerland

with biochemical and physico-chemical considerations suggest that most membrane proteins are globular, with an hydrophobic part of the molecule buried in a lipid bilayer and interacting with the lipid hydrocarbon chain, and an hydrophilic part (containing the carbohydrate components, when present) exposed to the aqueous environment (Singer 1971). Until recently membrane proteins were believed to be organized in a fixed, highly ordered array and this concept was apparently supported by the topographical distribution of specific protein or protein-associated components on the outer surface of the cell. This distribution was revealed by the use of specific antibodies, or other ligands such as plant lectins, capable of binding to different membrane glycoproteins. Membrane antigens have been mapped by measuring the extent to which the binding of an antibody to one specific antigen inhibits the binding of an antibody to a second antigen (Boyse *et al.* 1968) thus determining the 'closeness' of the two antigens. Moreover, antibody, suitably labelled with a visible marker so that the antigen distribution on the cell surface might be visualized directly by immunofluorescence or electron microscopy, showed that specific antigens were grouped in discrete 'patches' on the cell surface (Aoki *et al.* 1969). It was not clear, however, how these conclusions could be reconciled with the demonstration by Frye & Edidin (1970) that mouse and human antigens were free to move and to mix in the plane of the membrane, as in a fluid, in hybrid heterocaryons.

Studying the distribution of cell-bound immunoglobulin (Ig) and other antigens on the surface of lymphoid cells by electron microscopy and immunofluorescence, we have found that the binding of antibody may induce a temperature-dependent redistribution of specific surface antigens (Taylor *et al.* 1971; de Petris & Raff 1972). The same conclusion was reached independently by Taylor & Duffus using immunofluorescence (Taylor *et al.* 1971) and has been confirmed by others (Loor *et al.* 1972; Kourilsky *et al.* 1972). These studies provide evidence that Ig and other surface antigens are distributed essentially at random in the unperturbed membrane and that grouping of antigens in discrete 'patches' is a secondary phenomenon (de Petris & Raff 1973; Raff & de Petris 1973).

For the experiments, we used suspensions of washed mouse spleen lymphocytes or thymocytes. The cells were incubated with antibody or antibody conjugates, washed, fixed with glutaraldehyde followed by osmium and prepared for electron microscopy (de Petris & Raff 1972). In most experiments, the antibodies were rabbit-anti-mouse Ig (RaMIg) or their univalent Fab fragments conjugated with ferritin (RaMIg-FT and Fab-FT). The concanavalin A–ferritin conjugate (Con A-FT) was prepared in a similar way and used to study the distribution of concanavalin A receptors. Fluorescein-labelled RaMIg (RaMIg-

Fl) or their Fab fragments (Fab-Fl) were used for immunofluorescence studies.

Immunofluorescence and electron microscopy show that about 30–60% of mouse spleen lymphocytes carry immunoglobulin on their surface. These lymphocytes belong to the class of thymus-independent or B lymphocytes (Raff 1970). Membrane Ig, which is not removed by washing the cells but can be released by non-ionic detergents, is isolated mainly as monomeric (8S) IgM (Vitetta *et al.* 1971).

LIGHT MICROSCOPIC OBSERVATIONS: CAP FORMATION

On lymphocytes labelled and washed at 0–4 °C with bivalent fluorescein-labelled RaMIg, the label is distributed as a granular or spotty fluorescent *ring* around the circumference of the cell. In 80–95% of the cells labelled and washed at 20–37 °C (or labelled at 0 °C, washed and then incubated at or above 20 °C in antibody-free medium) the label moves over one pole of the cell, forming a fluorescent 'cap'. The rate of this cap transformation depends on the antibody concentration, but under optimal conditions it takes less than two minutes at 37 °C. Between 20 and 37 °C the formation has a Q_{10} of 2–3 and requires active metabolism, for it is completely and reversibly inhibited by sodium azide (20–30mM at 37 °C) or dinitrophenol (Taylor *et al.* 1971).

The cap is formed only when bivalent antibody [complete IgG molecules or bivalent $F(ab)'_2$] fragments are used to label the cell (Taylor *et al.* 1971; Loor *et al.* 1972). Cells labelled with univalent Fab-Fl appear as smooth fluorescent rings at all temperatures. However, if a second layer of antibody [e.g. unlabelled goat-anti-rabbit IgG (GaRIg)] capable of cross-linking the Fab molecules is added, all the label moves into a cap.

Thus the distribution of Ig on the lymphocyte membrane is not static, but can be drastically altered by the binding of bivalent antibody. Ig cap formation appears to be irreversible, since the label is not redispersed if the cells are cooled to 0–4 °C.

ULTRASTRUCTURAL DISTRIBUTION OF LABEL

Experiments with univalent Fab-FT and bivalent RaMIg-FT have shown three distributions of surface Ig which parallel the patterns observed in immunofluorescence studies (de Petris & Raff 1972, 1973).

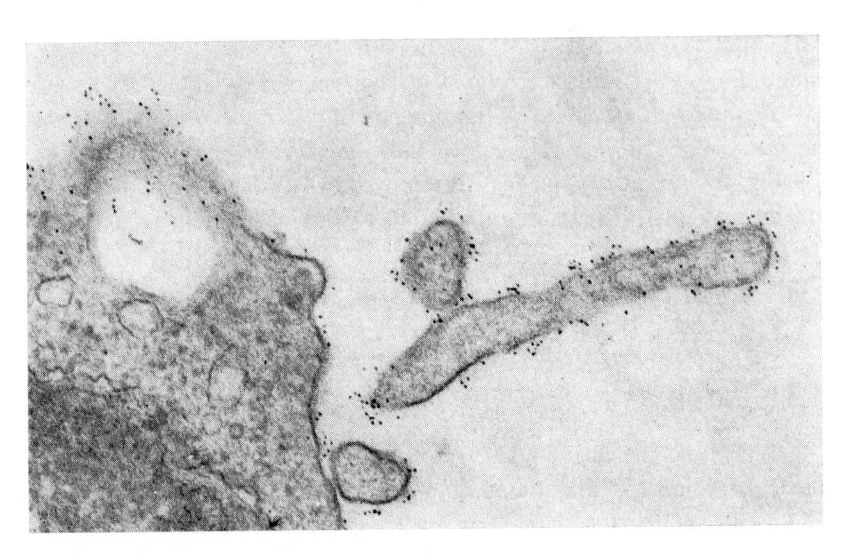

FIG. 1. Detail of the surface of a lymphocyte incubated for 140 min at 37 °C with Fab-FT. Ferritin molecules, isolated or in small groups, are distributed diffusely over the surface, including microvilli and an annular peripheral evagination (cut tangentially in this section, on the left upper side of the photograph). Unstained section. × 71 000.

(1) *Uniform distribution*

When cells are incubated with Fab-FT, the ferritin (FT) molecules, either singly or in small groups of 2–4 molecules (rarely larger clusters), are uniformly dispersed without any recognizable order over the entire lymphocyte surface (Fig. 1). The small groups of Fab-FT molecules probably correspond to antibodies bound to different antigenic determinants of the same Ig molecule. However, no cross-linking of different Ig molecules is possible. This pattern is independent of the temperature and is observed both at 0–4 °C and at 37 °C [even after more than 2 h incubation (see Fig. 1)]. At 37 °C, however, several ferritin molecules appear inside pinocytic vesicles, usually concentrated around the centrosomal–Golgi area of the cell.

(2) *Patches*

When cells are incubated with bivalent RaMIg-FT at 0–4 °C, ferritin molecules are still distributed over the entire surface, but now the majority of them are grouped in compact clusters of variable size ('patches') scattered without any order over the surface. Most clusters contain less than ten molecules if the

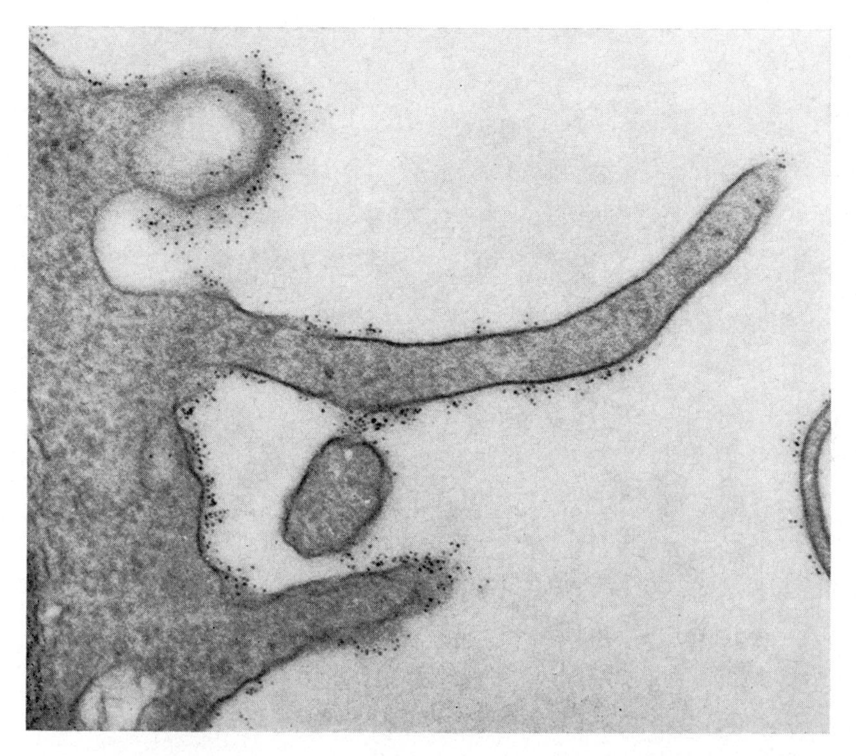

FIG. 2. Detail of a lymphocyte incubated with RaMIg-FT for 80 min at 0 °C, showing structures similar to those shown in Fig. 1. The ferritin molecules are clustered in patches separated by unlabelled gaps. Unstained section. × 71 000.

cells are incubated for only 20 min at 0 °C and washed at or below 2 °C. After 80 min or more at 0 °C (and washing at or below 2 °C) most clusters have coalesced into larger patches (Fig. 2). At 4 °C large patches are formed after 20 min. The formation of patches is unaffected by 30mM-sodium azide.

(3) *Caps*

When cells are incubated at 20 or 37 °C (or incubated at 0 °C, washed and then warmed) for 15 min or more, virtually all the isolated molecules or clusters of molecules move to one pole of the cell where they coalesce into larger patches or a single uninterrupted layer. This pole contains the centrioles, the Golgi apparatus and most of the cellular organelles. Occasionally ferritin aggregates (probably antigen–antibody complexes), labelled vesicles and other particulate

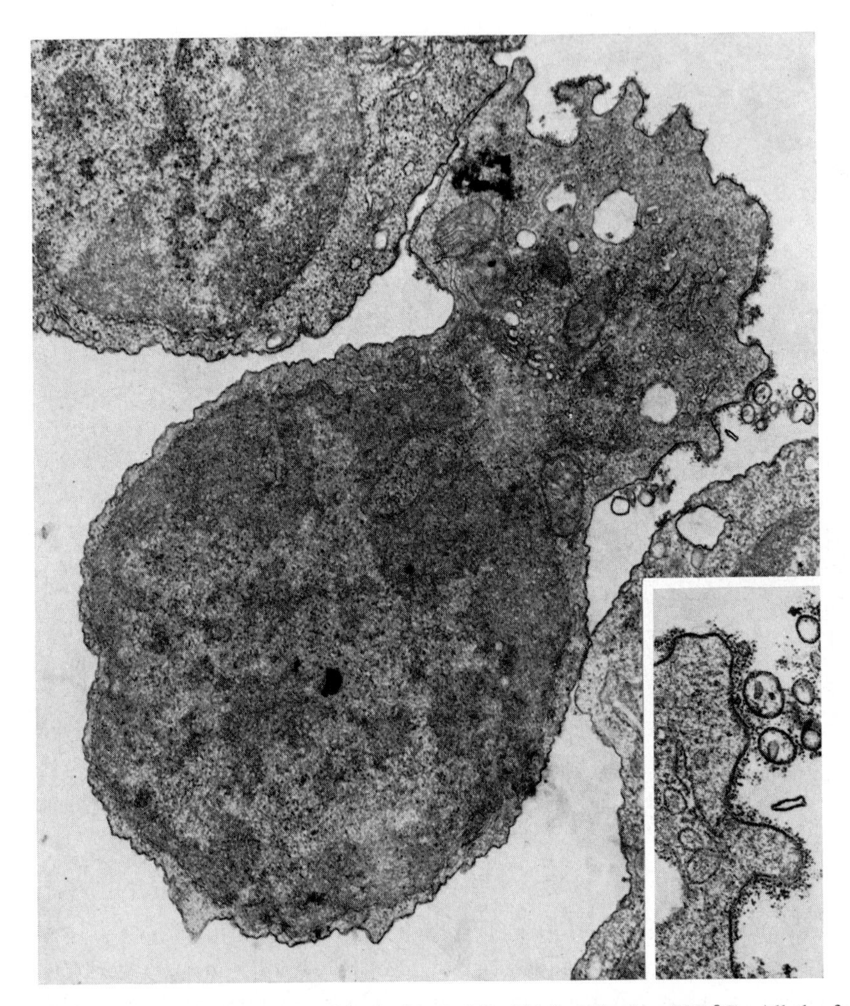

FIG. 3. Small lymphocyte incubated with RaMIg-FT for 45 min at 20 °C. All the ferritin molecules are clustered over the uropod, while the surface of the cell 'head', containing the nucleus, is completely free of ferritin. × 19 500 (inset : detail; × 36 000).

cellular debris, which all tend to stick to the labelled patches without any specific cellular localization at 0–4 °C, also move and concentrate over this pole of the cell (cf. Figs. 3 and 5). This pole corresponds to the trailing part of a lymphocyte moving over a substratum (e.g. de Bruyn 1944; Bessis & de Boisfleury 1971). When this part is elongated, as when lymphocytes interact with other cells, it has also been termed the 'uropod' (McFarland & Schechter 1970). In a few labelled cells with a uropod, all the ferritin is concentrated over this structure (Fig. 3). When, rarely, thin microvillous projections (probably retraction fibres) extend from the uropod, they are also covered by ferritin.

Cap formation is normally accompanied by pinocytosis of the labelled membrane (and some unlabelled membrane). Relatively large vesicles (usually of 0.2–0.4 μm diameter) lined with ferritin molecules accumulate around the Golgi area in the cytoplasm under the cap. There is, however, no strict correlation between the rate of cap formation and pinocytosis, although both increase with the temperature. With surface Ig, pinocytosis usually becomes conspicuous only when cap formation is completed. At room temperature it is slow (less than 20 % of the ferritin is ingested after 2 h), while at 37 °C it is considerable after only 5 min. When only some of the cells form caps (e.g. at low concentrations of antibody) pinocytosis is seen both in cells with caps and in cells with ferritin still distributed over the entire surface. After incubation with RaMIg-Fl or RaMIg-FT for 2–3 h, in about 50 % of the cells all the label becomes intracellular and on relabelling virtually no Ig can be detected on the surface. Thus pinocytosis is an irreversible phenomenon which can lead to the disappearance of the surface Ig [antigenic modulation (Taylor *et al.* 1971)]. The pinocytic vesicles are eventually transformed into secondary lysosomes (Biberfeld 1971).

If the lymphocytes are incubated with RaMIg-FT at or above 20 °C with sodium azide, the ferritin remains distributed over the whole surface in large patches similar to those observed at 4 °C but which do not coalesce in a single compact cap even after several hours. Also pinocytosis is absent. Clearly, azide inhibits cap formation and pinocytosis, but not patch formation.

These observations demonstrate that surface Ig molecules can move relative to each other in the plane of the membrane, behaving as if they were immersed in a two-dimensional fluid. This and other evidence (Pinto da Silva 1972; Brown 1972; Cone 1972) support the fluid-mosaic model of the plasma membrane proposed by Singer and Nicolson (1972) and originally suggested by the experiments of Frye & Edidin (1970). The experiments with Fab-FT indicate that the surface Ig molecules are distributed essentially at random over the cell surface at any temperature as single 7–8S molecules or small groups of a few such molecules (occupying an area of less than 400 nm^2). If the membrane is fluid and surface Ig molecules [or some other components to which they are probably attached (Vitetta *et al.* 1971)] are mobile in its plane, then it is clear how antibody could cross-link the mobile molecules into a two-dimensional antigen–antibody lattice, which would correspond to the patches observed at 0–4 °C, or at or above 20 °C in the presence of metabolic inhibitors. Thus, *patch formation* is probably a passive phenomenon due to diffusion by thermal agitation and cross-linking by bivalent antibody, which can occur when the membrane is fluid, namely at any temperature above 0 °C. *Cap formation,*

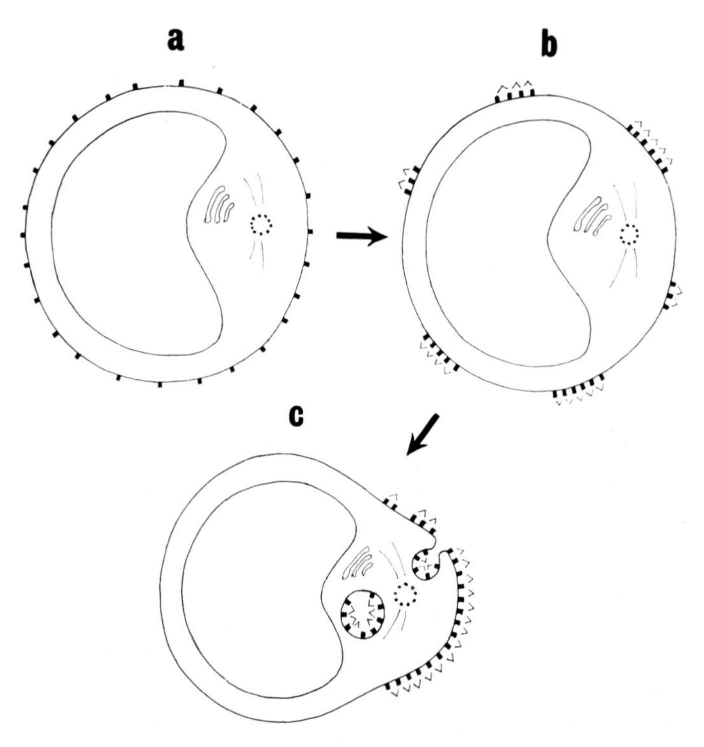

FIG. 4. Schematic representation of the changes in the distribution of surface Ig induced by bivalent antibody: (a) unperturbed distribution (as revealed by the binding of univalent antibody); (b) patches of surface Ig cross-linked by bivalent antibody in the cold or in metabolically inactive cells; (c) cap formation by active transport of the patches towards the pole of the cell containing the centrosomal (Golgi) area.

however, appears to be an active phenomenon which apparently follows patch formation and is essentially the transport of the patches to one pole of the cell and their coalescence at that pole, possibly through the interaction of membrane elements with cytoplasmic structures (see later and Fig. 4). That mobility of surface Ig molecules and cross-linking by bivalent anti-Ig are implicated in Ig redistribution was shown by experiments in which surface Ig was labelled at 0 °C with Fab-FT and the Fab-FT molecules were then cross-linked with a second layer of unlabelled GaRIg. If this second step was at 4 °C, all the diffusely distributed Fab-FT molecules were clustered in large patches which remained scattered over the entire surface (de Petris & Raff 1973), but if the temperature was 20 °C or higher all the ferritin moved into a practically uninterrupted layer at the 'tail' of the cell (Figs. 5 and 6).

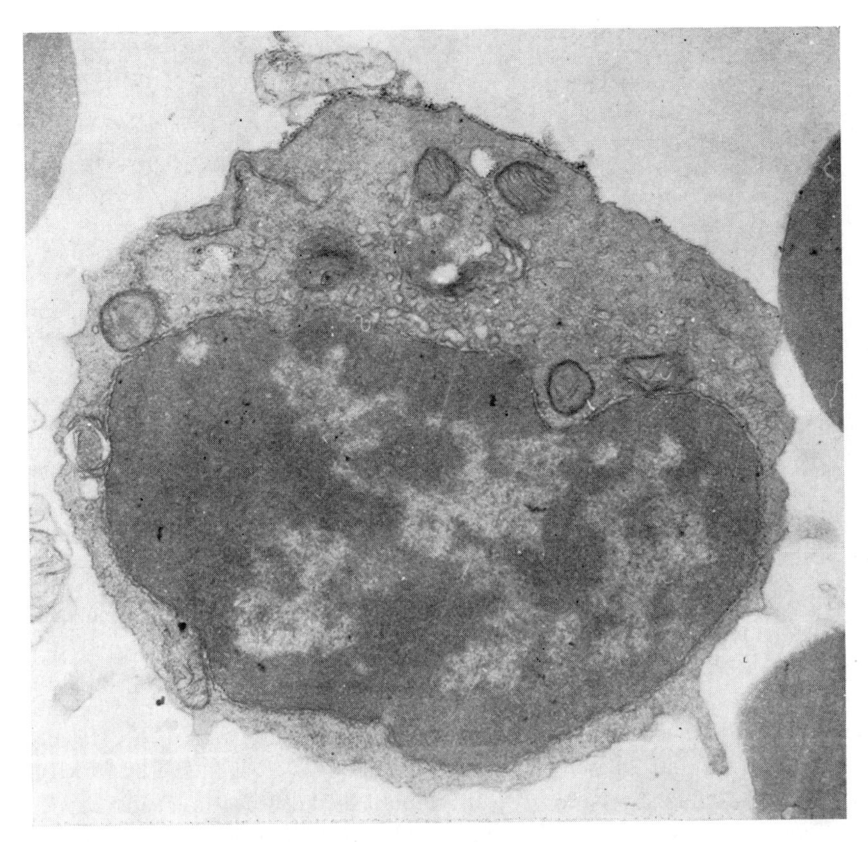

FIG. 5. Cell labelled with Fab-FT (cf. Fig. 1) in the cold, washed and incubated for 15 min at 30 °C with unlabelled anti-Fab antibody (GaRIg). All the cross-linked Fab-FT molecules have coalesced into a typical cap. Some cellular debris and aggregates have also accumulated over the cap. Unstained section. × 25 000.

Observations on other surface components

The random distribution of surface Ig in the unperturbed membrane, patching and capping induced by cross-linking are probably valid for most, if not all, lymphocyte surface components. For example, receptors for concanavalin A (ConA) labelled with ConA-FT appear to be diffusely distributed over the entire lymphocyte surface at 0 or at 20 °C (S. de Petris, unpublished observations). When ConA-FT is cross-linked by anti-ferritin antibody, it forms patches at 4 °C and caps at 20 °C. Under different conditions concanavalin A, which is multivalent, can induce patching and capping directly (Smith & Hollers 1970; Stobo & Rosenthal 1972). Similarly, the θ alloantigen on the thymocyte surface

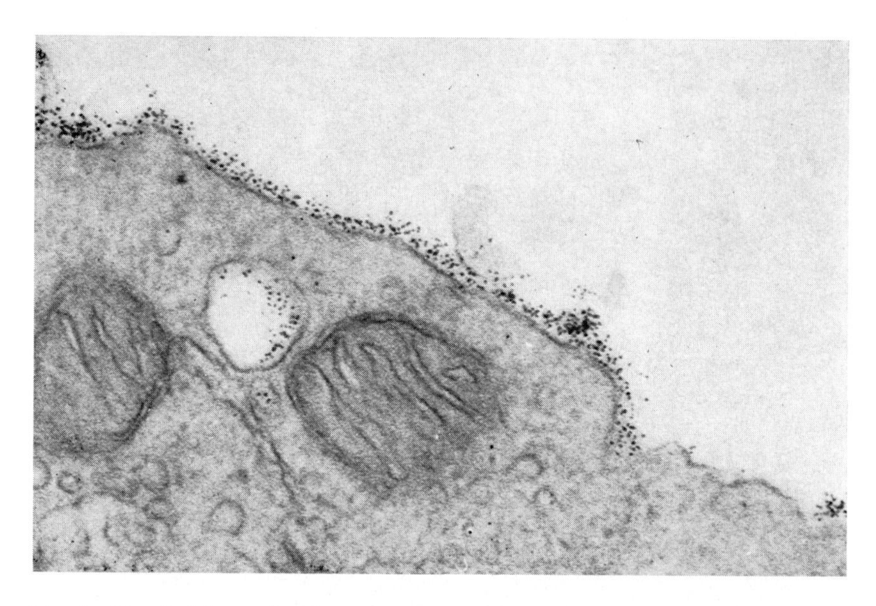

FIG. 6. Detail of the cap in Fig. 5, formed by a layer of closely-packed ferritin molecules (compare with Fig. 1). × 84 000.

also appears dispersed when labelled with anti-θ antibody raised in congenic mice and Fab-FT, but is patchy when anti-θ antibody is cross-linked by RaMIg-FT at 0–4 °C (S. de Petris & M. C. Raff, unpublished observations) and capped when cross-linked by RaMIg at or above 20 °C (Taylor *et al.* 1971). Similarly, HL-A antigen (Kourilsky *et al.* 1972) on human lymphocytes and H-2 antigens on mouse lymphocytes and thymocytes (our unpublished observations), which give a ring distribution when labelled at 0–4 °C with fluorescent antibody, can form caps at 37 °C. Phytohaemagglutinin receptors, labelled with phytohaemagglutinin and fluorescent anti-phytohaemagglutinin antibody, can also form caps in metabolically active cells (Osunkoya *et al.* 1970).

MECHANISM OF CAP FORMATION

The characteristic feature of cap formation is that elements of the plasma membrane, previously scattered over the whole cell surface, can move independently towards one pole of the cell and eventually segregate from the rest of the membrane. In this way, for example, surface Ig can segregate from H-2 antigens or from antigens recognized by an anti-lymphocyte serum, with which they were previously mixed over the surface (Taylor *et al.* 1971). Similarly, the

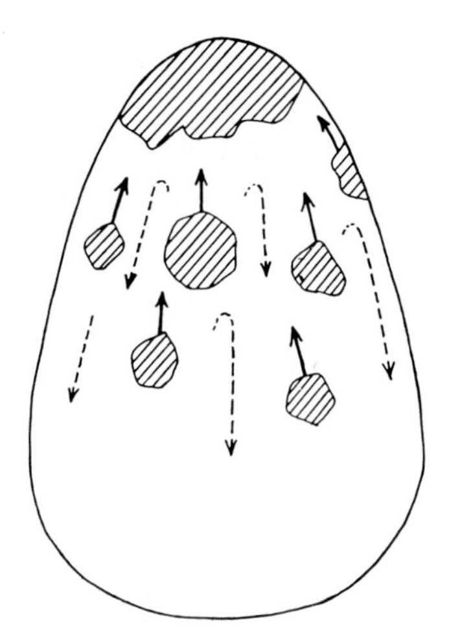

FIG. 7. Schematic representation of membrane movement during cap formation. An overall backward movement (solid arrows) of the labelled cross-linked patches (hatched areas) is accompanied by an overall countercurrent forward flow of the fluid unlabelled membrane components (broken arrows).

products of the D and K end of the H-2 locus on mouse lymphocytes, can also segregate independently (F. Kourilsky & F. Lilly, personal communication). These observations can be easily explained only if the overall backward flow of the cross-linked membrane components is accompanied by a simultaneous, countercurrent forward flow of the unlabelled components (Fig. 7). This possibility is implicit in the concept of a fluid membrane. Pinocytosis does not seem to be an essential part of the process. The cap forms without significant pinocytosis and under optimal conditions pinocytosis generally lags behind the rapid formation of the cap (the labelled membrane may move over the surface at 5 μm/min at least). Protein synthesis and normal membrane turnover are also too slow to be of any importance in cap formation [reappearance of surface Ig 'modulated' by antibody requires 4 h or more (M. C. Raff & S. de Petris, unpublished data; Loor *et al.* 1972)] and cap formation is unaffected by inhibition of protein synthesis with cycloheximide or puromycin (S. de Petris & M. C. Raff, unpublished observations; Kourilsky *et al.* 1972).

Since cap formation is an active process, apparently associated with morphological structures (such as the uropod) which are characteristic of moving lymphocytes, it seems extremely likely that the same mechanism of membrane movement applies in both cases. However, 'capping' does not appear to require adhesion to a substratum and actual cell movement, since caps can form in a few minutes in cells in contact with solid surfaces, in cells kept in continuous suspension with a magnetic stirrer, and in drops of cell suspensions floating in an immiscible oil (our unpublished observations). Thus membrane movement need not be considered with reference to a solid substratum but only to internal cellular structures.

Relative movement of membrane elements is possible only if the membrane has fluid characteristics; it is clear, however, that a countercurrent flow requires that some membrane elements are anchored to some cytoplasmic structures, while other parts of the membrane are free to move relative to them. By analogy with normally moving lymphocytes which have a rigid 'tail' as a probable site of contraction and a flexible advancing front (de Bruyn 1946; Norberg 1971), we suggest that during cap formation the cross-linked patches are interacting with a cytoplasmic contractile system, and are actively transported backwards, while the non-interacting, unlabelled parts of the membrane can retain their fluid properties and are displaced passively in the opposite direction. This suggestion does not exclude the possibility that the interaction with the contractile system is intermittent and also involves unlabelled membrane components which, however, would be able to return to an unconstrained state and more readily diffuse forward than the large, cross-linked patches can. The hypothesis is consistent with the observation that cytochalasin B partially inhibits cap formation (Taylor *et al.* 1971), and with the presence of a layer of microfilaments below the plasma membrane along the entire periphery of lymphocytes but often particularly thick in the region of the cap. This layer, which often appears as a tangled network, is more clearly visible in glycerinated lymphocytes (S. de Petris, unpublished observations). Preliminary observations indicate that these filaments bind heavy meromyosin, which implies that the filaments are actin-like and similar to those observed in many other cell types (Ishikawa *et al.* 1969).

It is not known whether the polarity of membrane movement during cap formation is connected with the layer of microfilaments itself or is determined by other structures, such as microtubules or 10 nm filaments. Treatment with Colcemid does not prevent the coalescence of the labelled membrane into a cap, but whether the normal polarity of the cell is preserved in these conditions is still under investigation. The 10 nm filaments are rare in labelled lymphoctyes.

SIGNIFICANCE OF CAP FORMATION FOR CELL MOVEMENT

Although cap formation can happen without cell locomotion, the mechanism responsible for membrane movement is likely to be the same in both cases. Thus, if a patch, moving backwards with respect to an internal cellular frame, is adherent to a solid substratum, the membrane and the cell as a whole would move forwards with respect to the substratum. Whatever the specific mechanism of cap formation, the properties and behaviour of the membrane described here have important implications for the function of the membrane during cell movement. The most important point is the fact that adjacent parts of the membrane can slide past each other and segregate independently. If this is a general property of the cell membrane, as now seems likely, then the backward movement of the membrane demonstrated in several cases during locomotion of amoebae and tissue cells [Wolpert & Gingell 1968 (a review); Ingram 1969; Abercrombie *et al.* 1970] could occur, as in the case of cap formation, without the 'resorption' of membrane at the tail of the cell, and 'formation' of new membrane at the front, postulated by several authors (e.g. Goldacre 1961; Shaffer 1963; Abercrombie *et al.* 1970). This conclusion is in agreement with the observations and the interpretation of others who concluded that the membrane is essentially conserved during cell movement (Griffin & Allen 1960; Wolpert & Gingell 1968). It does not exclude the possibility that the backward movement of the membrane is coordinated over a large area of the membrane (Abercrombie *et al.* 1970), but suggests that countercurrent movement of other sections of the membrane can occur simultaneously, resulting in an overall recirculation of the membrane. The sink of membrane at the rear of the cell would essentially correspond to the point(s) where the constrained backwards-pulled membrane returns to a liquid-like state and becomes able to flow forwards again. Formation and insertion of newly synthesized membrane (apart from normal membrane turnover) probably represents an important process only in cells in which the total surface area is progressively increasing, as, for example, in the growth cone of nerve fibres (Bray 1970).

This concept may also explain contradictory observations on the movement of particles on the cell surface [e.g. in amoebae (Mast 1926; Wolpert & Gingell 1968)], since adjacent parts of the membrane can move with different velocities in different directions. It is also clear that bivalent antibodies or markers capable of cross-linking membrane components, or any particulate matter, can be used to study the direction of flow of a particular region of the membrane (which, in most cases, would probably correspond to the 'constrained' or cytoplasm-linked regions, as in the model proposed above), but not to draw conclusions on the behaviour of the membrane as a whole. In particular, the appearance of unlabelled areas, in labelled cells, cannot be taken as unambiguous evidence for the formation of new membrane, especially at the level of resolution of the light microscope. Thus, in our view the 'new membrane' formed at the tip of pseudopods in amoebae (Jeon & Bell 1964) more likely represents fluid unlabelled molecules which have slipped through the majority of the molecules immobilized by the added marker. In principle, less ambiguous results would be obtained with the use of labelled univalent antibody or ligands. Conditions approaching those with univalent ligand could be obtained with bivalent antibody or ligands in certain cases in which the degree of cross-linking is minimized (e.g. when the surface antigen is greatly in excess of the added antibody or *vice versa*, or if the surface antigen is univalent). The mainly diffuse staining of amoebae with fluorescein-labelled antibody observed by Wolpert and his colleagues (see Wolpert & Gingell 1968; Garrod & Wolpert 1968) could represent examples of one such situation.

ACKNOWLEDGEMENTS

S. de P. was supported by a short-term fellowship of the European Molecular Biology Organization (EMBO).

References

ABERCROMBIE, M., HEAYSMAN, J. E. M. & PEGRUM, S. M. (1970). *Exp. Cell Res.* **62**, 389-398
AOKI, T., HAMMERLING, V., DE HARVEN, E., BOYSE, E. A. & OLD, L. J. (1969). *J. Exp. Med.* **130**, 979-1001
BESSIS, M. & DE BOISFLEURY, A. (1971). *Nouv. Rev. Fr. Hématol.* **11**, 377-400
BIBERFELD, P. (1971). *J. Ultrastruct. Res.* **37**, 41-68
BOYSE, E. A., STOCKERT, E. & OLD, L. J. (1968). *Proc. Natl. Acad. Sci. U.S.A.* **60**, 886-893
BRAY, D. (1970). *Proc. Natl. Acad. Sci. U.S.A.* **65**, 905-910
BROWN, P. K. (1972). *Nat. New Biol.* **236**, 35-38
CONE, R. A. (1972). *Nat. New Biol.* **236**, 39-43
DE BRUYN, P. P. H. (1946). *Anat. Rec.* **95**, 177-191
DE PETRIS, S. & RAFF, M. C. (1972). *Eur. J. Immunol.* **2**, 523-535
DE PETRIS, S. & RAFF, M. C. (1973). *Nat. New Biol.* **241**, 257-259
FRYE, L. D. & EDIDIN, M. (1970). *J. Cell. Sci.* **7**, 319-335
GARROD, D. R. & WOLPERT, L. (1968). *J. Cell Sci.* **3**, 365-372
GRIFFIN, J. L. & ALLEN, R. D. (1960). *Exp. Cell Res.* **20**, 619-622
GOLDACRE, R. J. (1961). *Exp. Cell Res.* **8** (Suppl.), 1-16
INGRAM, V. M. (1969). *Nature (Lond.)* **222**, 641-644
ISHIKAWA, H., BISHOP, R. & HOLTZER, R. (1969). *J. Cell Biol.* **43**, 312-328
JEON, K. W. & BELL, L. G. E. (1964). *Exp. Cell Res.* **33**, 531-539
KOURILSKY, F. M., SILVESTRE, D., NEAUPORT-SAUTES, C., LOOSFELT, V. & DAUSSET, J. (1972). *Eur. J. Immunol.* **2**, 249-257
LOOR, F., FORNI, L. & PERNIS, B. (1972). *Eur. J. Immunol.* **2**, 203-212
MAST, S. O. (1926). *J. Morphol.* **41**, 347-425
MCFARLAND, E. & SCHECHTER, G. P. (1970). *Blood J. Hematol.* **35**, 683-688
NORBERG, B. (1971). *Scand. J. Haematol.* **8**, 75-80
OLDFIELD, E. & CHAPMAN, D. (1972). *FEBS (Fed. Eur. Biochem. Soc.) Lett.* **23**, 285-297
OSUNKOYA, B. O., WILLIAMS, A. I. O., ADLER, W. H. & SMITH, R. T. (1970). *Afr. J. Med. Sci.* **1**, 3-16
PINTO DA SILVA, P. (1972). *J. Cell Biol.* **53**, 777-787
RAFF, M. C. (1970). *Immunology* **19**, 637-650
RAFF, M. C. & DE PETRIS, S. (1973). *Fed. Proc.* **32**, 48-54
SHAFFER, B. M. (1963). *Exp. Cell Res.* **32**, 603-606

SINGER, S. J. (1971) in *Structure and Function of Biological Membranes* (Rothfield, L.I., ed.), pp. 145-222, Academic Press, New York
SINGER, S. J. & NICOLSON, G. L. (1972). *(Science Wash. D.C.)* **175**, 720-731
SMITH, C. W. & HOLLERS, J. C. (1970). *J. Reticuloendothel. Soc.* **8**, 458-464
STOBO, J. D. & ROSENTHAL, A. S. (1972). *Exp. Cell Res.* **70**, 443-447
TAYLOR, R. B., DUFFUS, W. P. H., RAFF, M. C. & DE PETRIS, S. (1971). *Nat. New Biol.* **233**, 225-229
VITETTA, E. S., BAUR, S. & UHR, J. (1971). *J. Exp. Med.* **134**, 242-264
WILKINS, M. H. F., BLAUROCK, A. E. & ENGELMANN, D. M. (1971). *Nat. New Biol.* **230**, 72-76
WOLPERT, L. & GINGELL, D. (1968) in *Aspects of Cell Motility* (Miller, P. L., ed.), (Symp. Soc. Exp. Biol. no. 22), pp. 168-198, Cambridge University Press, London

Discussion

Allison: Dr de Petris has just confirmed what I said (p. 23); that attachment to either a substratum or a particle could provide cross-linking of membrane constituents, probably glycoproteins, which could move back differentially leaving the rest of the membrane in some kind of counter-flow.

Middleton: I find it difficult to reconcile Dr de Petris' data with those of Dr Harris. The former suggest that the membrane is basically a fluid structure in which isolated elements can move freely and independently, whereas the latter imply a more rigid structure for the membrane, capable of exerting considerable tension, with relatively large areas able to move in a coordinated and interdependent fashion. There seems to me to be a major difference between the movement of components within the membrane and the movement of large areas of the intact membrane.

de Petris: I am not saying that the membrane is an entirely fluid structure. I think it can pass from a state in which it is attached to cytoplasmic structures to one in which it is unattached and fluid. Normally the attachment to cytoplasmic structures is reversible, but in the presence of antibody the membrane elements which are cross-linked remain associated and cannot revert to an entirely fluid state. In normal cell movement it is possible that the cytoplasm-linked, backwards-moving membrane becomes fluid again and flows forward when it comes to the back of the cell (in the case of lymphocytes) or to the middle of the cell [as in fibroblasts (cf. Edidin & Weiss 1972)].

Ambrose: Does this not imply that fluidity lies in the surface layer, that is, within the plasma membrane in which surface antigens and other macromolecules are free to migrate? But the mechanical properties of cells, particularly their ability to withstand stretching forces, are dependent on the plasma gel lying immediately below the plasma membrane. This gel is an elastic solid, capable of maintaining tension.

de Petris: Although little is known about the organization of the inner layers, I think basically you are right; the fluidity is essentially in the membrane and in its fluid state the membrane could slide over the cytoplasm beneath. The contractile system is in the cytoplasm, but is intermittently attached to the membrane, which thereupon becomes gel-like to a certain extent.

Steinberg: There is another possible way in which fluid and solid properties can coexist in one and the same structure. If the cell envelope were an elastico-viscous body, the duration of application of a stress would determine whether it behaved as a solid or a liquid. Under stresses of short duration it would appear as an elastic solid, whereas under those of long duration it would appear as a viscous liquid. We have recently found that cell *aggregates* behave in precisely this way (Phillips *et al.* 1973).

Curtis: Dr de Petris, your results could be artifactual, since Fischer *et al.* (1967) have shown that attachment of immunoglobulin to cells stimulates production of lysolecithin in the plasmalemma. Chapman (1968) has postulated, probably correctly, that this makes the plasmalemma fluid.

de Petris: The plasmalemma does not become completely fluid. Patch formation requires the plasmalemma to be fluid, but cap formation necessitates the membrane being partially gel-like, partially fluid; the parts of the membrane which are fluid are essentially those which are free of antibody. Moreover capping is not occasioned only with antibody, but also with plant lectins and with multivalent antigens specific for surface immunoglobulin receptors. Thus it is unlikely to be due simply to an artifactual alteration of the membrane structure induced by antibody binding.

Allison: But, Professor Curtis, the fact that Fab, the univalent component of antibody, does not induce patching and capping means that attachment by itself is not enough. What you need is cross-linking.

Curtis: We do not know what Fab does to lysolecithin production.

Wohlfarth-Bottermann: Frame-by-frame analysis of time lapse films of *Amoeba proteus* shows that the moving amoeba is attached to the substrate. This can be seen by watching the moving cell from below with interference reflection microscopy. The plasmalemma is only attached to the substrate at a few points (Haberey 1971). Experiments with marker particles reveal that the plasmalemma is moving forwards between these points of attachment. Whether this means a shearing movement between neighbouring parts of the plasmalemma or perhaps only between proximal parts of the mucous layer is a difficult question. Strictly speaking, the analysis of the movement of the marker particles tells us only about the behaviour of the mucous layer to which the particles are attached. We do not think that the mucous layer and the plasmalemma behave differently, but we cannot be sure.

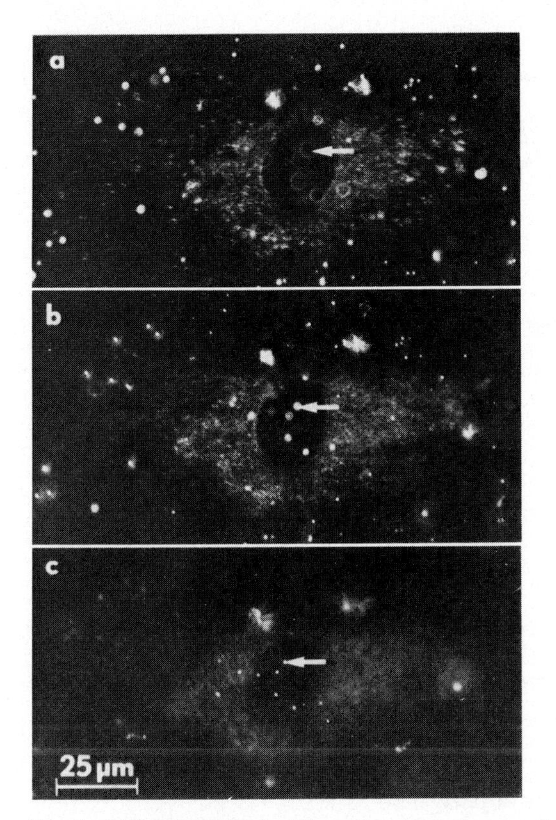

FIG. 1 (Albrecht-Bühler). Ultramicroscopic image of a 3T3 mouse fibroblast after treatment with gold particles. Arrows point to one of the gold particles attached to the surface of the cell. From a—c the focus is raised in order to show that the particles are located on the dorsal side of the cell [reproduced from Albrecht-Bühler & Yarnell (1973), with kind permission of Academic Press].

What do you know about the regeneration of the plasmalemma in endocytosis at the rear pole of the cell?

de Petris: Immunoglobulins are completely cleared from the surface after incubation for two hours at 37 °C, but if incubation is continued in antibody-free medium, immunoglobulins reappear within 4–8 h (Loor *et al.* 1972 and our unpublished work). So it would seem that turnover of the membrane, with regard to plasma immunoglobulin, takes 4–8 h.

Trinkaus: Professor Wohlfarth-Bottermann, are the points of adhesion of these amoebae confined to the marginal regions of the cell?

Wohlfarth-Bottermann: No; points of adhesion are found all over the lower surface of the cell, mostly in the front part of the amoeba. The uroid, the rear part, is not attached to the substratum (Haberey 1971).

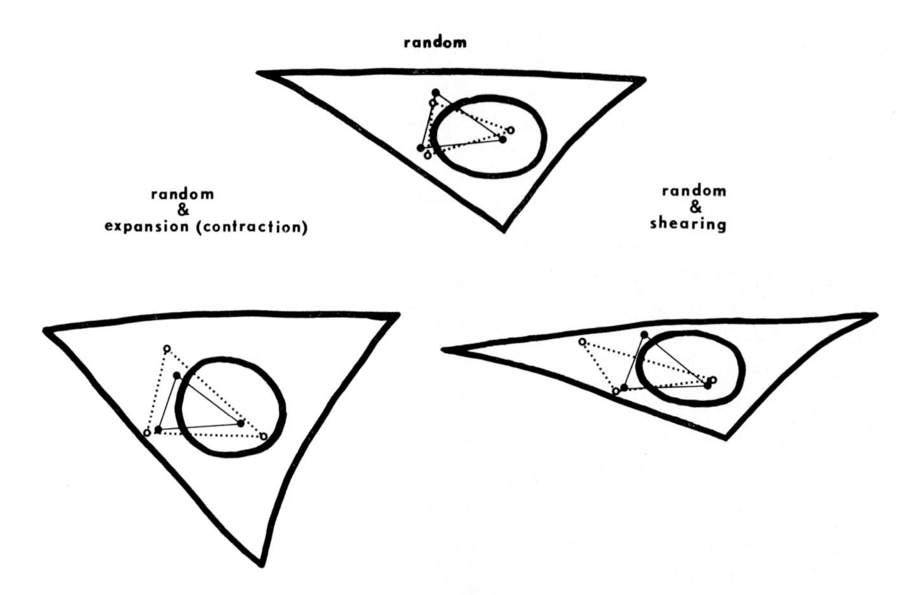

FIG. 2 (Albrecht-Bühler). Possible types of relative movement of three particles on the surface of a cell as described by changes of the triangle which is formed by the particles. The dotted lines indicate the triangles which developed from the initial triangles (solid lines) during a given time interval.

Trinkaus: What about the central part?

Wohlfarth-Bottermann: Both the front and the central parts are attached to the substratum (Haberey *et al.* 1969; Stockem *et al.* 1969).

Albrecht-Bühler: I shall try to avoid the term 'membrane movement' using 'particle movement' instead, and I shall restrict my comments to particle movement on the cell surface of mouse fibroblasts just above the nucleus.

We put large gold particles (200–400 nm in diameter) on cells which were grown on a glass coverslip and observed the particle movement in an ultramicroscope (Siedentopf & Zsigmondy 1903) (Fig. 1). Three particles in the central part of the cell were arbitrarily selected and regarded as vertices of a triangle. By following changes of this triangle during an experiment we measured the particle movement on the cell surface.

We observed three types of movement (Fig. 2). The main movement is irregular: the three vertices of the triangle move an unpredictable distance in unpredictable directions. There can also be a systematic shearing movement superimposed on this irregular movement: over a longer time the changes of perimeter are greater than the changes of area of the triangle compared to the case of random movement. Finally, a systematic expansion or contraction can

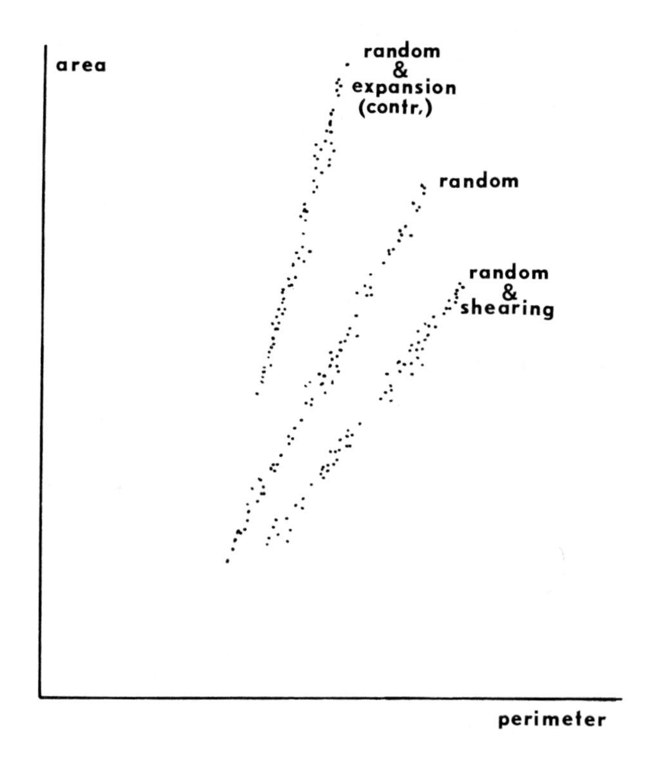

FIG. 3 (Albrecht-Bühler). Plot of perimeter against area of successive triangles formed by three particles on a cell surface for the three types of movement shown in Fig. 2. The slope gives the index of systematic movement.

be superimposed on the irregular movement: the changes of area exceed the changes of perimeter compared to random movement over a longer time.

We recorded the positions of the vertices of the same triangles and measured the perimeter and the area at three-minute intervals. From these measurements we calculated two quantities, one of which gives an index of systematic movement and the other of irregular movement. By plotting perimeter against area for successive triangles, we obtained straight lines (Fig. 3). This surprising result can be simulated by Monte Carlo methods (see Hammersley & Handscromb 1964) which show that it is due to the irregularity of the movement. An additional systematic movement would change the slope. From the deviation of the observed slope from the value predicted for random movement, we derived an index of systematic movement. The sign of this index designates the type of movement and its numerical value represents its extent.

The second quantity is similar to the augmented diffusion constant, D^*, of

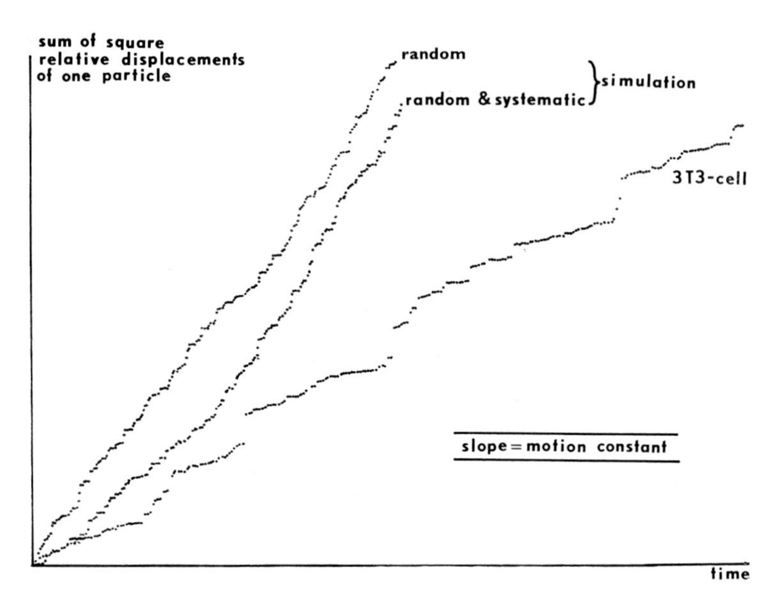

FIG. 4 (Albrecht-Bühler). Plot of the sum of the square displacements of one particle relative to another against time (calculated from the square changes of perimeter of the observed triangle). The curves for random movement and superimposition of random and systematic movement were simulated by Monte Carlo experiments. The curve obtained from an experiment with a 3T3 cell resembles neither of them.

Gail & Boone (1970). It is derived from a plot of the sum of the square displacements of one particle relative to another against time (for further details see Albrecht-Bühler & Yarnell 1973). The square relative displacements are calculated from the square changes of perimeter of the observed triangles. In this way net movements of the whole cell are eliminated from the measurements, because shifting or rotating a triangle does not affect its perimeter.

Plots of the sum of the square relative displacements of one particle against time are shown in Fig. 4. The mean square relative displacement of one particle is obtained from the slope, provided it is a straight line. We call the slope the 'motion constant'. The plots we obtained with gold particles on cells did not yield straight lines. However, their staircase character enabled the graphs to be decomposed into intervals of flat and steep slope, which can be plotted separately (Fig. 5). The slopes of the two resulting straight lines represent fast and slow particle movement in the cell surface. Hence, the particle movement in cell surface is described by a fast and slow motion constant and by the index of systematic movement.

Table 1 compares the particle movements on 3T3 mouse fibroblasts with those

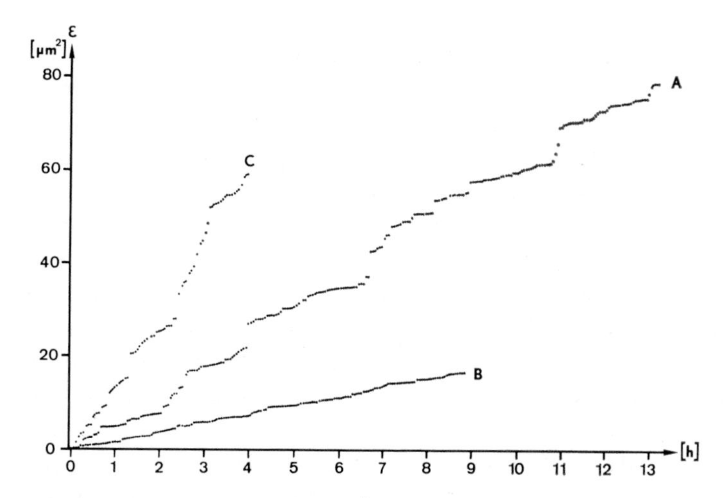

FIG. 5 (Albrecht-Bühler). Division of the sum of the square relative displacements of one particle into intervals of steep and flat slope (A). Separating the two slopes, two straight lines are obtained which represent the slow (B) and the fast (C) particle movement [reproduced from Albrecht-Bühler (1973), with kind permission of Academic Press].

TABLE 1

Comparison of the particle movement on the surface of 3T3 mouse fibroblasts and their polyoma transformants. The values obtained with SV3T3 cells are similar to the values for Py3T3 cells.

	Motion constant (*μm²/min*)		*Index of systematic movement*
	Fast	*Slow*	
3T3	0.9	0.07	0.8
Py3T3	0.13	0.012	0.4

on polyoma-transformed 3T3 cells. In the central part of the cells we found much slower particle movement for transformed than for normal cells. About 60% of the time between cell divisions the particles were in the state of slow movement. To judge by the sign of the index of systematic movement, shearing was about four times as frequent as both contraction and expansion.

Wessells: Have you investigated the variation of the index of persistent movement with temperature? Frye & Edidin (1970) found a sharp cut off in redistribution of surface materials at 15 °C. It would certainly be worth looking for a similar relationship in your system if the ideas gained from the polar capping phenomenon are to be extrapolated to surface movements during cell locomotion.

Albrecht-Bühler: No, we have not done any temperature-dependence experiments.

Wolpert: I would like to make a general point about the forces exerted by the cells. On the whole these forces are small and difficult to estimate, but seem to be of the order of 10^{-8} N (10^{-3} dyn). I don't think there is any question of the membranes that we are dealing with being unable to withstand such forces.

Porter: Assuming there is some turnover, I would like to hear some comments on the origin of the membrane. I recall that you, Mr Abercrombie, found some vesicles adjacent to the surface of the cell in electron micrographs of the lamellipodium in fibroblasts. I am reminded also of some work, which may not be familiar to some of you, on the development of the pollen tube (Rosen 1968; Rosen *et al.* 1964). Pollen tube grows rapidly—several mm/h. According to Rosen, membrane is added at the advancing tip by the imposition of vesicles into the tip. In other words, vesicles generated within the cell have a membrane of appropriate design to fit into the advancing edge. Is that a general phenomenon at the margin of advancing cells? I have the impression that these vesicles are derived from the Golgi apparatus.

Abercrombie: The sporadic appearance of these vesicles at the advancing edge of the fibroblast puzzles us. It seems that cells undergoing active ruffling, which according to our interpretation would mean active production of new surface at the front end of the cell, are the ones which have these vesicles. However, we do not know whether the vesicles originate from the surface or are discharging onto it.

Wessells: We have tried to investigate that problem with glial cells and elongating axons by applying Thorotrast (colloidal thorium dioxide) to cultures for periods of 15 min, 5 min or 45 s. Even in the group treated for 45 s only, we found a considerable number of vesicles containing Thorotrast. The vesicles must be rapidly forming at the cell surface during locomotion and axon elongation. This surprised us, since we expected the vesicles to fuse with the cell surface and thus generate new surface area, although that is not excluded by the observations: first, not all the cells have Thorotrast in vesicles and secondly, even in the cells exposed to Thorotrast for 15 min, there is a substantial population of vesicles both in nerve growth cones and near the front of ruffled membrane organelles that contains no Thorotrast at all. Perhaps there are two populations of vesicles in these parts of the cells; one group acting in surface addition, the other in micropinocytosis.

de Petris: There are several smooth vesicles also in lymphocytes but usually they are absent from the region of the pseudopodium. One of the reasons we fail to observe discharge of new membrane in lymphocytes might be that membrane turnover is slow in comparison with cap formation. The morphological evidence would suggest that the vesicles are discharged at the back of the cell and not at the front.

Goldman: This has also bothered us. We find large pinocytic vesicles forming at the leading, ruffling edge of cells and moving back towards the nucleus. Thus there is active surface membrane uptake in a region where the cell is supposedly increasing its surface area by addition of new membrane. We also see many small vesicles in this region and although we do not have any evidence about the direction of movement of these vesicles, it is possible that the smaller vesicles are fusing with the cell surface at its leading edge (Wang & Goldman, unpublished observations). We have also suggested that changes in surface topography such as the folding up of the surface in the form of microvilli and the subsequent unfolding or flattening of microvilli, account for localized changes in surface area (Follett & Goldman 1970).

Curtis: Dr de Petris, if the Harris theory is right would you not expect the lymphocytes to start moving when you cross-link them at surfaces?

de Petris: It does seem likely that cross-linking the membrane with antibody triggers the cell into moving more quickly. Some evidence for this is that few (2–3%) uropods are observed in lymphocytes stained with Fab (de Petris and Raff, unpublished results). In cells incubated with bivalent antibody about 30% of the cells have uropods. However, this might mean that either more lymphocytes are triggered into moving or the adhesiveness of the lymphocyte to the substratum is changed.

Curtis: So it is conceivable that your experiments support the Harris hypothesis?

de Petris: No. I don't see how!

Curtis: I imagine that when the cell is covered with antibody it would be totally immobilized. But before this immobilization, you might stimulate movements of an abnormal type by the cross-linking.

de Petris: With regard to surface immunoglobulin not more than one third of the surface is covered by antibody and there is no indication that the cells are immobilized. I am not claiming, however, that the movement of the cross-linked membrane during capping is identical to membrane movement during locomotion, but that their basic features and mechanism are the same. During cell locomotion there could be a more extensive coordinated movement of large areas of the membrane, as in fibroblasts. The principle of membrane movement which I suggest is basically different from that suggested by Dr Harris. I think that there is a countercurrent movement of membrane elements over the surface.

Harris: Particles attached to polymorphonuclear leucocytes also collect at the uropod, which suggests to me that this is the membrane sink or area of disassembly.

de Petris: We found that particulate debris, labels etc., tended to stick to the patches of antibody, and then the patches with the attached particles would move together towards the back of the cell, where they would accumulate on the cap. This shows that there is a relative movement of separate parts of the membrane.

Harris: Why do you think that the different attachment sides are separate parts of the membrane?

de Petris: Initially they are distributed at random all over the membrane, and then they fuse together over one pole.

Harris: If the membrane continually flowed to the uropod, was disassembled there and returned to the front through the cytoplasm and if the marker material stayed on the surface, then label would accumulate at the uropod and my theory would explain your observations.

de Petris: But how does all this happen in one minute?

Harris: The speed of membrane movement is not surprising since the cells themselves move rapidly.

de Petris: More than half the membrane has to be transferred from the back to the front, and this transfer would be restricted to the unlabelled parts of the membrane. There is no evidence of dramatic pinocytosis or transport of unlabelled vesicles towards the front of the cell.

Harris: The 'disassembled' membrane need not move in the form of vesicles.

Allison: The problem is whether the rearrangement is within the membrane itself or whether you have to postulate movement through the cytoplasm. We maintain that most of the rearrangement is in the membrane and not in the cytoplasm.

Harris: That is perfectly plausible but I see no evidence for it.

Allison: The evidence in favour is the general information about the fluid–mosaic behaviour and turnover of membrane constituents. From the evidence of Warren and others already discussed (pp. 15 and 24) it is unlikely that membranes can turn over sufficiently rapidly to account for movement in your model.

Harris: Warren & Glick (1968) studied rates of turnover of incorporated material, that is, rates of synthesis of molecules. I am talking about recycling of these molecules, so there is no contradiction.

Allison: I know of no evidence that material can cross from the membrane into the cytoplasm and out again.

de Petris: Labelled and some adjacent unlabelled membrane are pinocytized together and remain inside the cytoplasm.

Porter: Should we not define the membrane? Some investigators consider the antigenic portion to be the glycocalyx on the cell, which might be independent of the unit membrane; the two might move separately.

Wolpert: Are you suggesting that Harris' particles are binding to the unit membrane?

Porter: I am suggesting that they are binding to the polysaccharide-rich layer on the outside, and that this might move independently of the membrane beneath.

Dunn: It is difficult to reconcile Dr Harris' observation that blebs and other surface features of complex shape are transported backward on the dorsal cell surface with Professor Wolpert's hypothesis that only binding sites move backward.

Abercrombie: Many observations suggest that backward movement of the upper surface of the cell is independent of any kind of marker. The ruffles themselves go backwards.

Porter: The membrane could flow over the ruffles. Surely it does not have to move with the ruffle?

Abercrombie: True, though the shape of ruffles does not suggest that they are waves, and they move at the same speed as the particles. I think we have here two hypotheses that we cannot at present sort out: (i) the hypothesis of counter-currents and (ii) the hypothesis of circulation through the cytoplasm. As regards the former, I suppose our marking experiments with concanavalin A (Abercrombie *et al.* 1972) could still allow counter-flow of unmarked molecules.

Wolpert: It is vital to show that the membrane actually goes into the cytoplasm at a plausible rate for Harris' theory to be acceptable. All the evidence is against the rapid entry of membrane during cell movement (Wolpert 1971).

Curtis: Fischer *et al.* (1967) reported that some components turn over rapidly. Our own experiments (unpublished results) have proved that the rate of turnover depends on the culture conditions. A slight change results in either the shutdown of the whole process or rapid degradation of the membrane. That, of course, is the typical consequence of putting the cells in inadequate media.

Steinberg: I thought Dr de Petris' observations indicated that there was a countercurrent movement in the membrane. If you add a linking agent, such as an antibody which goes onto the surface, the cross-linked parts of the surface move back. On subsequent addition of a different marker, such as concanavalin A, are not the target sites found to have been shifted forward?

Harris: Also, if there were counter-currents and you added two markers simultaneously, would you not expect two separate caps, or one marker accumulated in a cap and the other dispersed?

de Petris: Yes, that would be the important experiment: to see if the two markers can segregate after addition, because only one of the two is able to form a cap. We have not done this yet, but only for technical reasons—mainly lack of suitable markers; however, this difficulty does not appear insuperable.

References

Abercrombie, M., Heaysman, J. E. M. & Pegrum, S. M. (1972) The locomotion of fibroblasts in culture. V. Surface marking with concanavalin A. *Exp. Cell Res.* **73**, 536-539

Albrecht-Bühler, G. (1973) A quantitative difference in the movement of marker particles in the plasma membrane of 3T3 mouse fibroblasts and their polyoma transformants. *Exp. Cell Res.* **78**, 67-70

Albrecht-Bühler, G. & Yarnell, M. M. (1973) A quantitation of movement of marker particles in the plasma membrane of 3T3 mouse fibroblasts. *Exp. Cell Res.* **78**, 59-66

Chapman, D. (1968) *Biological Membranes*, Academic Press, New York

Edidin, M. & Weiss, A. (1972) Antigen cap formation in cultured fibroblasts: a reflection of membrane fluidity and cell motility. *Proc. Natl. Acad. Sci. U.S.A.* **69**, 2456-2459

Fischer, H., Ferber, E., Haupt, I., Kohlschütter, A., Modolell, M., Munder, P. G. & Sonak, R. (1967) Lysophosphatides and cell membranes. *Protides Biol. Fluids Proc. Colloq. Bruges* **15**, 175-184

Follett, E. A. C. & Goldman, R. D. (1970) The occurrence of microvilli during spreading and growth of BHK-21/C13 fibroblasts. *Exp. Cell Res.* **59**, 124-136

Frye, L. D. & Edidin, M. (1970) The rapid mixing of cell surface antigens after formation of mouse-human heterocaryons. *J. Cell Sci.* **7**, 319-335

Gail, M. H. & Boone, C. W. (1970) Locomotion of mouse fibroblasts in tissue culture *Biophys. J.* **10**, 980-993

Haberey, M. (1971) Bewegungsverhalten und Untergrundkontakt von *Amoeba proteus*. *Mikroskopie* **27**, 226-234

Haberey, M., Wohlfarth-Bottermann, K. E. & Stockem, W. (1969) Pinocytose und Bewegung von Amöben. VI. Kinematographische Untersuchungen über das Bewegungsverhalten der Zelloberfläche von *Amoeba proteus*. *Cytobiologie* **1**, 70-84

Hammersley, J. M. & Handscromb, D. C. (1969) *Monte Carlo Methods*, Wiley, New York

Loor, F., Forni, L. & Bernis, B. (1972) The dynamic state of the lymphocyte membrane. Factors affecting the distribution and turnover of surface immunoglobulin. *Eur. J. Immunol.* **2**, 203-212

Phillips, H. M., Steinberg, M. S. & Lipton, B. H. (1973) Elasticoviscous morphogenetic behavior of centrifuged chick cell aggregates. *Am. Zool.* in press

Rosen, W. G. (1968) *Annu. Rev. Plant Physiol.* **19**, 435-462

Rosen, W. G., Gawlik, S. R., Dashek, W. V. & Siegesmund, K. A. (1964) *Am. J. Bot.* **51**. 61-71

Siedentopf, H. & Zsigmondy, R. (1903) *Drudes Ann. Phys.* **10**, 1-39

Stockem, W., Wohlfarth-Bottermann, K. E. & Haberey, M. (1969) Pinocytose und Bewegung von Amöben. V. Konturveränderungen und Faltungsgrad der Zelloberfläche von *Amoeba proteus*. *Cytobiologie* **1**, 37-57

Warren, K. & Glick, M. A. (1968) Membranes of animal cells. II. The metabolism and turnover of the surface membrane. *J. Cell Biol.* **37**, 729-746

Wolpert, L. (1971) Cell movement and cell contact. *Sci. Basis Med. Annu. Rev.* 81-98

Surface movements, microfilaments and cell locomotion

NORMAN K. WESSELLS, BRIAN S. SPOONER and MARILYN A. LUDUEÑA

Department of Biological Sciences, Stanford University, California, and Division of Biology, Kansas State University

Abstract The movements carried out by the surfaces of migratory glial cells and elongating nerve cells are described. The distribution of linearly arrayed microfilaments (sheath, α-filaments) and of lattice (network) microfilaments in the two cell types are discussed. Lattice filaments are associated with the cell surface in regions where extension and surface movement go on. Sheath filaments, which bind heavy meromyosin, can apparently be rearranged within the cell after fresh plating, colchicine treatment and, probably, change in migratory direction. Sheath filaments have not been found in nerve cell bodies, axons or growth cones. It is proposed that (1) the extension phase of cell locomotion and surface addition in the growth cone (see Bray) characterize nerve axon elongation and (2) glial cell locomotion also includes the extension and surface addition (as proposed by Abercrombie) phases as well as a sheath-mediated contractile event (as proposed by Ingram) to displace the cell soma forward over the substratum.

The subject of this paper is difficult to treat because very few experiments have been performed on generalized plasma membrane–filament interactions. It is clear that certain classes of filaments insert into specialized structures associated with the cell surface. Desmosomal plaques and portions of the zonular regions of junctional complexes are two kinds of filament–membrane junctions. Terminal sarcomere insertion areas of muscle cells and the tips of microvilli are other specialized sites where intracellular filaments may exert effects on the cell surface. In all these instances, electron-dense material lines the internal side of the plasma membrane and individual filaments appear to move into or pass through that material. To the extent that such filaments engage in contractile or skeletal activities, such insertions must possess sufficient tensile strength to transmit force between filament and cell surface.

Such filament–membrane relationships are found in specific types of differentiated cells or in epithelia where the cell is in large part subordinated to the tissue in terms of activities and perhaps structure. We shall address ourselves to the more general question of how the membrane is related to underlying cytoplasm in vertebrate cells not specialized for massive contractility or function in tissue, for example a single cell in suspension or a single migratory cell. How is the cell surface moved in space during locomotion, cytokinesis, or other processes?

Wolpert & Gingell (1968) have summarized the possible sources of force generated at the cell surface. The two extreme propositions are that (i) the plasma membrane itself possesses the necessary elastic and contractile elements or (ii) cytoplasmic elements are responsible for movements of the membrane.

Little evidence favours the first possibility. Most 'contractile' events associated with large embryonic cells apparently depend on cortical materials located beneath the plasma membrane. The calcium-dependent 'cortical contractions' in *Xenopus* eggs, for instance, appear to be linked with filamentous materials which concentrate beneath the sites of calcium injection (Gingell 1970; J. Ash, personal communication). Plasma membrane at such sites is thrown into tortuous folds (Figs. 1 and 2). Such folds would not be expected if the calcium ion caused contraction in the membrane itself or caused localized withdrawal of membrane to reduce surface area. Wolpert & Gingell (1968) have marshalled other evidence against the autonomy hypothesis.

Moreover, the Singer–Nicolson (1972) concept of the plasma membrane as a fluid mosaic (see de Petris & Raff, pp. 27-40; Albrecht-Bühler, pp. 44-47) suggests that there is no single 'structural' protein associated with the plasma membrane that might act as a contractile system according to the autonomy hypothesis. The apparent ease with which particulate materials 'flow' laterally in the plane of the membrane (Frye & Edidin 1970; Pinto da Silva 1972) implies fluidity, at least with respect to proteins within and on the outer surface of the lipid bilayer. Relatively little information is available yet about proteins on the inner surface of the plasma membrane, though certain enzymes (sodium-, potassium- and magnesium-dependent ATPases) (Marchesi & Palade 1967; see also Marchesi & Steers 1968) are found only on that surface.

This inner surface is of paramount interest, since it is there that forces generated from cytoplasmic structures must be transmitted to membrane *per se*. It is likely that regularly spaced molecules (presumably proteins) influence cell surface movements associated with ion transport and protein uptake (for review, Kallio *et al.* 1971; also Allison, this volume). Thus 'bristle-like' structures are present beneath the surface of an osteoclast engaged in such activity. It has been proposed that such structures are the mechanochemical transducers responsible for movement of cell surface so that the physiological exchanges can proceed. What is intriguing about the situation in osteoclasts (Kallio *et al.* 1971) is that the cell cortex underlying these surface bristles is identical in appearance to the active ruffling areas of migratory cells and axonal tips (as

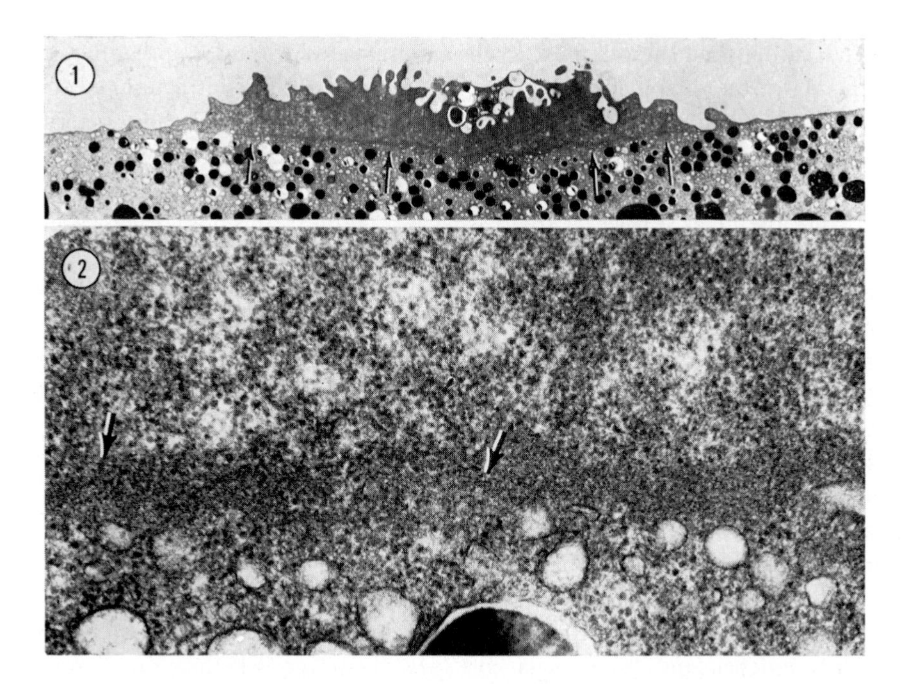

FIG. 1. A low-power electron micrograph of the surface of an early blastomere of a *Xenopus laevis* embryo. Calcium ions were injected through the surface iontophoretically with a microelectrode. Surface materials moved rapidly to the injection site (monitored visually in the living state). The area around the electrode is marked by densely packed cortical cytoplasm and considerable folding of the cell surface, which contrasts with the normal smooth cell surface seen to the right and the left. Note the electron-dense straight line (arrows) that runs beneath the site; vesicles and cytoplasmic granules are usually located beneath the line. Electron-dense lines are found only in those parts of the cortex beneath the sites of injection of calcium ions. × 2 580.

FIG. 2. A typical view of the electron-dense line at a site of calcium injection. Short segments of filaments (arrows) can be seen in most thin sections through such areas of the cell. × 45 000.

described later). Hence, the hypotheses concerning formation of coated vesicles, 'extension-dependent' macropinocytosis (Allison, pp. 122-125), and equivalent types of surface activity should be taken into account in discussing the basis of surface movements during locomotion.

We come, then, to the second possibility, namely that cytoplasmic agents cause movements of the cell surface. What are these movements? We shall describe the basic movements, since they may be the manifestation of organelle function. Seravin (1971) has discussed surface movements with emphasis on the underlying processes (see also Harris, pp. 3-19, and Bray, pp. 366-369).

First, the cell surface may move in response to gross contraction. The ends of skeletal, cardiac and smooth muscle cells are drawn towards one another by the shortening of the sarcomere. Similarly, it is proposed that contractions of the core filaments, which bind heavy meromyosin, cause microvilli to shorten periodically (Tilney & Mooseker 1971). Certain specialized cells, such as epidermal cells on the tails of metamorphosing ascidian tadpoles, possess filaments which insert into the inner surface of the plasma membrane and whose orientation suggests that they move the cell surface by contractility (see Cloney 1966, 1969). These and other cases [e.g. the heavy meromyosin-binding filaments in some morphogenetically acting epithelia (Spooner *et al.* 1973)] might represent instances of intracellular contractility moving the cell surface 'inward'.

A second type of surface movement, the formation of the cleavage furrow, is associated with mitosis. Studied most in large zygotes and blastomere cells, this process is associated with an increase in total surface area of the cell over a relatively brief period. Furthermore, furrowing may proceed in isolated portions of cell surface (underlying plasma membrane and cortical cytoplasm) removed from sea urchin zygotes (Rappaport 1969). It is not clear whether furrowing should be viewed as the result of a contractile process in which the cell surface in the furrow is pulled inward (hence, furrowing would be a subclass of the contraction-dependent phenomena just described), or whether membrane addition or other undefined processes generate the change in shape of the cell surface (Schroeder 1972; Bluemink 1971).

A third distinctive surface movement is the formation of ruffles at the edge of migratory cells or at the tips of elongating nerve axons. Here rather flat, plate-like extensions, 'ruffles' (see Boyde *et al.* 1972 for scanning electron microscopic views of ruffles), extend outward from the cell surface into the surrounding fluid environment. Cylindrical microspikes on nerve cells do the same (Ludueña & Wessells 1973). Two important characteristics of these extensions are their speed—they frequently extend fully in seconds or tens of seconds—and their small radius of curvature. Microspikes frequently have diameters of only 0.2–0.3 μm, though they can reach 20 μm in length. From the few available cross sections of ruffles, the widths appear to be 0.15–0.18 μm. The geometry of these highly asymmetric shapes forces one to conclude that substantial rearrangement of molecules in the cell surface must take place during extension. Such rearrangement is compatible with a fluid–mosaic membrane (Singer & Nicolson 1972).

Movement of surface extensions forms another class of cell surface movement. From this we exclude here cilia, sperm tails and other surface structures in which microtubules are the predominant cytoplasmic element, but rather consider: (i) microspike movement, either along the surface of a substratum or in a fluid; (ii) movement of ruffles that sweep backwards over the dorsal surface of single locomoting cells (cf. Harris, pp. 3-19); (iii) peristaltic wave activity along cylindrical volumes of cytoplasm (Byers & Abramson 1968); and (iv) localized extensions associated with ion or protein uptake by cells (see Kallio *et al.* 1971 for review; also Allison, pp. 124, 125). The wave movement (iii) implies propagation of the causal activities for surface movement through space. Similarly, one interpretation of the process by which vertical plate-like ruffles and cylindrical microspikes bend backward as their bases apparently are transported posteriorly over the surface of the cell or growth cone raises the possibility of a propagated alteration in the cortical systems of the cell.

In summary, cell surface movements fall into the three classes: extension, movement of extended regions, and regression (Seravin 1971). Possibly, the distinctive movement in the plane of the cell membrane, either at the molecular level or associated with addition–deletion of cell surface, is a component of these movements.

Any hypotheses about movement of cell surfaces must take into account the adhesive forces between a cell surface and surrounding substratum. Vertebrate cells in fluid suspension become spherical, showing little ability to retain the flattened morphology characteristic of the same cells on a suitable solid substratum (Wolpert & Gingell 1968). Nor can highly asymmetric processes such as nerve axons be maintained in fluid suspension, despite the fact that such processes contain the supposed 'skeletal' microtubules. When a cell is presented with a suitable solid substratum under appropriate conditions, the edges of the cell can spread down upon the substratum. The degree of spreading can be altered by extracellular factors (such as serum levels for nerve cell bodies) or intracellular factors, since during the last stage of the mitotic cycle cells commonly 'round up' before cytokinesis (a phenomenon that is also observed in intact tissues, even epithelia).

There is no obvious explanation of these cell surface movements *solely* in terms of altered adhesions between a cell and its substratum. The most obvious process that could be adhesion-dependent is the extension of cell surface. However, Ingram's observation (1969) that ruffles extend upward into the fluid medium, as do elongating microspikes (Ludueña & Wessels 1973), as well as the blebbing of many cell surfaces (Middleton and Trinkaus, this volume) clearly show that these extensions must be caused by cellular activity and *not* by interaction between the cell and the substratum. Similarly, the movement of extended cell surfaces, microspikes, ruffles, or perhaps peristaltic waves must be attributed to cellular processes because the movements are at cell surfaces not in direct contact with solid substrata.

Nevertheless, adhesion to a solid substratum is essential to and can affect various parameters of cell locomotion [Carter 1965; Curtis 1967 (review)].

Since the subject will be dealt with elsewhere (pp. 357–361) we shall concentrate upon the membrane and associated filament systems in relation to cell surface movement during locomotion.

LINEAR MICROFILAMENTS AND CELL LOCOMOTION

The rationale for now discussing the distribution of certain filament systems within migratory single cells and nerves rests on the assumptions that surface movements result from mechanochemical processes (Seravin 1971), that polymeric systems in the cell cortex are the sites of such processes and that those polymeric systems can be fixed and visualized in the electron microscope (Behnke *et al.* 1971). Microtubules and intermediate-sized filaments (about 11 nm, tonofilaments, neurofilaments) are discussed later.

In a single migratory glial cell, the linearly arrayed microfilaments, called sheath filaments (Spooner *et al.* 1971), α-filaments (McNutt *et al.* 1971) or stress filaments (Buckley & Porter 1967), are located in bundles immediately beneath the lateral, immobile cell surfaces (Figs. 3–5). The outermost microfilaments in such bundles can be found within 6 nm of the inner surface of the plasma membrane (Figs. 5 and 6), in other words only the diameter of a filament away (the diameter varies between 4 and 8 nm depending on fixation, source of material, etc.). This small separation does not necessarily imply a fixed relationship, since the cell surface can be moved away from such filament bundles in a short time, when transient local ruffles appear along the sides of migratory cells.

The lateral filament bundles are portions of a more widely distributed system. On the inside of the lower plasma membrane of these elongate cells, the sheath filaments commonly form a complete layer, extending anteriorly from about the level of the nucleus to the proximal portion of the microfilamentous network or lattice (Spooner *et al.* 1971; Ludueña & Wessels 1973). The individual filaments lie parallel to the long axis of the cell. Towards the posterior, the sheath is commonly arranged in discrete bundles that extend in straight lines toward the narrow trailing tail of the cell (Figs. 4 and 5). These bundles are similar in appearance to those found at the lateral surfaces of the cell. Serial sections cut parallel to the substratum, and upward through the cell, establish that the sheath filaments are close to both the lower and upper surfaces. Sheath filaments near the upper surface also run parallel to the long axis of the cell, although they are sparser than those near the lower cell surface (Ludueña & Wessells 1973). Therefore, the sheath should not be viewed as a specialization

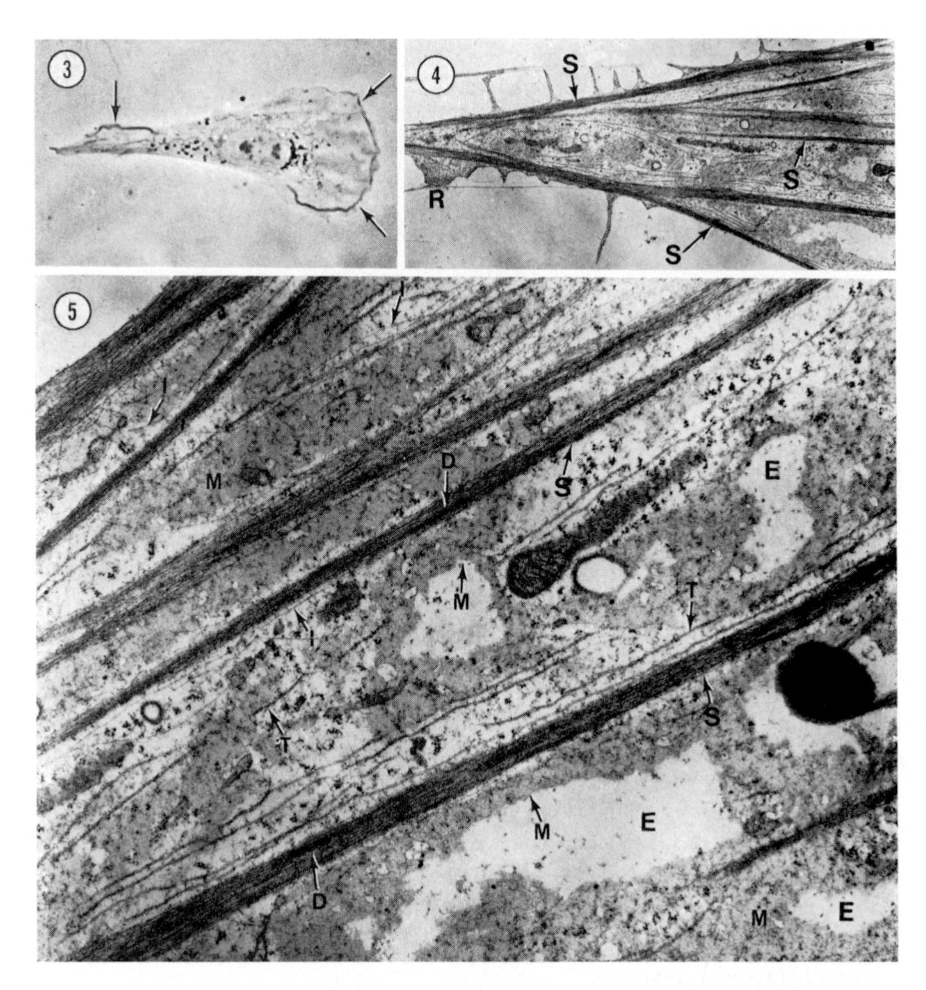

FIG. 3. A living migratory glial cell from a spinal ganglion of an eight-day-old chicken embryo. The predominant ruffling area is to the right, as is the direction of locomotion. Wave-like ruffles (arrows) are seen at the anterior surface as well as in the lateral ruffled membrane, near the tail of the cell. × 940.

FIG. 4. The tail of a cell similar to that shown in Fig. 3. The bundles of sheath filament (S) can be seen extending beneath the lateral surfaces as well as in lines parallel to the long axis of the cell (the direction of locomotion). Note the small lateral ruffle (R) at the lower left; at higher resolution a few sheath filaments splay out into that ruffle and merge with its lattice microfilaments. × 4 460.

FIG. 5. The right-hand portion of Fig. 4 at higher magnification. Extracellular space (E) beneath the cell, as well as tangentially sectioned lower cell surface (M) establish proximity of this view to the substratum side of the cell. Microtubules (T), and intermediate sized filaments (I) are seen in addition to the sheath bundles (S). Densities (D) are seen in the sheath. × 15 000.

FIG. 6. A sheath bundle beneath the lateral surface of a migratory cell with densities (D) visible (see Spooner *et al.* 1971 for spacing between densities). Higher magnification reveals distances of only 6–10 nm between sheath and plasma membrane. Areas of cell surface like this can extend outward to form lateral ruffled membrane organelles in short times, though during the process, the bulk of the sheath does not alter position (cf. Fig. 9). × 34 200.

found only in one portion of the cell, but as an organelle which may be present on the inside of all cell surfaces.

The length of individual sheath elements has not been established, although filaments as long as 1.3 μm have been measured. The great variation in length might be due to filaments traversing the plane of section. Numerous electron-dense elements perpendicular to the axis of the sheath filament bundles are seen in many preparations. It is premature to consider such structures to be cross bridges (Spooner *et al.* 1971; Goldman 1972), since the regularity of spacing characteristic of such bridges in skeletal muscle or microtubular systems has not been reported for the sheath. Since sheath microfilaments bind heavy meromyosin (Ishikawa *et al.* 1969; Ludueña & Wessells 1973; Spooner *et al.* 1973) prepared from mammalian skeletal muscle, the filaments can be presumed to be actin-like in at least this property (in the same cells, microtubules, intermediate-sized filaments and other cytoplasmic organelles show no hint of binding heavy meromyosin).

Electron-dense patches (Figs. 5 and 6) (Spooner *et al.* 1971; Goldman 1972) are sometimes periodically distributed along the bundles of sheath filaments at intervals varying between 0.6 and 1.2 μm. These patches are reminiscent of dense bodies in smooth muscle (Cooke & Fay 1972), and hence might be the equivalent of Z bands in skeletal or cardiac muscle by serving as insertion points for ends of sheath filaments. Their irregular distribution in the sheath of any one cell and apparent absence from most cells (as judged by ultrastructure) do not elucidate the function of the densities and suggest caution in using their presence as a criterion that the sheath is a contractile system.

Sheath filaments are not altered morphologically by colchicine, cytochalasin B, 10mM-cyanide, low-temperature fixation (2 °C), primary fixation in osmium tetroxide instead of glutaraldehyde, or by the glycerol extractions used for the

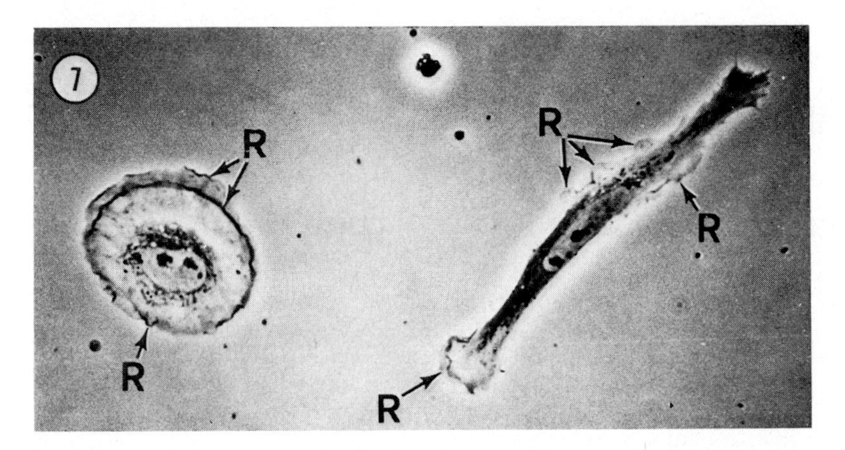

FIG. 7. Two glial cells cultured in the presence of colchicine (2.5μM) for 24 h. Electron micrographs do not reveal any intact cytoplasmic microtubules in either glial cell type. Nevertheless, the type on the right retains its refractile lateral surfaces, shows ruffling restricted to discrete areas (R), and continues normal migratory rates (Spooner *et al.* 1971). Note the wave-like ruffles (R) on the cell at the left. The ruffles originate at the edge of the cell and sweep inward toward the nucleus. At the top of this cell, a second ruffling wave is starting at the edge while an earlier one still moves inward. These sorts of 'extension' movements of cell surface are certainly not dependent upon intact microtubules. × 640.

heavy meromyosin assay (Spooner *et al.* 1973). This inertness of the sheath filaments, in contrast to the alteration induced in cytoplasmic microtubules by colchicine, makes analysis of the function of the sheath difficult. It is really only on the basis of binding with heavy meromyosin, structural analogies to smooth muscle filaments, and distribution within cells that explanatory hypotheses can even be constructed (see later). Nevertheless, the preliminary results of our work on the extraction and analysis of the sheath material promise more positive information.

In assessing the significance of the sheath, we must consider structural rigidity. As mentioned, sheath bundles are the predominant organelle located just beneath the immobile, highly refractile (when viewed with phase contrast microscopy) sides of migratory cells. In glial cells that remain elongate and continue net cell movement at normal rates despite prolonged colchicine treatment (2.5μM for 24–48 h) (Fig. 7) (Spooner *et al.* 1971), the bundles are found in their normal position. However, no intact cytoplasmic microtubules are observed but instead the 10 nm filaments typical of cells treated with colchicine are seen (Ishikawa *et al.* 1968; Spooner *et al.* 1971; Yamada *et al.* 1971). This suggests that, in the absence of ruffling, the immobility and normal

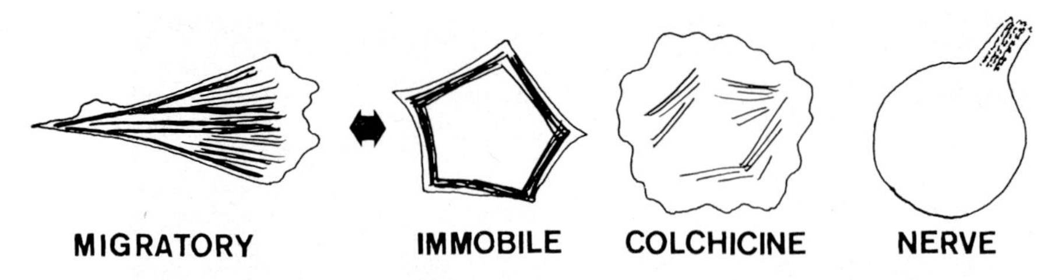

FIG. 8. Diagrammatic representation of the arrangements of sheath filament. Transition between the typical migratory and the immobile condition is reversible. Colchicine-treated cells may show active lateral ruffling (as in Fig. 7) and sheath distribution as depicted here; those that do not ruffle resemble the 'immobile' cell. In contrast to glial cells (and, in fact, other types of migratory cells), the nerve cell contains no sheath filament bundles in its body. Microtubules and neurofilaments (upper right) are seen in the axon.

morphology of lateral cell surfaces is retained where sheath filaments are present and intact microtubules are not.

Bundles of sheath filaments are the sole organelle always found within the long, peculiarly branched, and apparently rigid processes found on the surface of cytochalasin-treated single cells (Spooner *et al.* 1971). Microtubules are frequently absent from such structures and colchicine treatment has no effect upon the integrity of these processes. It is possible that they owe their stability to adhesion with the substratum; if so, the straightness and small number of sharp angles in the processes (as in Fig. 1 of Spooner *et al.* 1971) imply a heretofore unexpected arrangement of adhesive sites. For the moment, the possibility that bundles of sheath filaments have considerable rigidity must be kept in mind.

Any account of the sheath as a highly ordered contractile or skeletal system must allow for the cell's capacity for internal rearrangement (Fig. 8). Abundant observations establish that single migratory cells frequently change their direction of movement yet, when elongate migratory cells are sectioned, the predominant sheath orientation is parallel to the long axis of the cell. Hence, during change of direction due to contact inhibition or contact guidance, the sheath system must reorient. The time required for such reorientation is uncertain, but in some cases the anterior portion of a cell appears to move sideways for a short while as the shape gradually changes. The new axis of elongation can be established in 15–60 min. The reorientation of the sheath can be examined, at least indirectly, by ascertaining the arrangement of sheath filaments in colchicine-treated cells that have lost their elongate shape and in freshly plated single cells as they are beginning to spread upon the substratum and elongate before movement.

Glial cells treated with colchicine (2.5μM for 1–24 h) assume either of two gross morphologies (Fig. 7). Many cells lose elongate shape, the tail retracting into the cell body. Frequently, ruffling continues on *all* sides of such cells; ruffles have been observed (with phase contrast microscopy) around the entire periphery of the cell. Electron micrographs show no intact cytoplasmic microtubules in such cells, only typical colchicine-induced 10 nm filaments (as in Fig. 16 of Spooner *et al.* 1971). Thus the ruffling phase of cell locomotion is *not* dependent upon cytoplasmic microtubules (though restriction of ruffling to limited portions of the cell periphery may be microtubule-dependent) (Vasiliev *et al.* 1970).

In colchicine-treated cells ruffling over much of their lateral periphery, bundles of sheath filaments tend to extend tangentially and are located well beneath the ruffling regions. Serial sections have not revealed the extensive, mat-like arrays of sheath filaments such as are seen toward the anterior end of elongated migratory cells, nor is there one predominant orientation of the sheath system.

Similarly in freshly plated cells that have not yet elongated, bundles of sheath extend tangentially either beneath non-ruffling lateral cell surfaces or beneath regions containing numerous vesicles. No extensive mats of filaments are seen near the lower cell surface nor sets of parallel bundles characteristic of the tail of an elongated cell. Future observations will establish the intermediate steps between this arrangement of sheath filaments and that characteristic of elongated cells showing net movement. But it is noteworthy that every observation so far is consistent with the hypothesis that the sheath system is *not* a stable, sarcomere-like arrangement, which cannot be altered morphologically. The ability to reorient appears to be one of the basic characteristics of the sheath filament system. Further support is added by Heaysman's demonstration (pp. 187-194, this volume) that microfilaments identical to those of the sheath system appear within 20 s of the contact of an extended portion of the ruffled membrane organelle (a lamellipodium) with another cell (such contact initiates the withdrawal due to 'contact inhibition'). Moreover, the appearance of electron-dense materials at the junction of those microfilaments with the inner surface of the plasma membrane suggests that a system to link filaments to cell surface can be generated with equal or greater speed (see later).

LATTICE MICROFILAMENTS AND THE RUFFLED MEMBRANE

Microfilaments in oriented sheaths, intermediate-sized filaments and microtubules rarely extend into the anterior ruffled membrane region or into the

FIG. 9. A portion of a large lateral ruffle on a glial cell. Note that the lattice (L) filaments approach close to the inner surface of the plasma membrane (small arrows). Large arrows indicate areas where sheath (S) and lattice filaments merge together [lower plasma membrane (M); microtubules (T); intermediate filaments (I)]. × 20 000.

transient lateral ruffles found along the sides or near the tail of migratory cells. (Note that the term ruffled 'membrane' refers to the complete region of the cell where ruffles are seen, not the plasma membrane *per se* in such regions.) Instead, a network of filaments occupies the cytoplasmic volume of the ruffles (Spooner *et al.* 1971). The average diameter of individual elements of this lattice is about 5 nm and, as in the case of the sheath, the length is indeterminate. It is the rule rather than the exception that elements of the lattice do not extend far before changing direction and meeting other elements of the lattice; in this respect the lattice differs from the sheath. One can view the lattice as a three-dimensional scaffolding within the cytoplasm (Fig. 9).

The lattice filament system appears to approach close to the internal surface of the plasma membrane and to enter into it. Aggregations of electron-dense materials, such as are seen at desmosomal plaques or zonular filament insertion areas of epithelial cells, have not been reported. However, it is not safe to conclude that equivalent insertion substances are not present. The high density of tonofilament or zonular-associated microfilaments in the specialized insertion areas might be a reason for the presence of substantial quantities of junctional materials. Elements of the lattice probably approach the plasma membrane at much greater intervals, and thus might not be associated with sufficient filament–membrane junctional material to be detected by standard fixation and staining procedures. The lattice–membrane association is of critical importance to the problems of whether intracellular organelles can move the cell surface in space and, if so, how?

The disposition of the proximal portion of the lattice system within the cell is of equal importance if activities of more centrally located organelles (the sheath system?) are to be transmitted to the cell surface, or if the surface is to be anchored in any sense to intracellular structures such as microtubules, vesicles etc. It is of interest therefore to explore the lattice–sheath junction. As pointed out initially by Buckley & Porter (1967), the sheath appears to splay out as it approaches the cell surface. As seen in lateral ruffles of migratory cells, individual filaments in the sheath occasionally extend away from the main bundles and merge into the lattice region (Fig. 9). Similarly, at the proximal portion of the ruffling area in migratory cells, the sheath and the lattice merge into one another.

Lateral ruffles containing a normal complement of lattice filaments can grow in a short time (a few minutes). At a localized region along the side of a cell, the distance between the plasma membrane and the lateral sheath bundle may increase from only 6 nm to as much as several μm in such a short period. Where does the lattice material in that large intervening space originate? One possibility is that lattice components are moved into the expanded ruffle from elsewhere in the cell. Thus, as suggested by Behnke *et al.* (1971) for filamentous polymers in platelets, the lattice system may appear at localized sites in response to localized control factors. That is, the lattice in such ruffles might be derived from the transition from soluble monomers to filamentous polymers. Alternatively, the outer portion of the bundle itself might change from the more ordered sheath system to the less ordered lattice system. This change would correlate with the outward extension of the lateral cell surface. Whether such a change could generate the force required for extension outward is unknown.

LATTICE MICROFILAMENTS AND NERVE CELLS

The tip of an axon elongating on a solid substratum such as plastic or gelatin is the site of intensive ruffling and the formation, movement and disappearance of microspikes (Figs. 10–13) (Yamada *et al.* 1971). The remainder of the cell

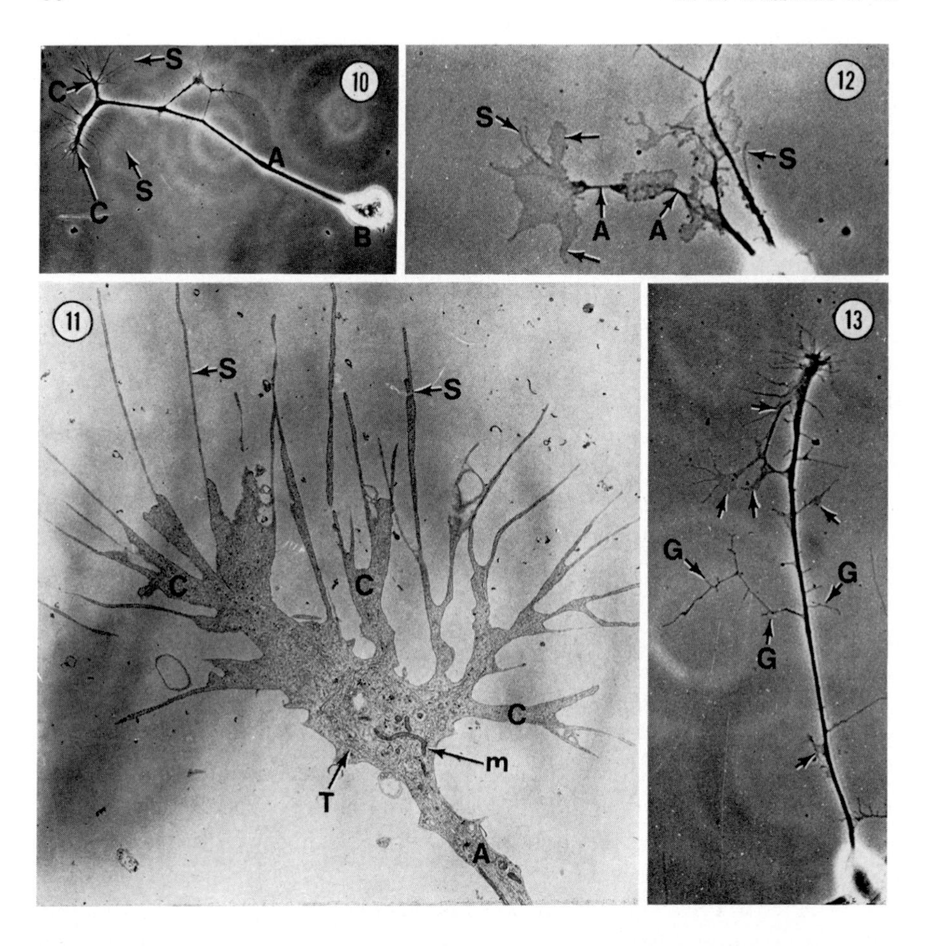

FIG. 10. A spinal ganglion nerve cell elongating on plastic in medium with 10% serum. The cell body (B) is highly refractile; the axon (A) can send out lateral microspikes (S) from any part of its surface, though most are seen here near the growth cone region (C). × 230.

FIG. 11. A typical growth cone at the tip of a long axon (A) in medium with no serum. Both the large complex microspikes (C) and smaller single ones (S) show the movements depicted in Fig. 14. Only lattice material and, occasionally, vesicles are located in these areas of surface extension and surface movement: microtubules (T); mitochondrion (m). × 10 400.

FIG. 12. Typical areas of flattening of microspikes or axonal endings in medium with 1.0% serum. Areas, as at the arrow, may rapidly form a phase-dense, cylindrical microspike: axon (A); microspikes (S). × 550.

FIG. 13. Typical microspikes on an elongating axon in medium with 1.0% serum. Flattened membrane areas (arrow) appear and disappear in short times, testifying to the dynamic nature of these cell surfaces. Common features (G) of more stable microspikes are the sharp angles along the axis of the microspike. Only plasma membrane and lattice material occupy such points of bending, suggesting that the restricted geometrical arrangement reflects some property of those systems. × 435.

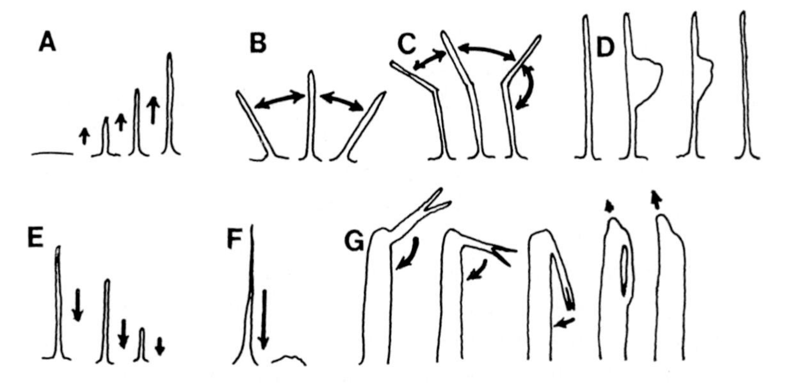

FIG. 14. Diagrams of microspike movements: A, lengthening as a stiff, straight cylinder; B, bending at the base as a stiff rod; C, bending of the tip with the proximal portion maintaining a fixed position relative to the growth cone; D, appearance and disappearance of a lateral flattened area; E, gradual shortening; F, violent retraction; G, possible fusion.

body of spinal ganglion, neural retina or motor-horn spinal-cord cells shows virtually no ruffling when the concentration of serum in the culture medium is about 5–10%. At most, microspikes protrude from such cell bodies or the sides of axons, wave about and retract into the cell body. The sides of the spinal ganglion cell are highly refractile in phase contrast microscopy, just as the immobile sides of migratory glial cells are (see later).

Microspikes along axons may be highly dynamic structures, extending outward, waving about in the medium or along the solid substratum and then retracting into the side of the axon (Fig. 14). They may also extend and apparently establish a relatively permanent adhesion with the substratum and by virtue of regular spacing along the axon serve as braces or anchoring points for the long cylindrical structure (see Fig. 10). The possibility that some lateral microspikes represent an early step in collateral axon formation should also be considered.

If the edge of the growth cone of an axon is observed, individual microspikes can be seen to grow outward as far as 25 μm in about 60 s; the time spent waving before retraction is variable. During elongation, the diameter of an individual microspike remains reasonably constant along its length at, for example, about 0.2 μm.

The extended microspikes exhibit several distinctive motions (Fig. 14). First, the microspike may bend at its base or at any position along its length, so that the distal portion can swing back and forth in space. When either the whole microspike or a portion of one moves, no undulations or sinuous waves are seen; the bodies move as rigid rods. A second type of movement is the sudden spreading of restricted portions to form thin ruffled membrane-type sheets upon the substratum (Figs. 12 and 13) within 2–5 s. This behaviour is reversible; small extended ruffling areas can rapidly become cylindrical microspikes. Such areas can retract into the microspike at any time leaving no evidence of the former condition. Finally, phase-dense particles may move outward or inward along microspikes (just as occurs more frequently within axons).

Microspikes regress in several ways (see Fig. 14). Some slowly sink back directly into the edge of the growth cone (about 6–15 μm/min). With others, the tip of the microspike appears to pull free from a firm adhesion whereupon the whole microspike collapses back into the growth cone in a few seconds (at about 70 μm/min). Another mode of retraction is fusion of the lateral surface either with other microspikes or, after bending of the microspike back toward its own cell body, with the side of the growth cone and distal axon (M. A. Ludueña, unpublished observations). These motions are particularly important in the interpretation of ultrastructure, since they imply a fluidity or plasticity of the cell cortex that must be reconciled with filaments, membrane or other organelles present at the sites of dynamic change.

Time lapse cinematography of such movements and regressions gives the impression of overall movement of surface material backwards over the surface of the growth cone. Most new microspikes are continuously generated near the front of the growth cone, bend back and disappear near the rear of the cone [precisely as described by Bray (pp. 195-209) for other types of nerve cells], similarly to the manner in which the surface of the ruffled membrane organelle of a migratory cell appears to move backward to a level over the nucleus (Harris, pp. 3-19). This movement in nerve cells must be interpreted in the light of Bray's demonstration (1970) that new axonal surface material is added in the growth cone region. Thus it can be proposed (see Bray, pp. 195-209) that during axonal elongation, a pool of surface material constantly circulates through the growth cone, moving from the rear toward the front, where it re-enters the surface itself. It then moves posteriorly in the form of microspike and plate-like ruffle surfaces. Net addition to the pool, perhaps from vesicles originating in the Golgi apparatus (Yamada *et al.* 1971), could compensate for the portion of surface continually 'left behind' in the more stable walls of the axonal cylinder.

These speculations re-emphasize the importance of analysis of the vesicular components of the growth cone. An intriguing feature of the growth cone–microspike complex is the presence of strings of flattened, elongate vesicles [identical to those described by Bray (pp. 195-209) and Bunge (1973)], which are often arranged in straight lines directly beneath the bases of adjacent microspikes or along the sides of the growth cone, as might be expected after fusion of the microspike with the cell surface. Furthermore, this flattened

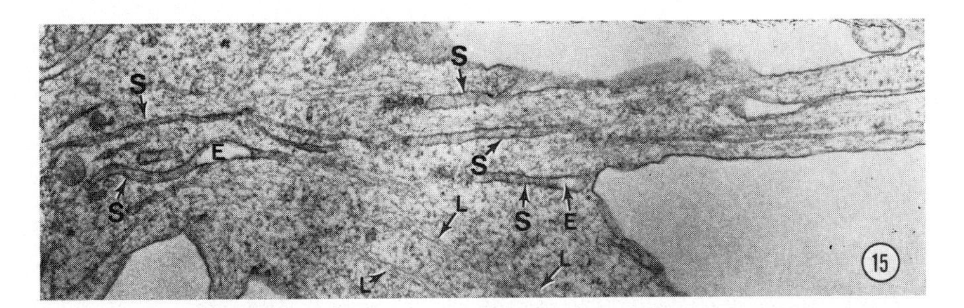

FIG. 15. An area suggestive of lateral fusion of microspikes. The flat vesicular sacs (S) extend parallel to the axes of the microspikes on the right. Possible extracellular space (E) is indicated. Asymmetric lattice areas (L) extend into the base of the very large, complex microspike that extends toward the lower right. × 22 500.

vesicular system is like that seen near the front of glial cells (Spooner *et al.* 1971), where Harris (pp. 3-19) hypothesizes addition of new surface material (see also Ingram 1969). Continuity of the lumen within such flat vesicles with extracellular spaces is indicated by the presence of Thorotrast (colloidal thorium dioxide) within the lumina after exposure of living cells for only 60 s to that electron-opaque substance (Wessells, unpublished results).

What ultrastructural features are found where microspikes are generated, move or retract? The periphery of the growth cone and core of microspikes are occupied by the same sort of microfilamentous lattice observed in the ruffled membrane organelle of migratory cells (Figs. 11, 15 and 16) (Yamada *et al.* 1971; Ludueña & Wessells 1973). No microtubules or neurofilaments extend into such regions. An invariant characteristic of microspikes is an asymmetric arrangement of lattice elements; the basic polygonal nature is deformed by the arrangement of the long axes parallel to the axis of the microspike (Fig. 16). Thus, in such areas the lattice superficially resembles the truly linear sheath filaments seen in glial cells. The regions of asymmetric lattice arrangement of each microspike can be traced for up to 2–3 μm into the cortically situated lattice of the growth cone. Frequently, elements of the lattice located at each side of the base of microspikes are closely placed, suggesting less distance between insertion points on the inner surface of the plasma membrane. The question of whether this close spacing relates to microspike generation, movement or retraction deserves investigation. Asymmetric arrangements of the lattice have been seen in the infrequent microspikes that extend from the surface of glial cells, and can also be followed for several μm into the ruffled membrane lattice. It should be emphasized that, in this respect, the oriented lattice elements of nerve growth cones do not resemble the sheath micro-

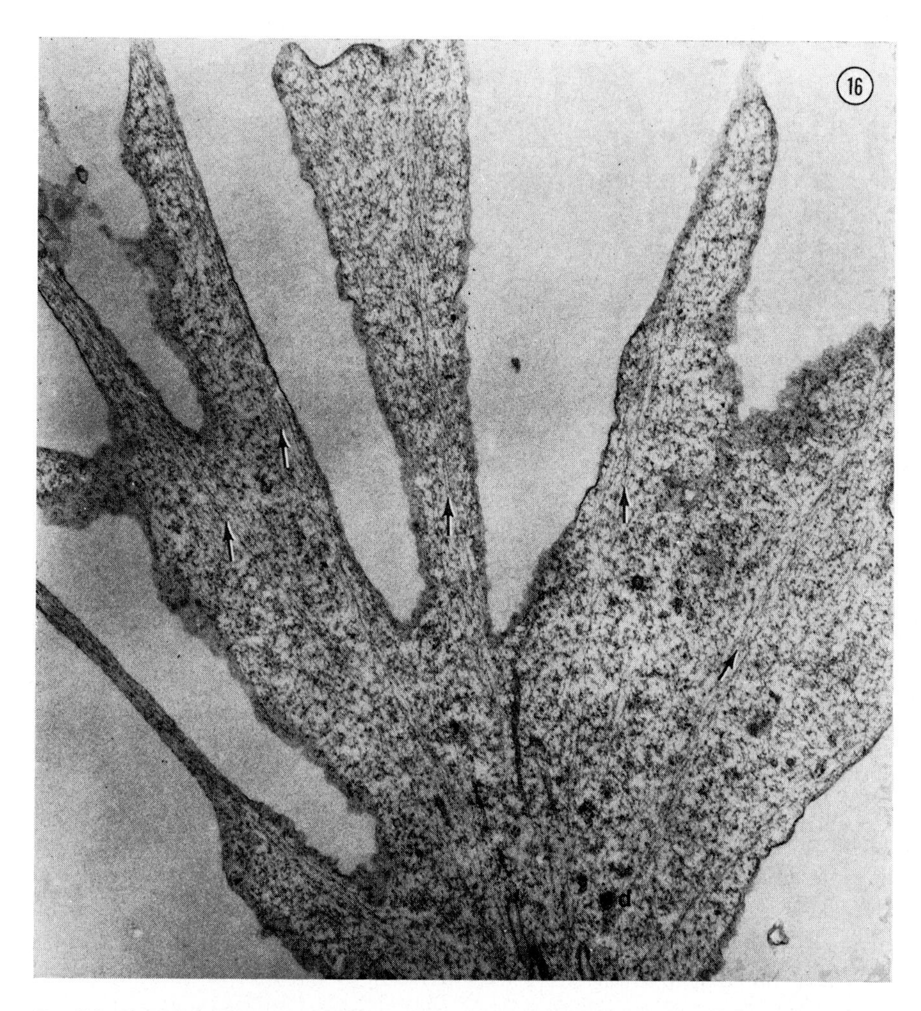

FIG. 16. The left-hand portion of the growth cone seen in Fig. 11. Asymmetries (arrows) in the lattice extend into every microspike. Note the absence of microtubules, neurofilaments and other organelles except a few flattened vesicles and a dense core (d) granule. Microspikes, as seen here, might be moving at their base, extending or retracting at the time of fixation. Few hints of the ultrastructural basic for such behaviour can be seen. × 56 250.

filaments of glial cells or those that form in leading lamellipodia of fibroblasts (Heaysman, pp. 187-194). Electron-dense amorphous materials similar to those described by Heaysman are not found beneath the plasma membrane of growth cones or microspikes. This argues against interconvertibility of sheath and lattice systems (see above). More important, the absence of sheath system and dense insertion areas from growth cones correlates with the general failure of growth cones to show 'contact inhibition'; that is to say, growth cones can move over the surface of glial cells and also of heart fibroblasts (M. A. Ludueña, unpublished observations.)

THE LATTICE: FACT OR ARTIFACT?

To what extent is the lattice an artifact? To what degree does the arrangement of its elements reflect a condition present before fixation?

The lattice material is the major fixable, electron-dense substance within the volume of the ruffled membrane and peripheral growth cone–microspike complex. If the lattice is wholly a precipitation artifact, then those volumes of cell periphery must be filled with soluble substances *in vivo* (as in Figs. 9 and 16). Whether the kinds of cell surface movement described here could be instigated by the plasma membrane and such a soluble phase is a moot point. The asymmetric array of lattice elements (Fig. 16) in all microspikes and localized areas of ruffled membranes argues that at least some differential properties in potentially fixable materials exist *in vivo;* otherwise, the whole lattice would appear homogeneous in arrangement after fixation. Furthermore, lattice material or similar structures are not observed elsewhere in the cell (within mitochondria, the lumen of the endoplasmic reticulum, etc.) where the same fixation conditions no doubt apply to soluble proteins. Finally, the highly condensed masses of material in cytochalasin-treated growth cones, or beneath the immobilized surfaces of migratory cells, raises the possibility that, owing to the action of cytochalasin, the original lattice can be fixed in an altered, collapsed form (Yamada *et al.* 1971; Spooner *et al.* 1971).

The limited tests currently available do not support the hypothesis that the lattice filament system is an artifact of the preparative procedures for electron microscopy. The interpretations of Behnke *et al.* (1971) that actin-like and myosin-like elements in blood platelets are monomeric in circulating platelets but polymeric and filamentous in platelets subjected to various physiological stresses can be applied to the lattice system. Such an extrapolation demands caution, since fixation might precipitate the soluble monomers in a pattern unrelated to their polymeric configuration in response to normal stimuli. Yet, one could argue that the monomeric system is not preserved at all by fixation, and that the lattice does indeed represent a polymeric arrangement characteristic of active cell cortices.

NERVE CELLS AND THE SHEATH SYSTEM

A striking feature of spinal ganglion cell bodies, axons, growth cones and microspikes is absence of sheath filaments (Ludueña & Wessells 1973). Serial sections parallel to the substratum and upward through the cell body, for instance, show only tangential sections of lower plasma membrane, scattered elements of the lattice system and the typical internal cytoplasmic organelles (ribosomes, endoplasmic reticulum, microtubules, etc.). The same situation is seen at the lateral and upper cell surfaces. It is of note that the lateral periphery of nerve cell bodies, which appears highly refractile in phase contrast microscopy (as in Figs. 10, 12 and 13) reflecting the characteristic 'rounded-up' condition of the soma, *lacks* the thick sheath bundles characteristic of the equally refractile, immobile sides of glial cells. Furthermore, in nerves treated with colchicine so that axons have retracted, the sides of the cell soma remain typically refractile, despite the fact that no intact cytoplasmic microtubules or sheath bundles are present. Therefore, it cannot be concluded that the refractile, immobile condition of the nerve cell periphery necessarily correlates with, or results from, the presence of either microtubules or sheath bundles. Nor is the disruption of cytoplasmic microtubules in nerve cells followed by activation of cell surface ruffling as in many migratory cells. The contrast between the nerve cell and migratory cell, with respect to microtubules, microfilaments and the degree of surface mobility, implies different means of stabilizing and moving the surfaces in the two cells.

Lowering the serum concentration of the culture media to 0–0.5% induces the lateral surfaces of nerve cell bodies to flatten onto the substratum and to extend abundant microspikes. This is observed not only with spinal ganglia, but also with cells of the neural retina and central nervous system. Thus there is the ability to change from the immobile refractile condition to the thin active one, as along the sides of elongate migratory cells (where transient local ruffles may appear or, owing to contact inhibition, whole major ruffling areas can be rapidly set up).

CELL LOCOMOTION AND AXON ELONGATION

The presence of both lattice and sheath microfilaments in migratory cells, compared with the presence of just lattice microfilaments in nerve cells, suggested a useful working hypothesis to us (Fig. 17). In culture, no net movement of nerve cell bodies occurs except by occasional 'dragging' of the cell body, which is not pertinent to the argument. Only the tip of the axon (or

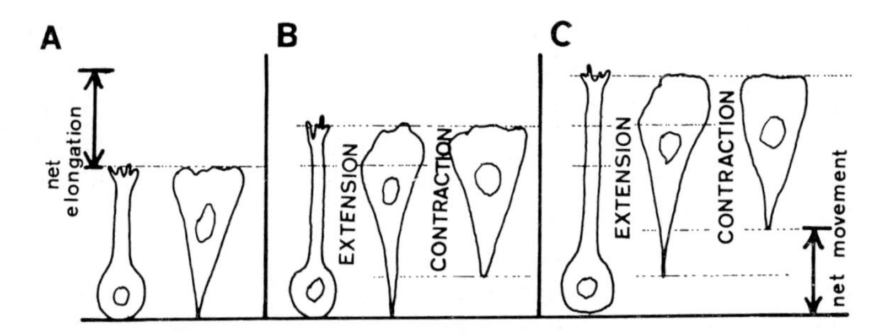

FIG. 17. The difference between axon (left) elongation and cell (right) migration is emphasized in these diagrams. The axon extends (A – C) and adds new surface material. The migratory cell also extends its anterior ruffling area, probably with membrane addition at the front too, but it is believed that a contractile step is required to gain net displacement forward. Thus the whole cell inches forward, whereas the nerve cell body remains in its original position. (The drawings cannot depict what is probably the real situation with a large number of minute extensions and surface additions in going from A to C.)

its branches) moves and functions as a locomotory organelle. New surface materials are apparently added near the tip of the axon (Bray 1970) to allow elongation. We can hypothesize that the extension phase of the locomotory cycle in such a cell is controlled by the plasma membrane–lattice microfilament complex and is concomitant with the addition of new surface material.

Extension and surface addition (Ingram 1969; Abercrombie *et al.* 1970) also are associated with the ruffled membrane organelle. But the whole cell soma is displaced forward over the substratum, either smoothly or by relatively violent movements (as when the tail seems to 'snap' free of the substratum and pulls into the cell body in a few seconds). Sheath microfilaments extend parallel to the long axis of such cells in a suitable orientation for a contractile system controlling such a displacement. The binding of heavy meromyosin by the sheath is consistent with contractility.

The basic difference between axonal elongation and single cell migration is emphasized by the behaviour of single cells from eight-day-old chick spinal ganglia on suspension in a soft agar matrix (0.33 %) prepared in normal culture medium (R. Strassman, unpublished results): none of the glial cells spread or assumed the typical flattened morphology with a single, predominant ruffled membrane. In contrast, many nerve cells produce long axons which branch and extend tortuously in three dimensions through the agar substratum. The tiny proportion of cells at the agar–cover slip interface behaves differently: the glial cells flatten and move about with typical morphology, while nerve axons extend in straight lines and show no hint of tortuous bending. This is a direct demonstration of axon elongation in a matrix that will not support locomotion of a single cell [as also seen by Weston & Kaplan (1973) for outgrowth from intact spinal ganglia]. This may be because agar is not sufficiently adhesive to permit the generation of forces required to move the cell body forward; agar is sufficiently adhesive to allow extension and surface addition in the growth cone so that axonal elongation can occur.

The idea of two-step locomotion (i.e. extension and surface addition) in nerves as distinct from three-step locomotion (i.e. extension, surface addition and contraction) in migratory cells is compatible with Seravin's belief (1971) that glycerinated cells *cannot* perform the extension phase of amoeboid movement, but only manage the contractile process, which is activated by ATP. Whether this means that the contractile system (the sheath?) is distinct chemically from the extension system (the lattice?) can only be established by further investigation.

It is not clear whether or how the term 'contractile' should be applied to the activity associated with moving the cell body forward behind the extending ruffles, though most reviews (for example Ingram 1969) conclude that some contractile event must occur during locomotion. As a result of our knowledge of contractile protein *systems* in differentiated muscle cells, contractility generally connotes sliding filaments. The recent demonstration of 'contractility' based on calcium-induced alterations of state in a polymeric system in ciliates (Weis-Fogh & Amos 1972) shows that we have much to learn about alternatives to sliding filament systems of skeletal muscle. On the one hand, the evolutionary specialization seen in molluscan 'catch' muscles provides a model pertinent to the sheath of mobile or immobile single cells as well as to the microfilament bundles of morphogenetically active epithelial cells. The contracted state in the non-striated catch muscle is stabilized by paramyosin (tropomyosin A) so that expenditure of energy is not required to maintain tension (Laki 1971; Millman & Elliott 1965). Thus a 'holding' condition characterized by considerable rigidity might be maintained by sheath acting as a catch muscle, at no great cost to the cell. On the other hand, nothing we know about the sheath filament system contradicts the hypothesis that it acts as an elastic structure, which when placed under strain due to extension forward and adhesion of the cell surface ultimately causes the rupture of more posterior points of adhesion with the substratum. Whatever the precise action of sheath filaments, if those structures help to move the cell body forward during locomotion, they must be coupled in some way to the cell surface. This is the reason that the observations of Heaysman (pp. 187-194) showing rapid generation of sheath–plasma membrane association in lamellipodia are so important.

The arguments concerning the concept of 'contractility' must also be applied to the other types of cell surface movement discussed in this paper. Extension of the surface outward, in particular, is difficult to reconcile with any simple sliding of filaments in the lattice. Localized polymerization and depolymerization of filaments [as proposed for platelets by Behnke *et al.* (1971)], change in state (for example, equivalent to adding a paramyosin-type catch mechanism) and change in relative arrangement of elements are all potential mechanisms of lattice action. Wohlfarth-Bottermann has demonstrated (p. 246) that an actomyosin gel capable of contracting to half its original length has an ultrastructure characterized by only scattered, seemingly disconnected, filaments, which observation emphasizes that 'contraction' can be generated by a system other than the linear sheath microfilaments. Therefore, the failure to demonstrate long, actin-like filaments (as the sheath system) in the leading edge of moving cells [as at the front of enveloping layer cells during epiboly (Trinkaus, pp. 363-365)] says nothing about the presence or absence of a system capable of contraction or other higher order processes. Furthermore, the absence of linear sheath-type microfilaments from ruffles, microspikes and areas of cell surface showing extension, coupled with the appearance of such microfilaments just before a massive retraction (Heaysman, pp. 187-194), suggests that the more highly ordered sheath arrangement is not required by cells in much of the locomotory cycle.

One can only conclude from studying the changing shapes of ruffles, microspikes and extended regions of the cell that molecular rearrangement in the plane of the cell surface is a basic feature of locomotion. The fluid–mosaic model of the plasma membrane is compatible with the sorts of relative movement that must exist in the plane of the membrane. Furthermore, the lattice–membrane junction cannot be composed of static, fixed points of 'insertion', but must be viewed as capable of rapid rearrangement of lattice elements relative to the membrane (as, for instance, when flat expanded ruffled membranes suddenly protrude from the sides of thin cylindrical microspikes). An advantage of the fluid–mosaic model is that it eliminates important difficulties encountered by models where the 'membrane' moves over cylindrical microspikes or flat, plate-like ruffles (cf. Bray, pp. 366-369). The possibility that the lipid bilayer is relatively immobile and that proteins (lattice insertion points?) move through it (see Huxley, pp. 349, 350) would explain how the 'membrane' so moves. Furthermore, if the presence of specialized junctions (cf. the electron-dense patches seen by Heaysman, pp. 187-194), inhibits movement of proteins through the lipid bilayer, then a simple explanation for 'contact inhibition' of locomotion can be advanced: contact of the leading lamellipodium of a cell with the surface of another cell results in the generation of such surface specializations. Consequently, flow in the plane of the membrane (a correlate of forward motion) is inhibited, and insertion sites are provided for the linear microfilaments that act to pull back the anterior lamellipodium away from its contact with the other cell.

CONCLUSION

Three distinct processes of cell locomotion can be conceived: (i) extension, movement and adhesion of cell surfaces; (ii) addition, deletion and re-arrangement of surface materials; and (iii) displacement of the cell body in migratory cells (but not nerves). Though the functions of the microfilamentous lattice and sheath in these events remain speculative, it seems most likely that those organelle systems are crucial in cell surface movements and locomotion. The dynamic nature of the cell surface implies a plasticity in the membrane and membrane-linked filament systems that must receive strong emphasis in any explanatory hypothesis of cell locomotion. All these processes must interact in the simple (for the cell) yet complex (for the biologist) activity of cell locomotion.

ACKNOWLEDGEMENT

We thank Kenneth M. Yamada for helpful discussion of this manscript. Work reported here was supported by grants HD04708 (NIH) and GB-31934 (NSF) (to N. K. W.) and GM19289 (NIH) (to B. S. S.).

References

ABERCROMBIE, M., HEAYSMAN, J. E. M. & PEGRUM, S. M. (1970). *Exp. Cell Res.* **62**, 389-398
BEHNKE, O., KRISTENSEN, B. I. & NIELSEN, L. E. (1971). *J. Ultrastruct. Res.* **37**, 351-369
BLUEMINK, J. G. (1971). *Z. Zellforsch. Mikrosk. Anat.* **121**, 102-126
BOYDE, A., WEISS, R. A. & VESELÝ, P. (1972). *Exp. Cell Res.* **71**, 313-326
BRAY, D. (1970). *Proc. Natl. Acad. Sci. U.S.A.* **65**, 905-910
BUCKLEY, I. K. & PORTER, K. R. (1967). *Protoplasma* **64**, 349-380
BUNGE, M. B. (1973). *J. Cell Biol.* **56**, 713-735
BYERS, B. & ABRAMSON, D. H. (1968). *Protoplasma* **66**, 417-435
CARTER, S. B. (1965). *Nature (Lond.)* **208**, 1183-1187
CLONEY, R. A. (1966). *J. Ultrastruct. Res.* **14**, 300-328
CLONEY, R. A. (1969). *Z. Zellforsch. Mikrosk. Anat.* **100**, 31-53
COOKE, P. H. & FAY, F. S. (1972). *J. Cell Biol.* **52**, 105-116
CURTIS, A. S. G. (1967) *The Cell Surface: its molecular role in morphogenesis*, Academic Press, New York
FRYE, L. D. & EDIDIN, M. (1970). *J. Cell. Sci.* **7**, 319-335
GINGELL, D. (1970). *J. Embryol. Exp. Morphol.* **23**, 583-609
GOLDMAN, R. D. (1972). *J. Cell Biol.* **52**, 246-254

INGRAM, V. M. (1969). *Nature (Lond.)* **222**, 641-644
ISHIKAWA, H., BISCHOFF, R. & HOLTZER, H. (1968). *J. Cell Biol.* **38**, 538-555
ISHIKAWA, H., BISCHOFF, R. & HOLTZER, H. (1969). *J. Cell Biol.* **43**, 312-328
KALLIO, D. M., GARANT, P. R. & MINKIN, C. (1971). *J. Ultrastruct. Res.* **37**, 169-177
LAKI, K. (1971) *Contractile Proteins and Muscle*, Marcel Dekker, New York
LUDUEÑA, M. A. & WESSELLS, N. K. (1973). *Dev. Biol.*, **30**, 427-440
MARCHESI, V. T. & PALADE, G. E. (1967). *J. Cell Biol.* **35**, 385-404
MARCHESI, V. T. & STEERS, E., JR. (1968). *Science (Wash. D.C.)* **159**, 203-204
MCNUTT, N. S., CULP, L. A. & BLACK, P. H. (1971). *J. Cell Biol.* **50**, 691-708
MILLMAN, B. M. & ELLIOTT, G. F. (1965). *Nature (Lond.)* **206**, 824-825
PINTO DA SILVA, P. (1972). *J. Cell Biol.* **53**, 777-787
RAPPAPORT, R. (1969). *Exp. Cell Res.* **56**, 87-91
SCHROEDER, T. E. (1972). *J. Cell Biol.* **53**, 419-434
SERAVIN, L. N. (1971). *Adv. Comp. Physiol. Biochem.* **4**, 37-111
SINGER, S. J. & NICOLSON, G. L. (1972). *Science (Wash. D.C.)* **175**, 720-731
SPOONER, B. S., YAMADA, K. M. & WESSELLS, N. K. (1971). *J. Cell Biol.* **49**, 595-613
SPOONER, B., ASH, J. F., WRENN, J. T., FRATER, R. B. & WESSELLS, N. K. (1973). *Tissue Cell* **5**, 37-46
TILNEY, L. G. & MOOSEKER, M. (1971). *Proc. Natl. Acad. Sci. U.S.A.* **68**, 2611-2615
VASILIEV, JU. M., GELFAND, I. M., DOMNINA, L. V., IVANOVA, O. YU., KOMM, S. G. & OLSHEVSKAJA, L. V. (1970). *J. Embryol. Exp. Morphol.* **24**, 625-640
WEIS-FOGH, T. & AMOS, W. B. (1972). *Nature (Lond.)* **236**, 301-304
WESTON, J. A. & KAPLAN, R. A. (1973). *Dev. Biol.*, in press
WOLPERT, L. & GINGELL, (1968). *Symp. Soc. Exp. Biol.* **22**, 169-198
YAMADA, K. M., SPOONER, B. S. & WESSELLS, N. K. (1971). *J. Cell Biol.* **49**, 614-635

Discussion

Abercrombie: Isn't it true that the bodies of nerve cells can be pulled along when their axons advance? In other words tension is developed in the axon.

Wessells: That is absolutely correct.

Abercrombie: Does this not indicate that the axon has some contractile system of the same sort as the fibroblast rather than merely a mechanism of protrusion?

Wessells: Permanent microspikes are frequently observed extending laterally from axons or from inactive growth cones. Time lapse films show that when the central portion of a microspike lifts free of the substratum, the microspike straightens out in seconds. This implies a great deal of tension between the base and the tip of a microspike. The major points of adhesion appear to be at the tips of certain microspikes, which electron micrographs reveal as expanded, round, disc-like areas. It is important to emphasize that when a microspike straightens out violently and rapidly, there is *no* reason to assume the tension involved is contractile. It could be due to elasticity. Analogous tension and elasticity could 'move' nerve cell bodies behind an elongating axon.

Trinkaus: DiPasquale (1972), studying epithelial cells in culture, has observed microspikes whose ultrastructure appears to be identical to that which you have described. He has also observed, from time lapse films, that the cells somehow use microspikes in the process of locomotion. Have you seen anything like this?

Wessells: Although we have never seen that in glial cells, we have seen what I believe is the equivalent in nerves (Ludueña & Wessells 1973) when the side of a cylindrical microspike spreads laterally to form a large flattened sheet (see pp. 67, 68 and Figs. 12–14). The cell surface can undergo fantastic deformations and rearrangements in association with microspike systems.

Huxley: In the microspikes in which you have the rather well-oriented filaments can you see anything on staining with heavy meromyosin?

Wessells: We have been extracting nerves with glycerol for over 18 months and have never made any satisfactory preparations in which filament–plasma membrane insertion points are preserved. Therefore, though aggregates of filaments binding heavy meromyosin are seen in nerve cells, we do not know to what those filaments correspond in an unextracted cell (Ludueña & Wessells 1973; Spooner *et al.* 1973).

Huxley: When you put heavy meromyosin on the sheath filaments, can you see any indication of 'arrowheads' with a preferred orientation?

Wessells: In glial cells, sheath filaments appear to be arranged randomly; there is no preferred orientation of the arrowheads. Ishikawa *et al.* (1969) have reported the same thing for chondrocytes.

Wohlfarth-Bottermann: Do you think that the fibrillar system of filaments, the 'sheath system', and the lattice system are different? You proved that the filaments on the sheath system are F-actin by binding with heavy meromyosin. In acellular slime moulds it has been shown that the 5–8 nm filaments, identical with F-actin (Alléra *et al.* 1971), are aggregated in the fibrils and are in the non-aggregated form in the groundplasm. While the function of the non-aggregated form is chiefly contractile, that of the fibrillar state might be to transmit tension over longer distances.

Wessells: I don't think there is any way yet of knowing whether they are one system in two forms or two distinctive systems.

Trinkaus: I think Dr Wessells' hesitancy about whether cell shortening is due to contraction is well advised. One constantly notices the use of the term 'contraction' in the literature when, I think, all we have a right to say is that the cell shortens. The shortening of a cell could as well be due to elastic recoil from a stretched state (see Francis & Allen 1971).

Wohlfarth-Botterman: I always thought that microtubules could also cause these shortenings.

Porter: Yes, I wondered why Dr Wessells didn't point out the microtubules—they were so obvious in the pictures! I am sure that some people would have interpreted the evidence in terms of microtubules providing a surface for the fibrillar system to move over as the cell advances. The thicker filaments are, of course, associated with the deeper parts of the cytoplasm.

Wessells: We are not certain what the microtubules do. There is no evidence to indicate that they or the deep-lying 10 nm diameter tonofilaments function in either ruffling or the activity of the growth cone. Very few microtubules are near the sheath system. Furthermore, ruffling continues for over 24 h in concentrations of colchicine (2.5μM) that disrupt all cytoplasmic microtubules. Thus one would suspect that net transport of membrane material through the cytoplasm to the front ruffling edge of the cell [as implied by Bray, Harris and others (see p. 75)] would not be dependent upon microtubules. I must re-emphasize that glycerination disrupts the lattice–plasma membrane junction. Therefore, we have no proof that the lattice filaments bind heavy meromyosin (for complete discussion see Spooner *et al.* 1973).

Allison: One of the big problems, it seems to me, in these primitive contractile systems is that, unlike the well-ordered arrangements of filaments in muscle, the constituent proteins are probably in a state of reversible transition from the globular to the filamentous form all the time. There are some striking examples of this. In thin sections of freshly prepared blood platelets, very few filaments are observed but if the platelets are activated, for example with ADP, enormous numbers of filaments develop. Pollard & Korn (1971), using preparations of amoeba cytoplasm stimulated by ATP, obtained similar results. These observations suggest that the proteins are in a metastable state and interconvert readily. Presumably this is important in adapting from the sheath system, which you have described, to the loose reticular system of microfilaments.

Wessells: From their results, Behnke *et al.* (1971) have suggested that platelet filament systems are easily modifiable so that linear ones appear only under limited conditions.

Goldman: Using an improved glycerination procedure with neuroblastoma cells we have been able to demonstrate a continuous submembranous network of F-actin–heavy meromyosin complexes (decorated thin filaments) which runs from the cell body and along axons to the nerve endings (Chang & Goldman, unpublished observations). However, using the same techniques on embryonic chick spinal ganglion cells, we have not yet seen such complexes.

Wessells: I don't think there is any doubt that they are there, but the areas of filaments which show the binding are not associated with plasma membrane.

Goldman: Originally you proposed that the sheath of microfilaments became thickened into microfilamental bundles in some regions of the cell. Is this what you see at the rear end of a moving cell or do you see bundles of microfilaments without the sheath?

Wessells: We believe that the bundles of filaments found posteriorly in a cell splay out towards the anterior end of the cell. Serial sections establish the presence of a sheath system just beneath the dorsal plasma membrane (see Ludueña & Wessells 1973).

Wohlfarth-Bottermann: I think the generation of the fibrils has something to do with the G- to F-actin transformation. In preliminary results, H. Hinssen (unpublished findings, 1972) showed that there is a factor in the groundplasm preventing the polymerization of G- to F-actin. This factor can be inactivated by high concentrations of magnesium chloride. Further, we looked for different degrees of contraction of the fibrils by trying to find distinct states of contraction on one side and of relaxation on the other (Fig. 1).

It seems that the fibril itself is in the contracted state and that the relaxed system is non-fibrillar (Wohlfarth-Bottermann, unpublished results, 1972). During relaxation the fibrils depolymerize. Whether a polymerization of F-actin is involved in relaxation is a matter of speculation.

Allison: It is unlikely that actin alone can cause contraction. Pollard & Korn (1971) have separated the two types of filaments in the cytoplasm of the amoeba. The polymerized form of F-actin does not contract by itself, but only when it is recombined with the myosin component.

Wohlfarth-Bottermann: Yes, it is clear that only actin *and* myosin can contract, but I was discussing only the role of actin. Of course, for a contraction, you must have the actomyosin system.

Wolpert: I did not understand the distinction you were drawing between the two types of cells, since you said that ultrastructurally the growth cone was the same as the ruffled membrane area of a glial cell.

Wessells: The reason I made the distinction is important. We have no experimental probe of the sheath system: it is resistant to several treatments (see pp. 60, 61). Therefore, we looked for cells lacking the sheath system. Such are sensory ganglion nerve cells, and the striking and unexpected finding is that such sheath-less cells *do not move* their cell bodies by gliding cell locomotion.

It is interesting that the soma of the nerve cell can respond to elevated calcium or lowered serum concentrations by flattening and forming microspikes (see p. 72). Thus, though such behaviour is never normally associated with the refractile, rounded cell bodies of nerve, such areas have the capacity to flatten and perhaps ruffle.

Dunn: Lamont & Vernon (1967) claim that, in the migration of nerve cell bodies from chick dorsal root ganglia *in vitro*, contraction of the leading nerve process often pulls the cell body from the explant.

Wessells: Yes, the rounded cell body of a nerve can apparently be dragged

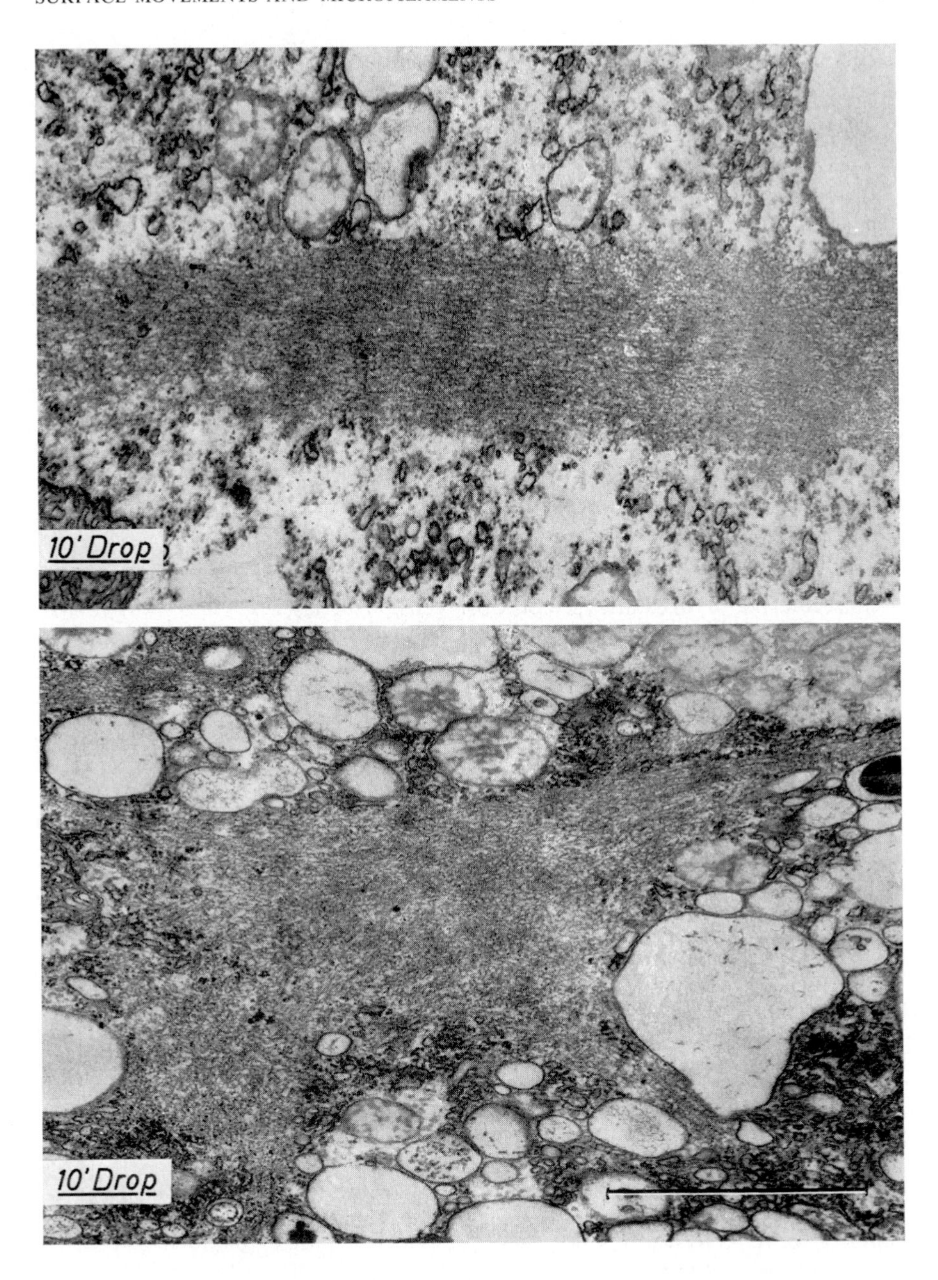

FIG. 1 (Wohlfarth-Bottermann). Fine structure of the actomyosin fibrils of *Physarum* after application of contraction solution (upper part) and relaxation solution (lower part) for 2.5 min (K. E. Wohlfarth-Bottermann, unpublished findings, 1972).

over a substratum. My point is that the body does not perform gliding movement itself (ruffling, contractions etc.).

Dunn: The nerve cell bodies can migrate for considerable distances by extension, adhesion and contraction of their processes.

Abercrombie: On this point of extension, there have been two proposals which are not necessarily mutually exclusive: one is extension by surface assembly, as outlined by Dr Harris, and the other is Dr Wessells' idea of extension by means of the lattice. Surface assembly necessitates a forward flow of material within the cell, which might be brought about by a squeezing (as a result of contracting the back end) or by microtubule transport. Such a forward flow of material, but including a flow of surface material at the surface without new surface assembly, is another possibility.

Wessells: I must emphasize again that extension in the form of active ruffling and microspike formation can go on for hours in the absence of intact cytoplasmic microtubules.

Trinkaus: I have often wondered whether extension might be caused in part by the internal hydrostatic pressure of cells. Harris (1973) has shown that blebbing can be prevented if internal hydrostatic pressure is reduced by adding sorbitol to the medium. It would be of interest to test the effect of sorbitol on the forward extension of the leading edge.

References

ALLÉRA, A., BECK, R. & WOHLFARTH-BOTTERMANN, K. E. (1971) Weitreichende fibrilläre Protoplasmadifferenzierungen und ihre Bedeutung für die Protoplasmaströmung. VIII. Identifizierung der Plasmafilamente von *Physarum polycephalum* als F-Actin durch Anlagerung von heavy meromyosin *in situ*. *Cytobiologie* **4**, 437-449

BEHNKE, O., KRISTENSEN, B. I. & NIELSEN, L. E. (1971) Electron microscopy of actinoid and myosinoid filaments in blood platelets. *Ultrastruct. Res.* **37**, 351-369

DIPASQUALE, A. (1972) *An Analysis of Contact Relations and Locomotion of Epithelial Cells*, Ph. D. Dissertation, Yale University, New Haven, Connecticut, USA

FRANCIS, D. W. & ALLEN, R. D. (1971) Induced birefringence as evidence of viscoelasticity in *Chaos carolinensis*. *J. Mechanochem. Cell Molitity* **1**, 1-6

HARRIS, A. K. (1973) Ruffling, blebbing, and surface flow in fibroblast locomotion., in press

ISHIKAWA, H., BISCHOFF, R. & HOLTZER, H. (1969) Formation of arrowhead complexes with heavy meromyosin in a variety of cell types. *J. Cell Biol.* **43**, 312-328

LAMONT, M. D. & VERNON, C. A. (1967) The migration of neurons from chick embryonic dorsal root ganglia in tissue culture. *Exp. Cell Res.* **47**, 661-662

LUDUEÑA, M. A. & WESSELLS, N. K. (1973) Cell locomotion, nerve elongation, and microfilaments. *Dev. Biol.* **30**, 427-440

POLLARD, T. D. & KORN, E. D. (1971) Filaments of *Amoeba proteus*. II. Binding of heavy meromyosin by thin filaments in motile cytoplasmic extracts. *J. Cell Biol.* **48**, 216-219

SPOONER, B. S., ASH, J. F., WRENN, J. T., FRATER, R. B. & WESSELLS, N. K. (1973) Heavy meromyosin binding to microfilaments involved in cell and morphogenetic movements. *Tissue Cell* **5**, 37-46

Fibrillar systems in cell motility

ROBERT D. GOLDMAN, GERMAINE BERG, ANNE BUSHNELL, CHENG-MING CHANG, LOIS DICKERMAN, NANCY HOPKINS*, MARY LOUISE MILLER, ROBERT POLLACK* and EUGENIA WANG

*Department of Biology, Case Western Reserve University, Cleveland, Ohio and * Cold Spring Harbor Laboratory, Cold Spring Harbor, New York*

Abstract Several types of cytoplasmic fibres are found in various cultured fibroblastic and epithelial-like cells (including the BHK-21 and BSC-1 cell lines): microtubules (about 25 nm in diameter), filaments (about 10 nm in diameter) and microfilaments (about 6 nm in diameter). The distribution, coordinated activities and the dynamic nature of these fibres suggest that they function in several aspects of cell motility, including cell spreading, locomotion and changes in cell shape during the normal life cycle.

Experiments with cycloheximide indicate that pools of precursors, which might be used to assemble fibres, are available in cells during cell spreading. Colchicine treatment demonstrates that microtubules are important in the formation and maintenance of the shape of fibroblastic cells, but not that of epithelial cells. The use of cytochalasin B and its interactions with the cell surface are also discussed.

One of the characteristics of the ultrastructure of cultured cells is the abundance of cytoplasmic fibres, such as microtubules, microfilaments and filaments (Goldman & Follett 1969; Spooner *et al.* 1971; Goldman & Knipe 1973). Microtubules (about 25 nm in diameter) are part of a cytoskeletal system participating in many motile processes, for example, in the formation and maintenance of the shape of fibroblastic cells (Goldman 1971; Vasiliev *et al.* 1970) and the elongation of cultured nerve axons (Yamada *et al.* 1971). Filaments (about 10 nm in diameter) are connected with the shaping of a cell during normal cell spreading (Goldman & Follett 1970; Goldman & Knipe 1973), with the intracellular transport of organelles in fibroblasts (Goldman & Follett 1969) and with axoplasmic transport in neurons (Huneeus & Davison 1970). Microfilaments (about 6 nm in diameter) are thought to operate directly in membrane ruffling, cell locomotion, pinocytosis and cytokinesis (Goldman & Follett 1969; Schroeder 1970; Wessells *et al.* 1971; Goldman & Knipe 1973).

Although the literature provides much information relating the presence of cytoplasmic fibres to cell motility, relatively little is known about the specific function of each type of fibre. A major obstacle to the furtherance of our knowledge is the scarcity of suitable tools to investigate the *in vivo* function of each type of fibre. Colchicine, which results in a reversible loss of microtubules in cultured cells (Goldman 1971; Vasiliev *et al.* 1970), and cytochalasin B, which reversibly inhibits many aspects of cell motility and acts either directly or indirectly at the level of microfilaments (Wessells *et al.* 1971; Goldman & Knipe 1973), are two of the more useful tools. We shall describe our experiments with colchicine and cytochalasin B and discuss several other approaches which we have made to elucidate the functions of the fibres. We shall also discuss the coordinated activities of the different fibres during the motility of cultured BHK-21 fibroblasts and BSC-1 epithelial cells.

CELL SPREADING AND SHAPE

BHK-21 and BSC-1 cultures are initiated with a suspension of single cells usually obtained from a growing culture by trypsinization (Follett & Goldman 1970). Suspended BHK-21 and BSC-1 cells appear spherical and indistinguishable from each other by phase contrast microscopy. In order that the cells may go through their normal growth cycle, they are placed on glass or plastic substrates in medium at 37 °C. After 15–30 min most cells adhere to the substrate and within a few hours BHK-21 cells are no longer spherical but assume the shape of a fibroblast (Goldman & Follett 1970; Goldman & Knipe 1973) and BSC-1 cells exchange their sphericality for an epithelial cell shape (Fig. 1a–c).

The distribution of all three fibres during attachment and spreading of trypsinized BHK-21 cells has been studied by light and electron microscopy (Goldman & Follett 1970; Follett & Goldman 1970; Goldman & Knipe 1973). Just after attachment to the substrate the cells contain a mass of filaments which appears as a birefringent sphere next to the nucleus. The very few microtubules are randomly scattered through the cytoplasm in these round cells, and the few microfilaments present are found in microvilli at the surface.

As the cells spread into a fibroblastic shape, the birefringent zone becomes smaller owing to the redistribution of filaments along the longitudinal axes of the asymmetric cellular processes (Goldman & Follett 1969, 1970). Simultaneously, large numbers of microtubules and submembranous bundles of microfilaments assemble along the length of these processes (Goldman & Knipe 1973).

A similar sequence of events is seen in BSC-1 cells. Shortly after attachment

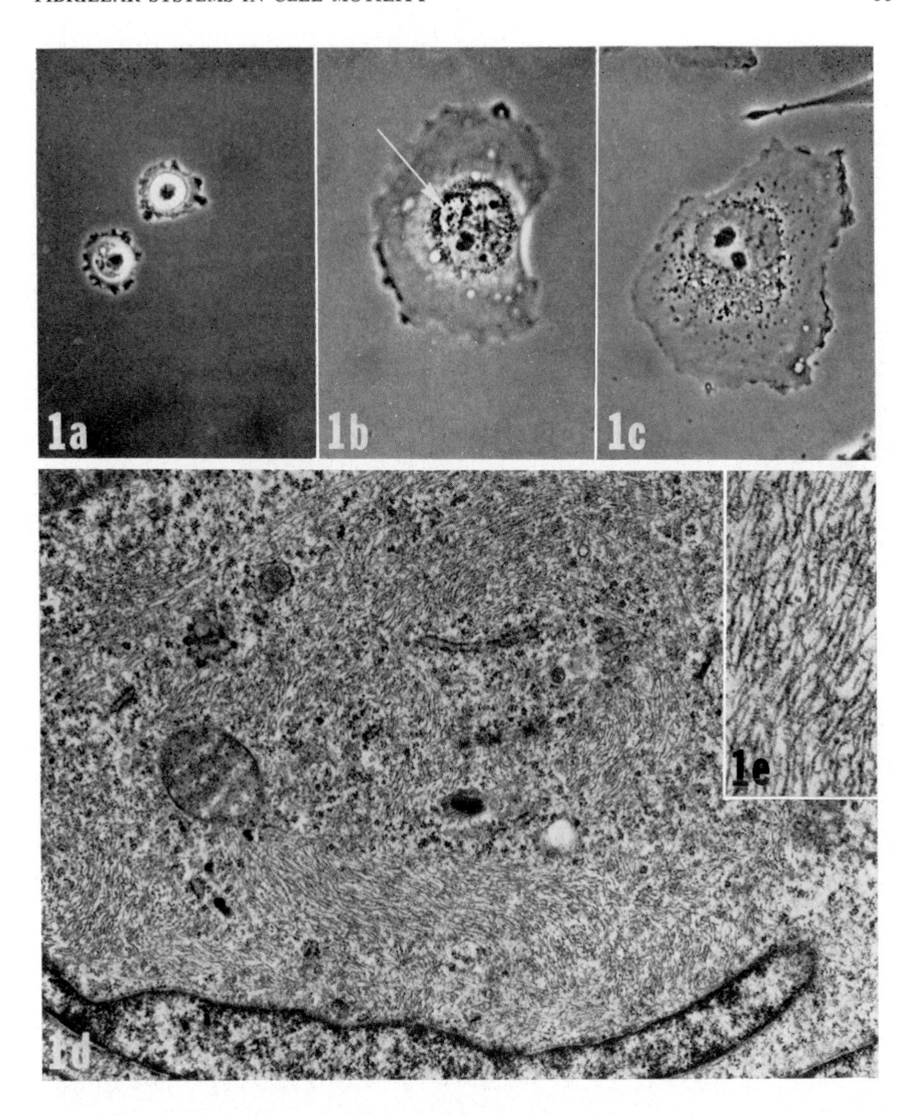

FIG. 1. (a) BSC-1 cell 30 min after being placed on a glass coverslip. Ruffling is beginning at the periphery of the cell; (b) BSC-1 cell after 1 h. Note clear region adjacent to nucleus (arrow); (c) BSC-1 cell after 3 h. Epithelial cell shape is obvious. (a)–(c) Phase contrast, × 480; (d) electron micrograph of a thin section through a clear region similar to the one seen in (b), masses of filaments are seen. × 16 000; (e) part of (d) at higher magnification showing 10 nm filaments. × 38 500.

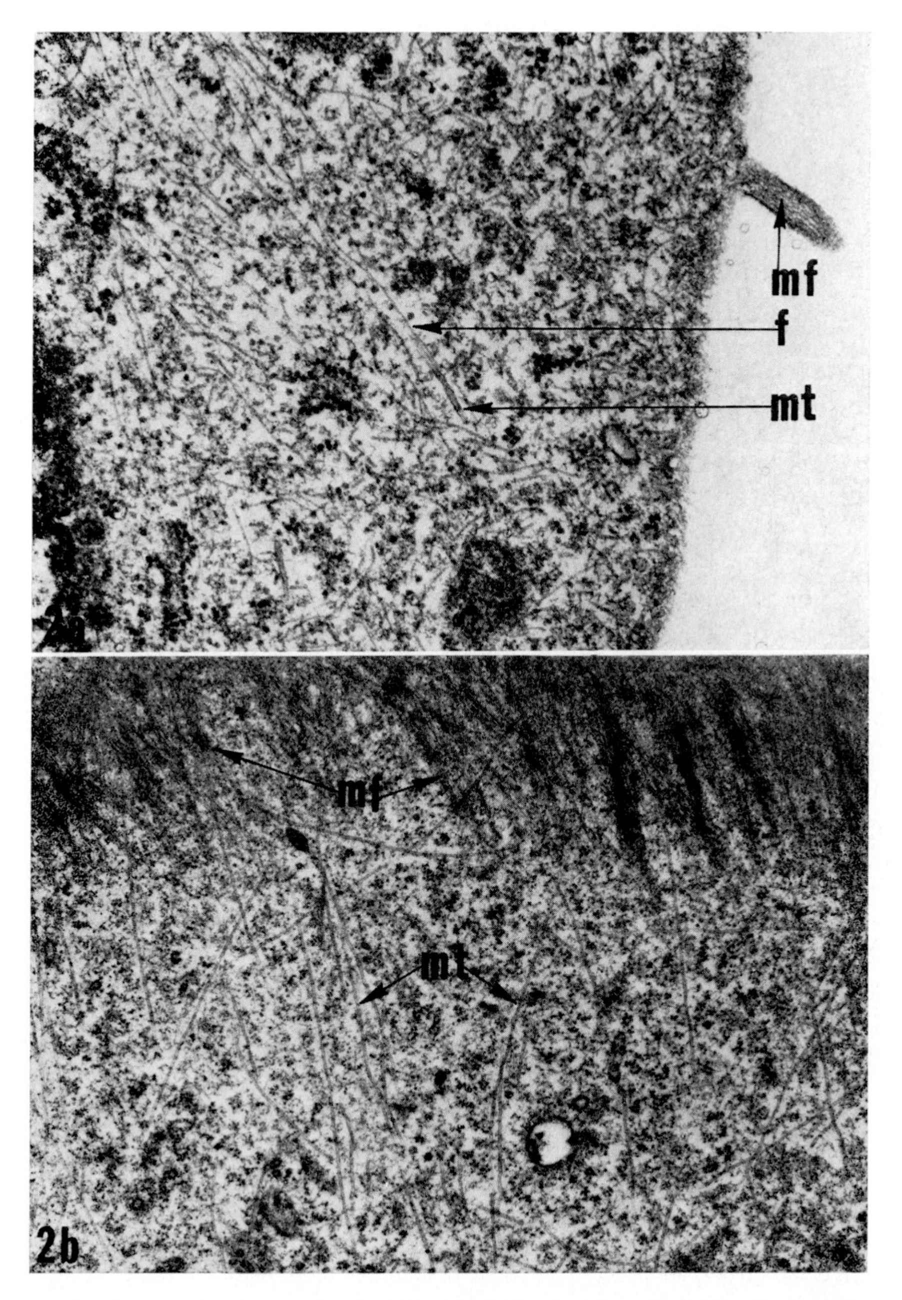

FIG. 2. (a) Electron micrograph of a BSC-1 cell 30 min after attachment to a substrate showing near the nucleus short, distinct microtubules (mt), filaments (f) and microfilaments (mf) localized in surface microvilli. × 32 000; (b) electron micrograph of a well-spread BSC-1 cell. Bundles of microfilaments (mf) are evident as well as long microtubules (mt). × 20 000.

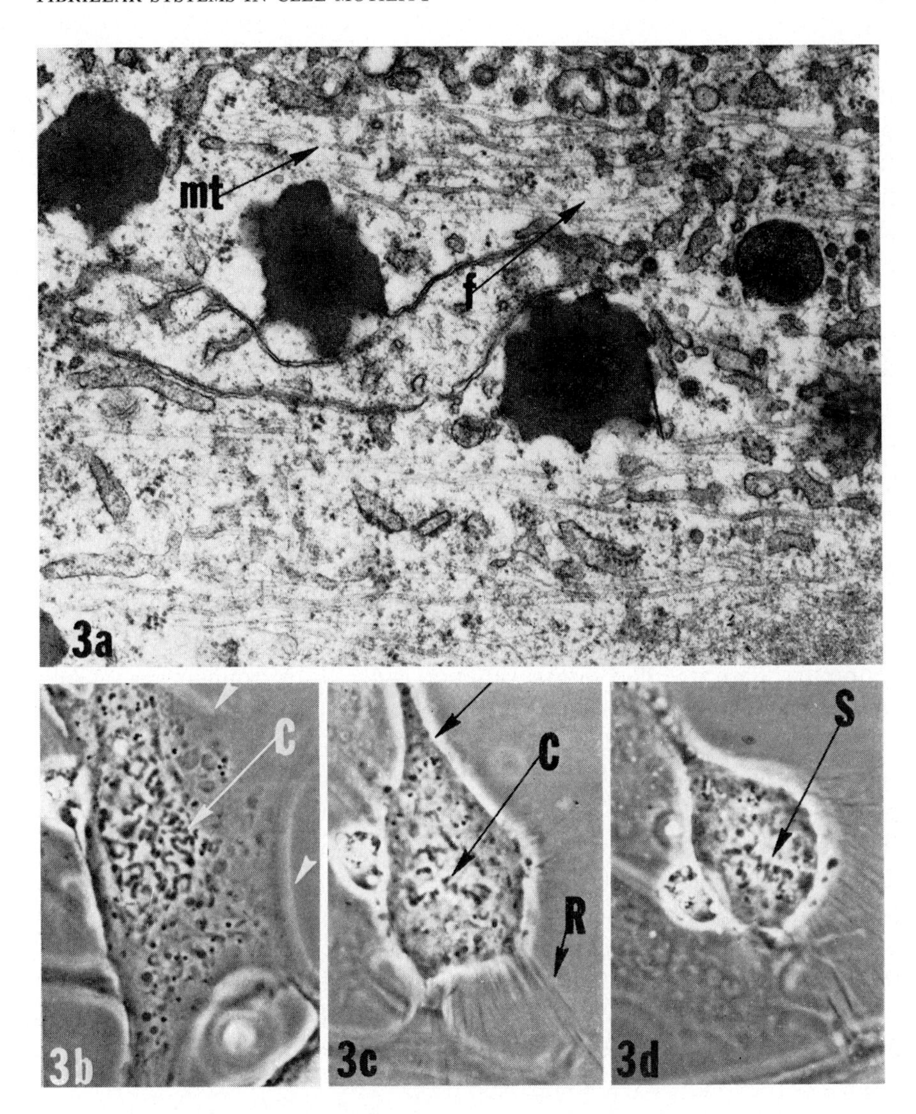

FIG. 3. (a) As Fig. 2b but a higher magnification, with filaments (f) and microtubules (mt) distributed throughout the cytoplasm. × 27 500; (b) BHK-21 cell rounding up for mitosis. Condensed chromosomes (C) are seen within the nucleus and the margins of the cell are still extended over the substrate (arrows); (c) as (b) but 10 min later. The cell has rounded up considerably with only fine retraction fibres (R) and a portion of a fibroblastic process (arrow) left extended over the substrate. Chromosomes (C) are now forming on the mitotic spindle; (d) as (c) but 5 min later. The cell is more rounded and chromosomes are lined up on the metaphase plate of the mitotic spindle (S). Phase contrast, × 1 000.

and the early phases of cell spreading, a clear region frequently appears next to the nucleus with phase contrast optics (Fig. 1b). This region is birefringent in polarized light, and electron microscopy reveals that it contains a large mass of filaments (Fig. 1d and e). A few microtubules are seen randomly scattered through the cytoplasm and small numbers of distinct microfilaments are located in microvilli at the cell surface (Fig. 2a). As the cell spreads into an epithelial configuration, large numbers of microtubules and submembranous microfilaments are formed (Fig. 2b), and filaments are distributed through the cytoplasm (Fig. 3a). These observations and those made on BHK-21 cells suggest that the assembly of microtubules and microfilaments and the redistribution of filaments are fundamental to the spreading of cells into their characteristic shapes.

CELL SHAPE AND THE DISTRIBUTION OF FIBRES DURING NORMAL CELL GROWTH

BHK-21 and BSC-1 cells in growing cultures show various degrees of spreading on the substrate. Part of this variation is due to the rounding up of cells which are preparing for mitosis (Fig. 3b–d). These rounded up cells divide, and each daughter cell respreads on the substrate (Fig. 4a–b). In the early stages of spreading each daughter cell contains a birefringent sphere next to the nucleus in polarized light, which also appears as a clear region in phase contrast optics (Fig. 4c). Electron microscopy shows that this region contains masses of 10 nm filaments and is identical to the clear region seen in spreading trypsinized cells (see Fig. 1d). Similarly, large numbers of microtubules and microfilament bundles are formed, and filaments are redistributed throughout the cytoplasm as the daughter cells spread. Thus the normal process of respreading after mitosis is similar to the spreading of cells which are induced to round up and detach from a substrate by trypsinization.

POOLS OF PRECURSORS FOR THE ASSEMBLY OF MICROFILAMENTS AND MICROTUBULES

Since we have implicated the assembly of fibres in the spreading of cells both in normal growing cultures and in cells induced to round up by trypsinization, it seems likely that pools of precursor subunits for the assembly of fibres are present in the cytoplasm. We tested this hypothesis for BHK-21 cells by allowing freshly trypsinized cells to attach and spread when protein synthesis was

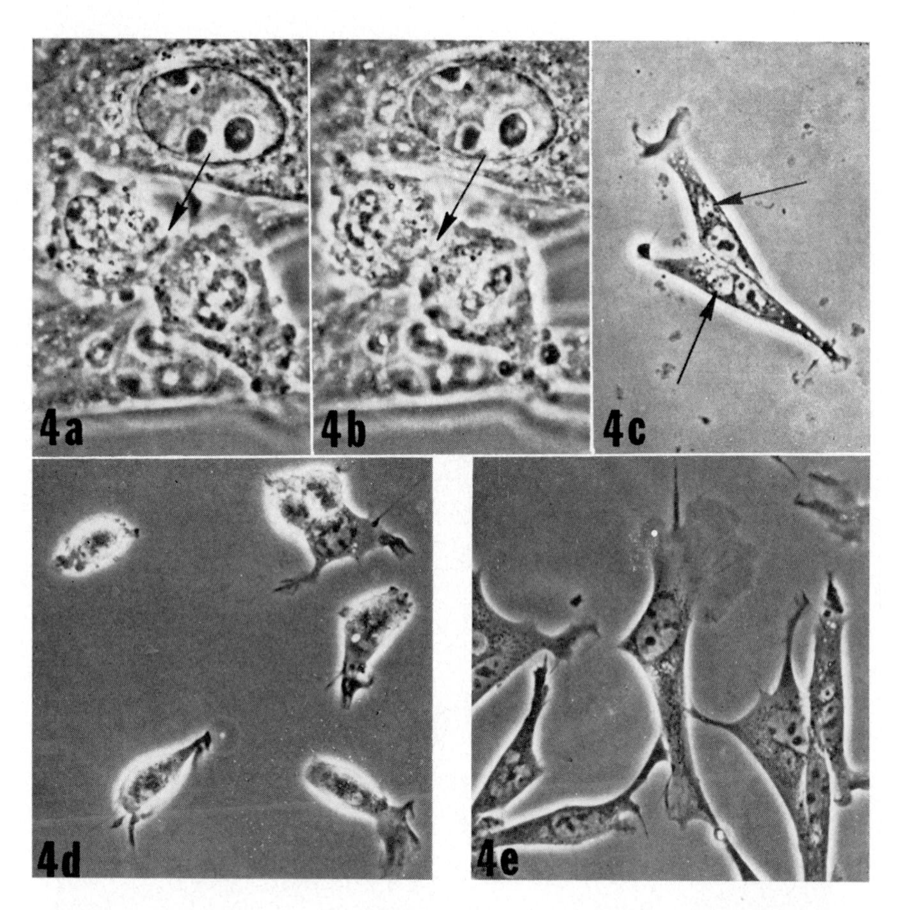

FIG. 4. Early (a) and late (b) stages in cytokinesis: in (a) arrow shows cleavage furrow and in (b) points to midbody. Phase contrast, × 1 000. (c) Daughter cells respreading on a glass coverslip 30 min after cytokinesis. Clear, spherical regions are evident in each cell (arrows). Phase contrast, × 400.

Early (30 min) (d) and late (3 h) (e) stages of BHK-21 cells spreading in cycloheximide. Phase contrast, × 480.

inhibited (Goldman & Knipe 1973). Cells in cycloheximide (20 μg/ml), which inhibits almost all (95%) protein synthesis, attach and spread normally (Fig. 4d and e). Electron microscopy shows that cells which have spread in cycloheximide contain a normal complement of microtubules, filaments and bundles of microfilaments (Fig. 5). Similar treatment of BSC-1 cells indicates that normal attachment, spreading and distribution of fibres are also seen in these cells when protein synthesis is inhibited.

We conclude from these observations that pools of precursors are available

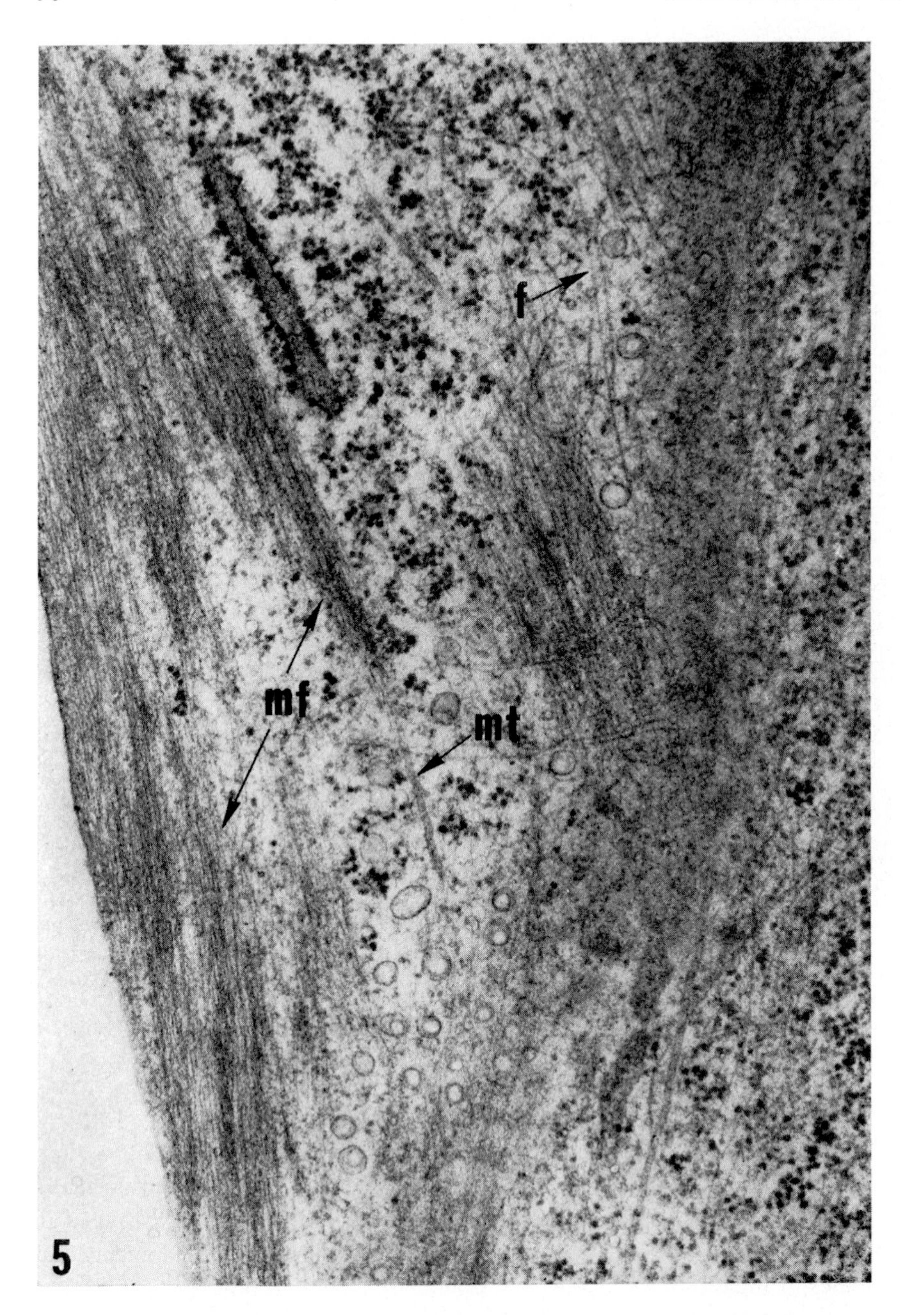

FIG. 5. Electron micrograph of a fibroblastic process of a BHK-21 cell which has spread in cycloheximide (20 μg/ml). Microtubules (mt), filaments (f) and microfilament bundles (mf) are visible just below the cell surface. × 38 500.

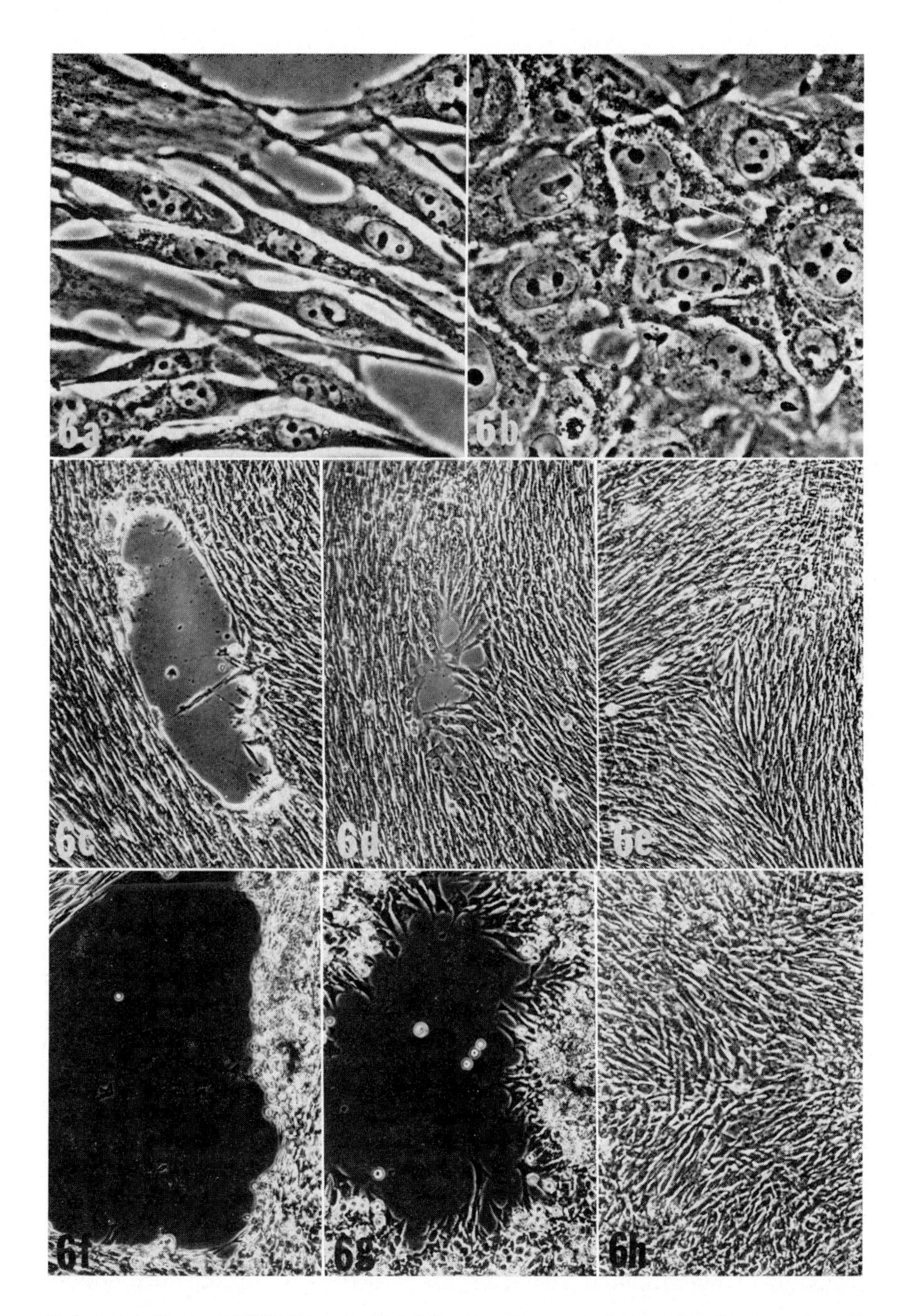

FIG. 6. (a) Control BHK-21 cells 18 h after attachment and spreading in normal medium. Phase contrast, × 480; (b) a similar field of BHK-21 cells 18 h after attachment and spreading in colchicine (40 μg/ml). Caps (arrows) next to the nucleus are evident. Phase contrast, × 480.

The same wound area in a monolayer of BHK-21 cells growing in normal medium: (c) 15 min after wounding; (d) after 6 h cells have moved into the available space; (e) after 12 h the wound area is filled with cells. Phase contrast, × 80.

The same wound area in a monolayer of BHK-21 cells in medium containing colchicine (40 μg/ml): (f) 15 min after wounding; (g) after 24 h; (h) 24 h after the replacement of colchicine with normal medium, cells have completely filled the space. Phase contrast, × 80.

for the formation of microtubules and microfilaments. Filaments, at least during the attachment and early spreading process, are stored as fibres in the birefringent regions next to the nucleus in both BSC-1 and BHK-21 cells.

POSSIBLE SPECIFIC FUNCTIONS OF MICROTUBULES

We have used colchicine to investigate the roles of microtubules in the attachment, spreading, shape formation and locomotion of BHK-21 cells. If these suspended cells are placed in a medium containing colchicine (40 μg/ml), they attach to and spread over the substrate (Goldman 1971; Goldman & Knipe 1973). The final shape, however, is epithelial-like (Fig. 6a, b). These cells contain no microtubules, but do have a normal distribution of microfilaments. Most of the filaments are localized in a cap near the nucleus and are not distributed through the cytoplasm as they are in normal spread BHK-21 cells (Fig. 6b). The effect of colchicine is completely reversible. From these observations we concluded that microtubules are necessary for the formation of the asymmetric cell processes of BHK-21 fibroblasts but are not necessary in the spreading of cells on the substrate. Further, it appears that the distribution of filaments throughout spread cells is dependent upon the formation and distribution of microtubules (Goldman 1971; Goldman & Knipe 1973).

We have also studied the locomotion of colchicine-treated BHK-21 cells and find that although membrane ruffling is apparently normal, locomotion is greatly inhibited (Goldman 1971). This may be demonstrated by wounding a confluent monolayer of cells (cf. Dulbecco & Stoker 1970) and watching cells move into the space created by the wound. In control cultures cells are seen to translocate towards the centre of a wound within 30–60 min; after 6–12 h, small wound areas are completely filled with cells (Fig. 6c–e). When wounds are made on monolayers in the presence of colchicine (40 μg/ml), cells along the periphery of the wound appear to spread into the space, but no locomotion to the centre of the space is seen after 24 h (Fig. 6f, g). When colchicine is removed, the cells regain their normal shape and fibre distribution and begin to locomote until the spaces are rapidly filled (Fig. 6h). Thus it is apparent that a normal distribution of microtubules and filaments is necessary for cell locomotion, although their specific functions are unknown.

Treatment of BSC-1 cells with colchicine (40 μg/ml) has also been investigated. When trypsinized BSC-1 cells are placed on a substrate with colchicine, the cells attach and spread normally into a shape which is similar to normal BSC-1 cells (Fig. 7a, b). The only obvious difference visible by light microscopy is a cap next to the nucleus which is similar to that seen in colchicine-treated

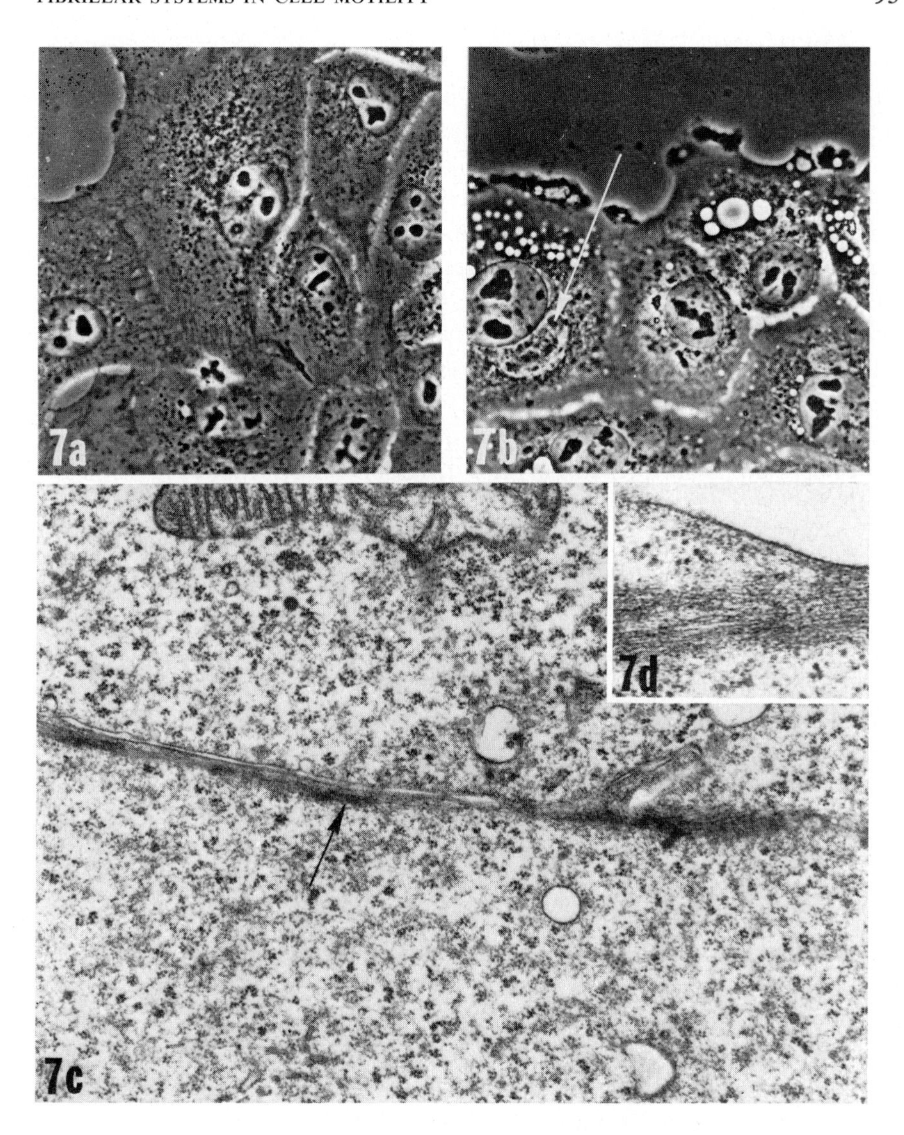

FIG. 7. (a) BSC-1 cells 18 h after attachment and spreading in normal medium. Phase contrast, × 440; (b) as (a) but in colchicine (40 μg/ml). Caps (arrow) next to the nucleus are evident. Phase contrast, × 440; (c) and (d) electron micrographs of a BSC-1 cell which has spread in colchicine (40 μg/ml). No microtubules are present, but submembranous microfilament bundles are evident [see arrow and (d)]. (c) × 18 000. (d) × 38 500.

BHK-21 cells. Electron microscopy reveals that the cap contains masses of 10 nm filaments, and that there are no microtubules visible in the cytoplasm (Fig. 7c). However, submembranous microfilament bundles are seen in BSC-1 cells which have spread in colchicine (Fig. 7c, d).

From these results, it appears that microtubules are not prominent in attachment, spreading or the formation of epithelial cell shape. The observation that colchicine-treated BHK-21 cells are similar in morphology to normal or colchicine-treated BSC-1 cells (compare Fig. 6b with Fig. 7a, b) suggests that one of the primary differences between the normal fibroblast and the epithelial cell is the manner in which microtubules are distributed through the cytoplasm. Since microfilament bundles are the only fibrous components which appear to be distributed normally in cells which have spread in colchicine, they are probably more important in the spreading of the cell than either microtubules or filaments.

POSSIBLE SPECIFIC FUNCTIONS OF MICROFILAMENTS

We originally suggested that bundles of microfilaments, owing to their submembranous localization and morphological similarity to skeletal muscle actin, help govern such activities of the cell surface as membrane ruffling, pinocytosis and cytokinesis (Goldman & Follett 1969). Similar bundles of fibres become 'decorated' with skeletal muscle heavy meromyosin in glycinerated models of BHK-21 (Goldman & Knipe 1973) and BSC-1 cells (Miller & Goldman, unpublished observations). Hence, it is probable that microfilaments represent an actin-like protein fibre in a contractile machinery for surface activities of BHK-21 and BSC-1 cells.

Another probe into microfilament function which has received much attention over the past few years is cytochalasin B (Carter 1967). Schroeder (1970) and Wessells *et al.* (1971) suggested a specific effect of the drug upon microfilaments and consequently upon many motile activities of cultured cells. Other workers have implied that cytochalasin B inhibits other aspects of cell surface activity, such as glucose transport (Zigmond & Hirsch 1972; Estensen & Plagemann 1972), mucopolysaccharide synthesis (Sanger & Holtzer 1972) or membrane fusion during cytokinesis (Hammer *et al.* 1971; Estensen *et al.* 1971).

We have looked at the effects of cytochalasin B on the attachment, spreading and formation of microfilament bundles in BHK-21 cells. When suspended cells are placed in a medium containing cytochalasin B (0.1–20 μg/ml), they attach to and spread over the substrate. The degree of spreading varies with the concentration of cytochalasin B (Goldman & Knipe 1973): at 0.1 μg/ml, BHK-21 cells attach and spread normally; at 0.5–1.0 μg/ml, the overall morphology appears normal after 18 h but both membrane ruffling and pinocytosis are inhibited (Fig. 8a). Also, the distribution of submembranous microfilament bundles, microtubules and filaments is normal when observed with the electron microscope. As the concentration is increased above 1 μg/ml, the cells show less spreading on the substrate after 18 h. At 5–20 μg/ml most of the cells send out only slender spikes of cytoplasm (Fig. 8b) over the substrate. These spiky processes abound with microtubules and filaments, but microfilament bundles are *extremely* difficult to locate. Within minutes of the removal of cytochalasin B, active membrane ruffling and surface expansion are seen together with a rapid return to normal morphology (Fig. 8c). This is accompanied by an equally rapid increase in the number of microfilament bundles.

Observations of wound areas made in monolayers of BHK-21 cells show that cell locomotion is halted at concentrations of cytochalasin B greater than 0.5 μg/ml. Figs. 8f and 8g show a wound area 15 min and 24 h, respectively, after wounding in the presence of cytochalasin B (1 μg/ml), and it is obvious that there is very little locomotion into the wound area. When cytochalasin B is removed, the cells rapidly fill in the space (Fig. 8h).

Analogous effects are seen with BSC-1 cells. At low concentrations of cytochalasin B (0.5–1.0 μg/ml) the cells are still well spread (Fig. 8d), and at higher concentrations (5 μg/ml or more) the cells are spiky (Fig. 8e). Ruffling and pinocytosis are inhibited at concentrations above 0.5 μg/ml. Cytokinesis in both BHK-21 and BSC-1 cells is inhibited at concentrations of 0.5–20 μg/ml and as a consequence large numbers of binucleated cells are seen 18–24 h after attachment and spreading (Fig. 8c, d).

The several interpretations of the effect of cytochalasin B fall into two general categories, in both of which cytochalasin B interacts with the cell surface: cytochalasin B interacts with or binds to submembranous microfilament proteins causing changes in their structure and function, or cytochalasin B acts directly at the cell membrane or at the surface coats of mucopolysaccharide and glycoprotein. In the latter explanation, the interactions could affect cell adhesion to substrates and, therefore, the ability of the cells to ruffle and spread over the substrate in a normal fashion. At the present time we do not favour either explanation, since both could interfere with the formation of bundles of microfilaments.

Several groups of workers have suggested that microfilament breakdown is an indirect consequence of treatment with cytochalasin B. Sanger & Holtzer (1972) have found that the incorporation of glucosamine into acid mucopolysaccharides and glycoproteins is rapidly inhibited in HeLa cells treated with cytochalasin B (10 μg/ml). This suggests that the cell surface coat, which is ultimately engaged in cell adhesion, is affected. We have looked for gross

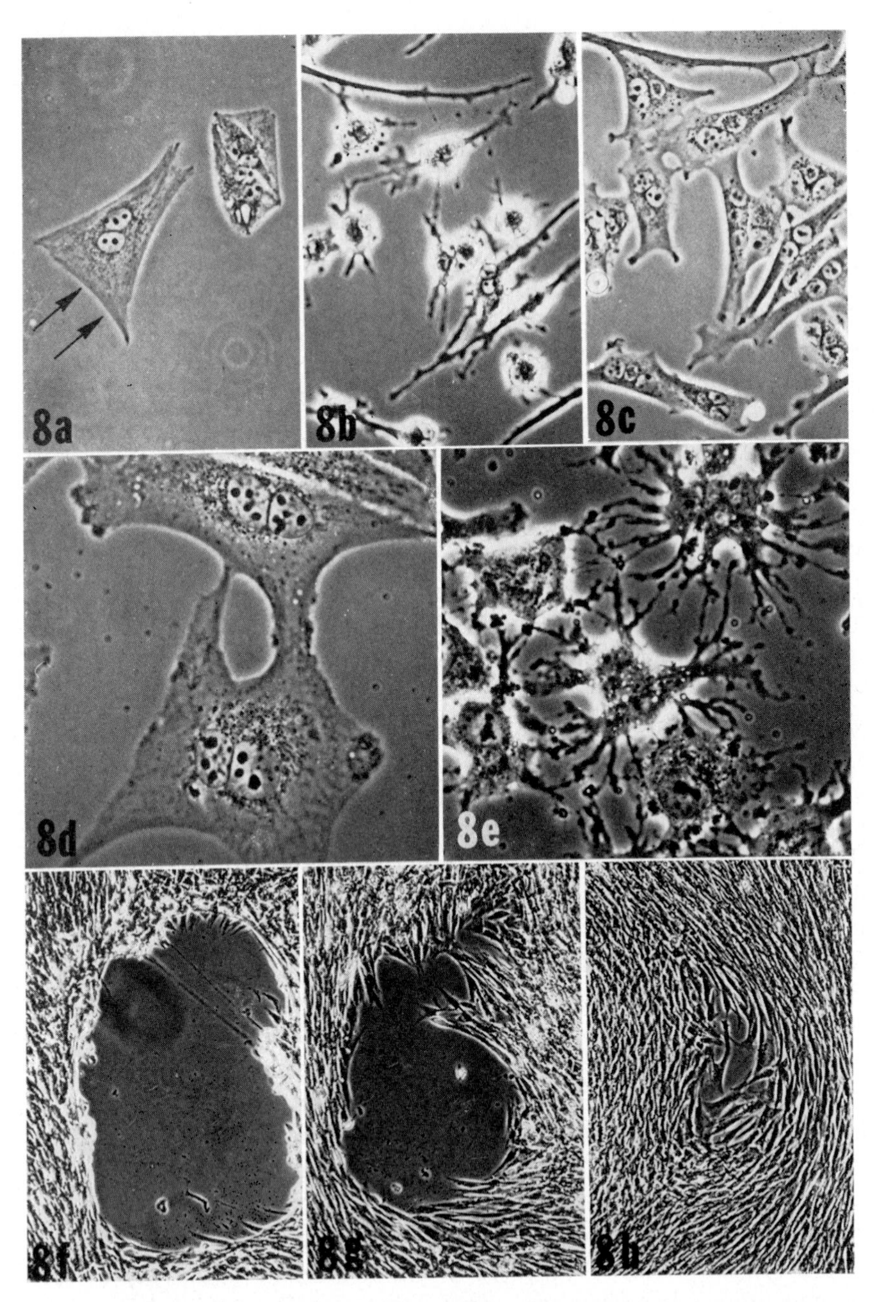

FIG. 8. (a) BHK-21 cells 18 h after attachment and spreading in medium containing cytochalasin B (1 μg/ml). No obvious ruffling or pinocytosis is seen at free edges of cells (arrows). Phase contrast, × 250; (b) as (a) but with 20 μg/ml cytochalasin B. Phase contrast, × 250; (c) same cells as in (b) 20 min after the removal of cytochalasin B. Phase contrast, × 250; (d) BSC-1 cells which have attached and spread in medium containing cytochalasin B (1 μg/ml). Pinocytosis, ruffling and cytokinesis are inhibited. Phase contrast, × 480; (e) as (d) but in 10 μg/ml cytochalasin B. Phase contrast, × 380; (f) wound area in a monolayer of BHK-21 cells showing inhibition of locomotion in medium containing cytochalasin B (1 μg/ml) 15 min after wounding; (g) as (f) but 24 h after wounding. Some cells have spread into the wound area but no locomotion to the centre of the space is seen; (h) as (g) but 6 h after removal of cytochalasin B. Phase contrast, × 80.

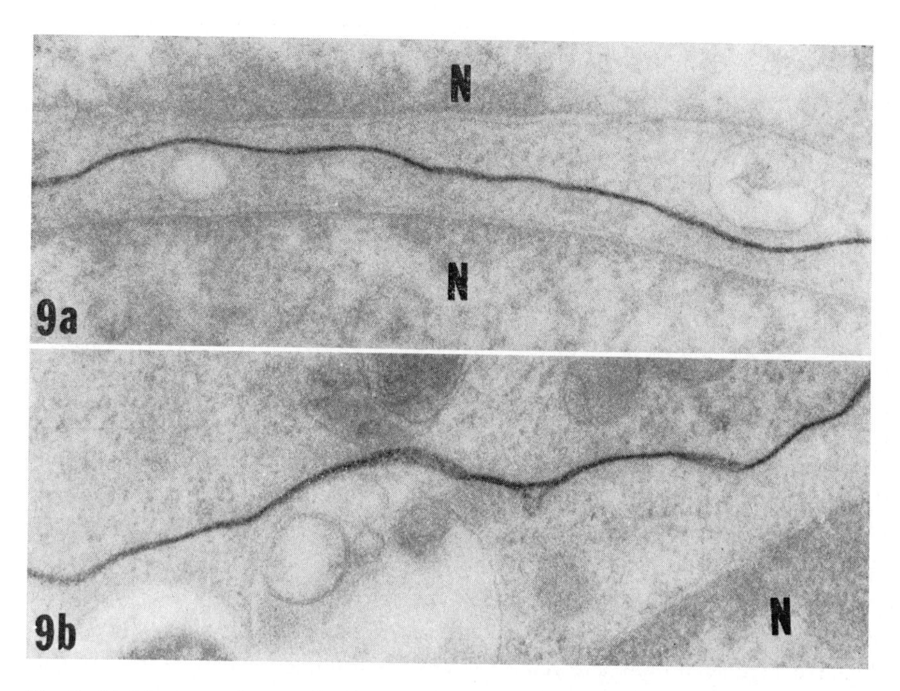

Fig. 9. (a) Electron micrograph of an unstained thin section of cells maintained in normal medium for 24 h and fixed in the presence of Ruthenium Red. The dense black line represents the Ruthenium Red positive layer of two spread BHK-21 cells. The nucleus of each cell can be seen (N). × 33 000; (b) an unstained thin section through a region of contact between two BHK-21 cells maintained in medium with cytochalasin B (10 μg/ml) for 24 h. The cells were fixed in the presence of Ruthenium Red and the dense black line represents the combined Ruthenium Red positive layers of each cell. × 33 000.

alterations in the (extracellular) surface coats of BHK-21 cells in the presence and absence of cytochalasin B. Cells were allowed to attach, spread and remain in cytochalasin B (10 μg/ml) for 18–24 h before being fixed and stained with Ruthenium Red [by the method of Luft (1971) which stains the extracellular coats of animal cells]. Since both control and cytochalasin-treated cells appear identical with regard to the Ruthenium Red positive layer (Fig. 9a, b), we concluded that there was no obvious alteration in surface coats on prolonged treatment with cytochalasin B. It must be emphasized, however, that small chemical changes due to interaction of cytochalasin B with the surface of cells probably would not be visible with this technique.

Hammer *et al.* (1971) have suggested that cytochalasin B interacts with the cell surface to prevent membrane fusion which takes place during amphibian egg cleavage. The fact that endo- and exo-cytosis are inhibited by cytochalasin B

(Wagner *et al.* 1971; Allison *et al.* 1971; Orr *et al.* 1972) also suggests that the drug acts on the surface to prevent membrane fusion, a prerequisite for both processes. Although we have noticed an extremely rapid inhibition of membrane ruffling and pinocytosis by phase contrast microscopy at concentrations above 0.1 μg/ml, we have seen indications in electron microscopic preparations of cells treated with up to 50 μg/ml of the drug that micropinocytic vesicles were still forming (Fig. 10a). To ascertain this, we used horseradish peroxidase as a tracer for electron microscopy (Graham & Karnovsky 1966). BHK-21 cells were treated with cytochalasin B (10 μg/ml) for one hour and then horseradish peroxidase (2 mg/ml) was added to the culture medium. Peroxidase reaction product was detected in small vesicles seen in electron micrographs of cells fixed within ten minutes of the addition of peroxidase (Fig. 10b). After longer intervals large vesicles containing reaction product were seen, indicating that the small vesicles had fused (Fig. 10c). Thus we conclude that micropinocytosis continues in the presence of high concentrations of cytochalasin B, and that membrane fusion is not completely inhibited.

From our work with cytochalasin B, we decided that the mode of action of the drug can only be understood by a more molecular approach to the problem. We have been unable to relate directly the total absence of microfilaments to the cessation of motile activity in BHK-21 cells. This is most evident at low concentrations of cytochalasin B, when the drug inhibits membrane ruffling, cytokinesis and locomotion, and yet microfilament bundles are formed. However, there is no reason to believe that a complete breakdown in the structure of microfilaments is necessary to alter their function. Studies such as those by Spudich (1973) on the interaction of cytochalasin B with skeletal muscle and platelet actin might resolve the controversy regarding the specific level of activity of the drug. The only conclusion we draw about the site of action at the cellular level is that the drug probably acts at the cell surface, but it is an open question as to whether it acts on the plasma membrane, the extracellular coats, and/or microfilaments.

CYTOCHALASIN B IN ENUCLEATION EXPERIMENTS

Even though we do not know exactly how cytochalasin B interacts with cultured cells, we have made use of the drug by taking advantage of the fact that it causes enucleation (Carter 1967). By modifying the procedure of Prescott *et al.* (1972) for enucleating cells we have obtained populations which are up to 99.5% enucleated (Goldman *et al.* 1973). We grew cells on coverslips, treated them with cytochalasin B (10 μg/ml) and then centrifuged the coverslips for

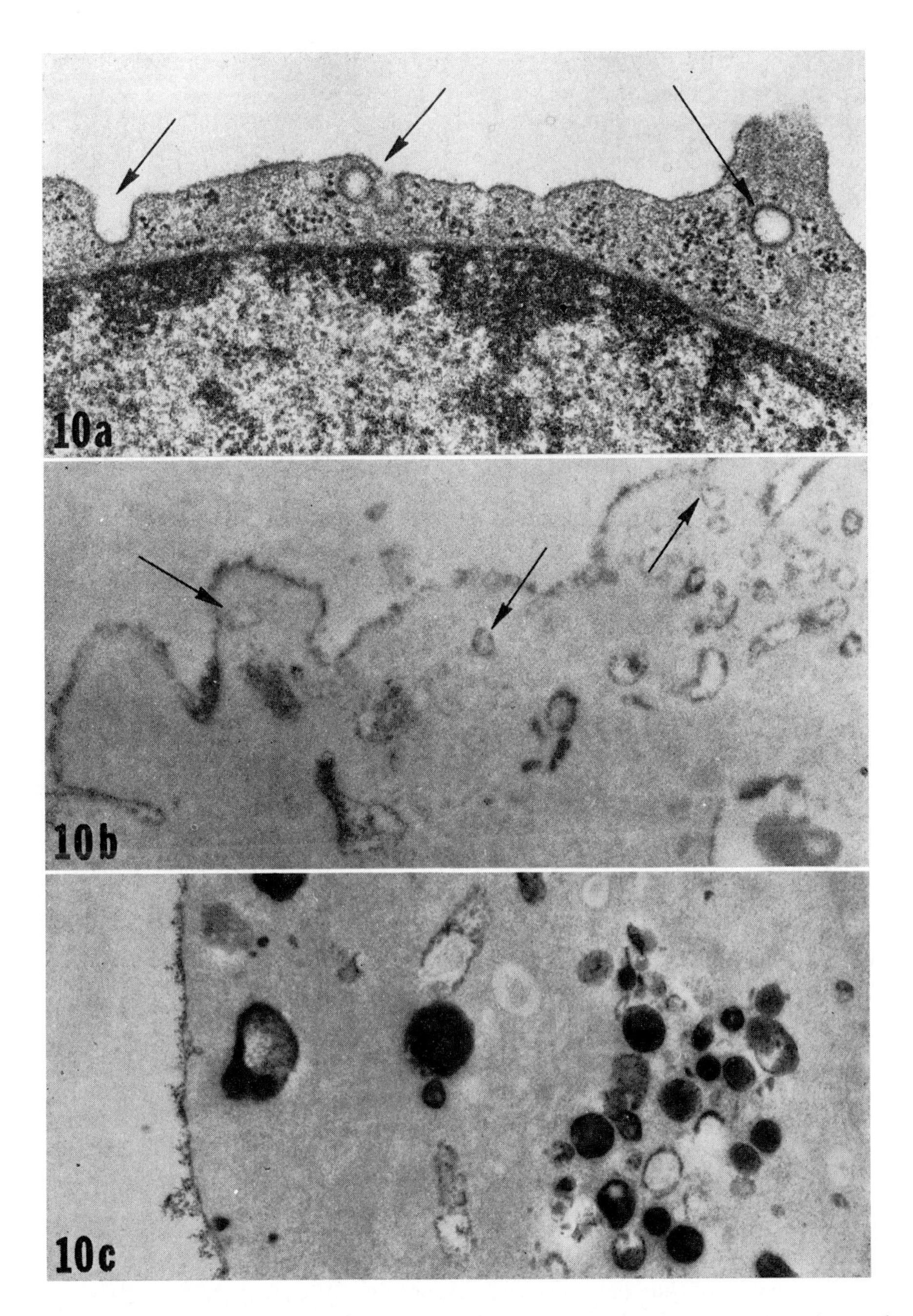

FIG. 10. (a) Electron micrograph of a section through the surface of a BHK-21 cell treated with cytochalasin B (10 μg/ml) for 24 h. Note the typical invaginations and probable stages in the pinching off of micropinocytic vesicles (arrows). × 32 000; (b) an unstained section showing horseradish peroxidase reaction product in small vesicles just beneath the cell membrane (arrows). This cell was treated for 1 h with cytochalasin B (10 μg/ml) followed by 10 min in medium containing cytochalasin B and peroxidase. × 33 000; (c) cytochalasin-treated cells containing large electron-dense vesicles after 4 h treatment with peroxidase in the presence of cytochalasin B (10 μg/ml). × 33 000.

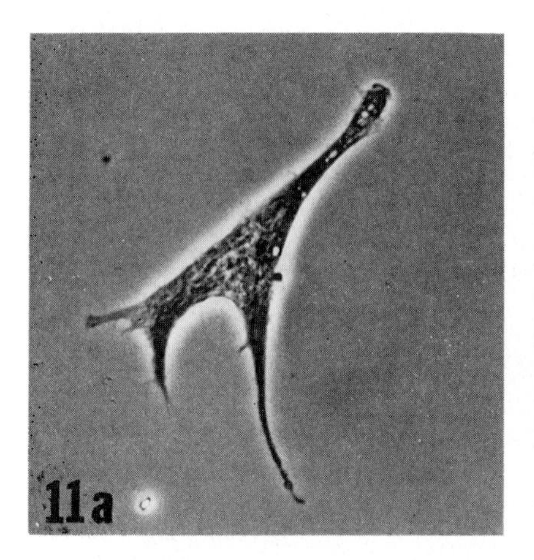

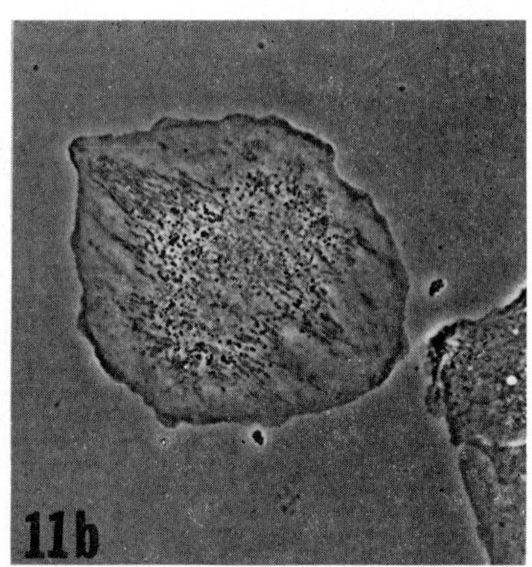

FIG. 11. (a) Enucleated BHK-21 cell 3 h after attachment to a glass coverslip. Phase contrast, × 520; (b) enucleated BSC-1 cell 3 h after attachment to a glass coverslip. Phase contrast, × 480.

about one hour. The coverslips still containing cells were removed from the centrifuge tubes and placed in normal medium. Within 1–2 h the morphology of the cells appeared normal except that most cells did not contain a recognizable nucleus. When the enucleates were removed from the glass coverslips by trypsinization, they appeared spherical, as did nucleated BHK-21 and BSC-1 cells. When these cells were replaced on substrates, they attached and spread in a fashion indistinguishable from normal cell spreading. The BHK-21 enucleates appeared fibroblastic (Fig. 11a) and the BSC-1 enucleates became epithelial-like within a few hours (Fig. 11b). Normal pinocytosis and membrane ruffling accompanied the spreading process. These experiments suggest that the information necessary for attachment and spreading into normal cell shape is present in the cytoplasm (Goldman *et al.* 1973).

DISCUSSION AND SUMMARY

From studies of the relationship of cytoplasmic fibres to aspects of BHK-21 and BSC-1 cell motility, it appears that the assembly and changes in distribution of all the cytoplasmic fibres are related to normal cell spreading, shape formation and locomotion. It is also likely that the processes of disassembly and redistribution of fibres are important in other aspects of cell motility. For example, microtubule disassembly apparently contributes to the rounding up of cells before mitosis. This is followed by the reassembly of microtubules in the mitotic apparatus (Dickerman & Goldman, unpublished observations). As chromosomes move towards the poles of the mitotic spindle, microtubules are disassembled and then after cytokinesis reassembly begins during spreading of the daughter cells.

The experiments with cycloheximide and enucleated cells demonstrate that the assembly and distribution of fibres during spreading continue despite the absence of protein synthesis or lack of a nucleus. Therefore, it is likely that the precursors for the assembly of fibres and the information for the distribution of the fibres are both present in the cytoplasm.

The 10 nm filaments are the least understood of all the known cytoplasmic fibres. In spread BHK-21 and BSC-1 cells, filaments and microtubules frequently lie parallel to each other. In the absence of microtubules (i.e. during colchicine treatment) filaments remain next to the nucleus as cells spread, indicating that they are dependent on the assembly of microtubules for their normal distribution through the cytoplasm of spread cells. Colchicine-treated BHK-21 cells exhibit active membrane ruffling at many points along their periphery and yet their locomotion is inhibited (Goldman 1971; Vasiliev *et al.* 1970). We have suggested (Goldman 1971) that the normal formation of the microtubule–filament association might determine which ruffling edge becomes the leading edge, so that the cell is capable of locomotion [see Abercrombie *et al.* (1970) for a discussion of the relation between ruffled membranes and cell locomotion].

The possible existence of a contractile system in cultured cells is, of course, another important factor in cell motility. Although there is much information available about the presence of contractile proteins in non-muscle cells, little is known about their interactions and localization. Microfilaments appear to be actin-like since they bind heavy meromyosin (Ishikawa *et al.* 1969; Goldman & Knipe 1973), and actin-like proteins account for as much as 10–20% of the total protein of non-muscle cells (Bray 1973). The recent finding of myosin-like proteins in cultured non-muscle cells (Adelstein & Conti 1973) supports the idea that the contractile system is of the actomyosin type. Such a contractile system is most likely localized in the microfilament bundles (Goldman & Knipe 1973). Pollard & Korn (1973) have presented evidence that actin-like protein fibres are attached to isolated fractions of the plasma membranes of *Acanthamoeba*, an idea which supports the notion of a contractile system being an integral part of the cytoplasmic side of the plasma membrane.

It is apparent that microtubules, filaments and microfilaments are important components of the molecular architecture of vertebrate cells. Furthermore, the fibres are dynamic structures and their assembly, disassembly and redistribution might be directly related to many aspects of cell motility.

ACKNOWLEDGEMENTS

This work was supported by grants from the American Cancer Society (E-639), the Damon Runyon Memorial Fund for Cancer Research, Inc. (DRG1083A) and the National Science Foundation (GB23185). R.D.G. is the recipent of an N.I.H. Career Development Award (1-K4-GM-32, 249-01).

References

ABERCROMBIE, M., HEAYSMAN, J. & PEGRUM, S. M. (1970) The locomotion of fibroblasts in culture. I. Movements of the leading edge. *Exp. Cell Res.* **59**, 393-398

ADELSTEIN, R. S. & CONTI, M. (1973) The characterization of contractile muscle protein from platelets and fibroblasts. *Cold Spring Harbor Symp. Quant. Biol.* **37**, 599

ALLISON, A. C., DAVIES, P. & DE PETRIS, S. (1971) Role of contractile microfilaments in macrophage movement and endocytosis. *Nat. New Biol.* **232**, 153-155

BRAY, D. (1973) Cytoplasmic actin; a comparative study. *Cold Spring Harbor Symp. Quant. Biol.* **37**, 567-571

CARTER, S. B. (1967) Effects of cytochalasins on mammalian cells *Nature (Lond.)* **213**, 261-264

DULBECCO, R. & STOKER, M. (1970) Conditions determining initiation of DNA synthesis in 3T3 cells. *Proc. Natl. Acad. Sci. U.S.A.* **66**, 204-210

ESTENSEN, R. D. & PLAGEMANN, P. G. W. (1972) Cytochalasin B1: inhibition of glucose and glucosamine transport. *Proc. Natl. Acad. Sci. U.S.A.* **69**, 1430-1434

ESTENSEN, R. D., ROSENBERG, M. & SHERIDAN, J. D. (1971) Cytochalasin B: microfilaments and 'contractile' processes. *Science (Wash. D.C.)* **173**, 356-357

FOLLETT, E. A. C. & GOLDMAN, R. D. (1970) The occurrence of microvilli during spreading and growth of BHK-21/C13 fibroblasts. *Exp. Cell Res.* **59**, 124-136

GOLDMAN, R. D. (1971) The role of three cytoplasmic fibers in BHK-21 cell motility. I. Microtubules and the effects of colchicine. *J. Cell Biol.* **51**, 752-762

GOLDMAN, R. D. & FOLLETT, E. A. C. (1969) The structure of the major cell processes of isolated BHK-21 fibroblasts. *Exp. Cell Res.* **57**, 263-276

GOLDMAN, R. D. & FOLLETT, E. A. C. (1970) Birefringent filamentous organelle in BHK-21 cells and its possible role in cell spreading and motility. *Science (Wash. D.C.)* **169**, 286-288

GOLDMAN, R. D. & KNIPE, D. M. (1973) The functions of cytoplasmic fibers in non-muscle cell motility. *Cold Spring Harbor Symp. Quant. Biol.* **37**, 523

GOLDMAN, R. D., POLLACK, R. & HOPKINS, N. (1973) Preservation of normal behavior by enucleated cells. *Proc. Natl. Acad. Sci. U.S.A.* **70**, 750-754

GRAHAM, R. C. & KARNOVSKY, M. J. (1966) The early stages of absorption of injected horseradish peroxidase in the proximal tubules of mouse kidney: ultrastructural cytochemistry by a new technique. *J. Histochem. Cytochem.* **14**, 291-302

HAMMER, M. G., SHERIDAN, J. D. & ESTENSEN, R. D. (1971) Cytochalasin B II. Inhibition of cytokinesis in *Xenopus laevis* eggs. *Proc. Soc. Exp. Biol. Med.* **136**, 1158-1162

HUNEEUS, F. C. & DAVISON, P. F. (1970) Fibrillar proteins from squid axons. *J. Mol. Biol.* **52**, 415-428

ISHIKAWA, H., BISCHOFF, R. & HOLTZER, H. (1969) Formation of arrowhead complexes with heavy meromyosin in a variety of cell types. *J. Cell Biol.* **43**, 312-328

LUFT, J. H. (1971) Ruthenium Red and Violet II. Fine structure localization in animal tissues. *Anat. Rec.* **171**, 369-392

ORR, T. S. C., HALL, D. E. & ALLISON, A. C. (1972) Role of contractile microfilaments in the release of histamine from mast cells. *Nature (Lond.)* **236**, 350-351

POLLARD, T. D. & KORN, E. D. (1973) The contractile proteins of *Acanthamoeba castellani. Cold Spring Harbor Symp. Quant. Biol.* **37**, 573

PRESCOTT, D. M., MYERSON, D. & WALLACE, L. (1972) Enucleation of mammalian cells with cytochalasin B. *Exp. Cell Res.* **71**, 480-485

SANGER, J. W. & HOLTZER, H. (1972) Cytochalasin B: effects on cell morphology, cell adhesion, and mucopolysaccharide synthesis. *Proc. Natl. Acad. Sci. U.S.A.* **69**, 253-257

SCHROEDER, T. E. (1970) The contractile ring I. Fine structure of dividing mammalian (HeLa) cells and the effects of cytochalasin B. *Z. Zellforsch. Mikrosk. Anat.* **109**, 431-449

SPOONER, B. S., YAMADA, K. M. & WESSELLS, N. K. (1971) Microfilaments and cell locomotion. *J. Cell Biol.* **49**, 595-613

SPUDICH, J. A. (1973) Effects of cytochalasin B on actin filaments. *Cold Spring Harbor Symp. Quant. Biol.* **37**, 585

VASILIEV, JU. M., GELFAND, I. M., DOMNINA, L. V., IVANOVA, O. YU., KOMM, S. G. & OLSHEVSKAJA, L. V. (1970) Effect of Colcemid on the locomotory behavior of fibroblasts. *J. Embryol. Exp. Morphol.* **24**, 625-640

WAGNER, R., ROSENBERG, M. & ESTENSEN, R. (1971) Endocytosis in Chang liver cells. Quantitation by sucrose-H^3 uptake and inhibition by cytochalasin B. *J. Cell Biol.* **50**, 804-817

WESSELLS, N. K., SPOONER, B. S., ASH, J. F., BRADLEY, M. O., LUDUEÑA, M. A., TAYLOR, E. L., WRENN, J. T. & YAMADA, K. M. (1971) Microfilaments in cellular and developmental processes. *Science (Wash. D.C.)* **171**, 135-143

YAMADA, K. M., SPOONER, B. S. & WESSELLS, N. K. (1971) Ultrastructure and function of growth cones and axons of cultured nerve cells. *J. Cell Biol.* **49**, 614-635

ZIGMOND, S. H. & HIRSCH, J. G. (1972) Cytochalasin B: inhibition of D-2-deoxyglucose transport into leucocytes and fibroblasts. *Science (Wash. D.C.)* **176**, 1432-1434

Discussion

Gail: I would like to comment on the effects of these agents on the motility of fibroblasts as measured in time lapse studies. I have measured motility in terms of a diffusion constant which measures how rapidly the cells spread over the surface. We found that Colcemid does inhibit motility as you suggested (Gail & Boone 1971*a*); this decrease in motility might be the consequence of the profound structural alterations which accompany Colcemid treatment. The cells lose their axial organization and become rounded, and the ruffling membrane becomes fairly uniformly disposed about the cytoplasmic periphery instead of being localized to the leading edge. Cytochalasin also reduces motility dramatically (Carter 1967). One can dissociate decreases in motility from inhibition of cytokinesis at low dose levels where motility is reduced but the proliferation rate is unchanged (Gail & Boone 1971*b*).

Goldman: We have also found that the inhibition of cytokinesis in BHK-21 cells is dependent on the concentration of cytochalasin B (Dickerman & Goldman, unpublished results and see p. 95).

Porter: Did you mean to suggest that the filaments and microtubules share the same protein?

Goldman: No. We do not think that the 10 nm filaments are a second form of microtubule (Goldman 1971). Similar filaments have been isolated from nerve axoplasm and preliminary biochemical analysis suggests that they contain unique proteins not found in microtubules (Huneeus & Davison 1970).

Porter: Is it valid to compare the number of microtubules in a rounded cell with that in an oriented cell? In one instance you can orient the sectioning with respect to some axis of the cell and in the other you cannot.

Goldman: It is true that we cannot be absolutely sure of the number of microtubules. However, after six years of looking at BHK-21 cells, we feel that we can state that there are fewer and certainly shorter microtubules in rounded up cells than in fully spread cells.

Porter: The description of the cell as 'epithelial-like' is crude in the sense that a cell which has been treated with colchicine flattens out and becomes round in outline; it loses its bipolarity. On examining these cells with the scanning microscope, we have noticed that the dimensions of a colchicine-treated cell are not the same as those of an epithelial cell (unpublished results). The colchicine-treated cell is so flat that it can hardly be defined under the scanning microscope, whereas a normal epithelial cell has depth and surface structure which are easily discernible.

Goldman: The longer you leave BHK-21 cells in colchicine the more they spread out over the substrate. After 21–48 h many of these cells completely fill, and sometimes are larger than, the microscope field (25×), whereas several normal BHK-21 cells can be seen in the identical microscope field.

Wessells: I want to comment on the frequency with which colchicine alters the locomotion of migratory cells. Fig. 7 in my paper (see p. 61) shows a cell, treated with colchicine, ruffling on all sides: the cell on the right, which has remained highly elongated, ruffles in one discrete area and shows net movement. About one quarter of treated glial cells behave in this way. Electron microscopy of such cells shows no microtubules but instead an abundance of the characteristic 10 nm filaments (see p. 61) (Ishikawa *et al.* 1968). This demonstrates clearly that colchicine is having a typical effect in the cytoplasm of the cells. We conclude that at least some cells obtained from normal embryos can move, and have immobile sides despite the absence of cytoplasmic microtubules (see also Trinkaus, this volume).

Goldman: We have analysed cell locomotion in the presence of colchicine (40 μg/ml), which breaks down all cytoplasmic microtubules in both BHK-21 cells (Goldman 1971) and BSC-1 cells (Miller & Goldman, unpublished results). We have not been able to detect significant locomotion in either of these cell lines in the absence of microtubules. This is especially evident in the wounded monolayer experiments (cf. Fig. 6).

Wessells: The wound healing system is complicated by the fact that the cells are encouraged not only to move in the presence of colchicine, but also to overcome the adhesive relations with neighbouring cells in the sheet. The same problem is encountered in the experiments of Vasiliev *et al.* (1970). Use of single cells in low density yields much cleaner results.

Goldman: We have taken several thousand feet of time lapse films of colchicine-treated, sub-confluent BHK-21 cells whose intercellular adhesions are minimal, and we are still unable to detect cell locomotion. Most cells show active ruffling at several points along their periphery, but are unable to translocate (Goldman 1971).

Porter: Vasiliev *et al.* (1970) have also described this.

Goldman: Yes, they demonstrated a cessation of directional cell locomotion in the presence of colchicine.

Gail: We have observed a decrease in the motility of Colcemid-treated fibroblasts in sparse culture (Gail & Boone 1971*a*). Even when the cells are essentially isolated from each other, their diffusion constants decrease from about 400 to 100 $\mu m^2/h$. Many such isolated cells do move during Colcemid-treatment, but motility is reduced in a quantitative sense.

You raised the possibility that cytochalasin alters cell–substrate adhesivity. We tried to gauge this adhesivity by measuring the extent to which cells were distracted by a calibrated air blast (Gail & Boone 1972). We found cells easier to distract after treatment with a high dose (10–100 μg/ml) of cytochalasin. Strangely, Colcemid-treated cells are also easier to distract, even though their apparent cell outline area is more than double the normal area.

Goldman: We have also attempted to measure the total force of adhesion of cells treated with cytochalasin B (5–20 μg/ml) by spinning BHK-21 cells on coverslips in a centrifuge. Many treated cells come off the glass at a force of 500–1 000 *g*. All control cells, treated with 0.01 % dimethyl sulphoxide, remain attached at forces as high as 7 000–8 000 *g* (unpublished observations).

Trinkaus: It is good to see the work of Vasiliev, Gelfand *et al.* (1970) and of Gail & Boone (1971*a*, *b*) confirmed. It occurs to me that if the cell is spreading all round its periphery, as is suggested by the ruffling, the inevitable result of this would be that the cell flattened. Also, I want to echo Dr Porter's objections to referring to flattened fibroblasts as 'epithelial-like'. One crucial difference between epithelial cells and fibroblasts is the strong lateral adhesions of epithelial cells. This is presumably due in part to the presence of desmosomes in epithelium. I do not believe fibroblasts have ever been shown to form desmosomes.

Goldman: By 'epithelial-like' we refer to no more than the overall morphology of BHK-21 cells treated with colchicine. It is difficult to distinguish these cells from normal BSC-1 cells by light microscopy (see Figs. 6 and 7).

Trinkaus: Has anyone seen desmosomes in fibroblasts?

Porter: I have observed that fibroblasts do form a continuous tissue and they do show close junctions at the points of contact (unpublished observations).

Steinberg: Dr Goldman, at the concentrations of cytochalasin which reduce the force required to pull the cells off the glass in the centrifuge, is the morphology of the cells changed?

Goldman: Yes, they become very 'spiky' (see Fig. 8b).

Steinberg: In that case, couldn't the change in the force of distraction be due to the change in the area of anchorage rather than to any change in the character of the individual anchorage regions?

Goldman: Yes, each anchorage point could have the same force of adhesion in cytochalasin B-treated cells as in normal cells. Perhaps fewer anchorage points are formed in cytochalasin-treated cells.

Wolpert: I was puzzled that at high concentrations of cytochalasin the cells were very elongated. How does this happen? Have you any time lapse studies on it? Do they ruffle?

Goldman: There is no obvious ruffling in the presence of high concentrations of cytochalasin B. The elongated cell processes contain microtubules and 10 nm filaments oriented along their long axes.

Ambrose: I have filmed (this is the first film to my knowledge) the movements of the internal membranes of a cell. Probably owing to the enormous size of the giant hypernephroma cell (this was the cell I studied) the movement of the internal membrane does not keep pace with the general locomotion, so that the membranes are made visible by the accumulation of material at their boundaries. In a film of the development of a smectic liquid crystal (Meyer & Jones 1972), the spreading of these liquid crystals was remarkably like the movements of the internal membranes of the hypernephroma cells, thereby suggesting that these exhibit liquid crystalline properties. By contrast the movement of the plasma membrane in my film is an active process with coordinated generation of lamellipodia etc., presumably due to the activity of the subsurface microfibrils (unpublished observations).

References

CARTER, S. B. (1967) Effects of cytochalasins on mammalian cells. *Nature (Lond.)* **213**, 261-264

GAIL, M. H. & BOONE, C. W. (1971*a*) Effect of Colcemid on fibroblast motility. *Exp. Cell Res.* **65**, 221-227

GAIL, M. H. & BOONE, C. W. (1971*b*) Cytochalasin effects on BALB/3T3 fibroblasts: dose dependent, reversible alteration of motility and cytoplasmic cleavage. *Exp. Cell Res.* **68**, 226-228

GAIL, M. H. & BOONE, C. W. (1972) Cell–substrate adhesivity: a determinant of cell motility. *Exp. Cell Res.* **70**, 33-40

GOLDMAN, R. D. (1971) The role of three cytoplasmic fibers in BHK-21 cell motility. I. Microtubules and the effects of colchicine. *J. Cell Biol.* **51**, 752-762

HUNEEUS, F. C. & DAVISON, P. F. (1970) Fibrillar proteins from squid axons I. Neurofilament protein. *J. Mol. Biol.* **52**, 415-428

ISHIKAWA, H., BISCHOFF, R. & HOLTZER, H. (1968) Mitosis and intermediate-sized filaments in developing skeletal muscle. *J. Cell Biol.* **38**, 538-555

MEYER, R. B. & JONES, F. (1972) Growth of smectic liquid crystals from dilute isotropic solution, *4th International Liquid Crystal Conference*, Kent State, Ohio, U.S.A., Abstract 38

VASILIEV, JU. M., GELFAND, I. M., DOMINA, L. V., IVANOVA, O. YU., KOMM, S. G. & OLSHEVSKAJA, L. V. (1970) Effect of Colcemid on the locomotory behavior of fibroblasts. *J. Embryol. Exp. Morphol.* **24**, 625-640

The role of microfilaments and microtubules in cell movement, endocytosis and exocytosis

A. C. ALLISON

Clinical Research Centre, Harrow, Middlesex

Abstract Many cell types contain actin-like microfilaments about 7 nm in diameter. These filaments can be identified morphologically by their distinctive pattern of heavy meromyosin binding and chemically by the presence of unusual amino acids and by their peptide map. Actins from different cell types are so far indistinguishable. Immunological studies show that leucocytes and other non-muscle cells contain myosin similar to that in smooth muscle but different from that in skeletal muscle. Cytochalasin probably inhibits directly the contraction of the microfilament system, ruffle membrane movement and cell motility. The formation of very small endocytic vesicles (micropinocytosis) is resistant to cytochalasin, but pinocytosis visible by light microscopy and phagocytosis in a variety of systems is inhibited by the compound. Cytochalasin also affects the discharge of packaged secretions in a variety of cell types exposed to appropriate agonists and it is suggested that microfilaments participate in the release, constituting an important effector system for hormone and drug action. Labile cytoplasmic microtubules probably help to maintain the normal shape and polarity of cells. Concentrations of calcium ions and availability of specific nucleating sites may determine whether microtubules are depolymerized or polymerized to form cytoplasmic or spindle microtubules. In some systems microtubules probably collaborate with microfilaments in the release of packaged secretions.

It has long been known that many cell types—from protozoa to mammalian leucocytes, fibroblasts and epithelial cells—can move, ingest particles of fluid and release packaged secretions when appropriately stimulated. However, the mechanisms underlying these movements have long remained obscure, and it is only during the past few years that plausible explanations could be offered. These explanations have followed the recognition of the widespread occurrence of microtubules and of microfilaments composed of actin- and myosin-like proteins. Microtubules have been observed not only in systems that are obviously contractile, such as flagella, cilia and the mitotic spindle, but also in a wide range of dispositions in the cytoplasm of other cells (see Porter 1966;

Margulis 1973). Moreover, evidence is accumulating that actin- and myosin-like proteins (and probably the related regulatory proteins, such as tropomyosin and troponin) are not confined to the several varieties of muscle cell but are present also in other cell types. The number of cell types in which appropriate identification has been performed is still limited, but already various leucocytes, blood platelets, fibroblasts and epithelial cells have been shown to contain microtubules as well as actomyosin microfilaments. Thus both these systems are probably widespread constituents of living tissues, and either might provide a contractile system in any cell. Probably their relative importance varies from cell to cell and in the same cell at different times, and the analysis of their interactions will require prolonged and detailed investigation.

The first attempts at such an analysis have depended on the use of compounds thought to interact more or less specifically with either microtubules or microfilaments, namely colchicine and vinca alkaloids on the one hand, the cytochalasins on the other. It is therefore necessary to review briefly the evidence for specificity or otherwise of the effects of these compounds.

COLCHICINE AND MICROTUBULES

That colchicine and vinblastine affect microtubular structure by binding with their tubulin subunits is well established (Weisenberg *et al.* 1968; Wilson 1970; Weisenberg & Timasheff 1970; Owellen *et al.* 1972). One molecule of colchicine or vinblastine is bound for each tubulin structural unit of 110 000 daltons, (i.e. a dimer of the individual tubulin proteins). Colchicine and vinblastine bind to different sites, so that neither competes with the other. Binding of radioactive colchicine is widely used as a means of identifying tubulin in fractionation of cell extracts. Moreover, it is generally agreed that in cells treated with colchicine or vinca alkaloids labile cytoplasmic microtubules are broken down. After vinblastine treatment tubulin is often deposited in crystal-like structures in the cytoplasm. Colchicine does not disrupt organized microtubular structures, such as cilia or flagella, but prevents their formation (Rosenbaum *et al.* 1969). It is thought that colchicine can bind the tubulin dimers, inhibiting their polymerization, but is prevented from gaining access to binding sites in organized microtubules, possibly by some additional component of flagella and cilia, such as dynein or nexin (see Gibbons & Fronk 1972). Hence not all microtubule-related functions are impaired by colchicine or vinblastine.

Nevertheless, there is good evidence that colchicine or vinblastine treatment breaks down microtubules in a great variety of cell systems (see Margulis 1973) and presumably functional impairment, such as metaphase arrest, follows. The problem in interpreting observed results of colchicine and vinblastine treatment is that the drugs, particularly when used in high concentrations, might affect systems other than microtubules. Thus, colchicine not only inhibits nucleotide transport across mammalian cell membranes (Mizel & Wilson 1972) and has a hypocalcaemic action (Heath *et al.* 1972), but also inhibits both pancreatic aldose reductase (Gabbay & Tze 1972) and protein synthesis (Creasey *et al.* 1971) and has anti-inflammatory activity which—judging from activity of structurally related compounds—is apparently independent of effects on microtubules (Fitzgerald *et al.* 1971). Doubtless other effects of colchicine and vinblastine will be discovered, so that interpretation of the observed results of treatment of cells must be cautious. It seems that colchicine and vinblastine, in concentrations that can disperse labile microtubules, do not significantly inhibit actomyosin systems, including contraction of smooth and skeletal muscle, cell movement and ruffle membrane activity, as discussed later.

CYTOCHALASIN AND MICROFILAMENTS

The evidence that cytochalasin disturbs certain microfilament functions is less direct than that for colchicine effects on microtubules. It is based on observations that cytochalasin inhibits functions of cells which on morphological and other grounds are thought likely to be related to an actomyosin-like microfilament system (Schroeder 1970; Wessells *et al.* 1971). There are also reports that treatment of cells with cytochalasin disrupts the regular arrangement of microfilaments associated with certain specialized functions. These include the contractile ring of microfilaments thought to be responsible for cleavage after mitosis in the sea-urchin egg (Schroeder 1970, 1972), the bundle of microfilaments observed in epithelial cells during resorption of the tail of ascidian larvae (Cloney 1966; Lash *et al.* 1970) and ordered microfilaments in other situations (Wessells *et al.* 1971). However, treatment with cytochalasin in low concentrations does not always demonstrably alter microfilament structure (see Goldman 1972). Forer *et al.* (1972) have reported that cytochalasin does not affect the appearance of actin-like filaments in blood platelets or their capacity to react with heavy meromyosin. Nevertheless, inhibition of microfilament-related functions would not necessarily be accompanied by morphological evidence of disorganization, as the effects of agents blocking neuromuscular transmission suffice to illustrate.

Evidence in support of a direct effect of cytochalasin B on actomyosin has been published by Spudich & Lin (1972). They reported that cytochalasin decreases the viscosity of the actomyosin complex of rabbit skeletal muscle by at least 60%. Cytochalasin does not affect the viscosity or ATPase activity of heavy meromyosin, which suggests that it interacts with the actin component of the complex; viscosity measurements showed a strong interaction with actin at nearly stoichiometric concentrations. This work is open to criticism on two grounds: first, the effective concentration of cytochalasin (250μM) is higher than that required to inhibit cell movement and secondly cytochalasin does not affect the contraction of striated muscle *in vivo* or in culture (Sanger *et al.* 1971). It would be of interest to repeat the work of Spudich & Lin with preparations of actin and myosin from cells that are highly sensitive to cytochalasin. One possibility is that cytochalasin affects purified actin, particularly the transition from G- to F-actin, but that when actin in intact striated muscle is polymerized and complexed with other materials such as troponin and tropomyosin, cytochalasin has little effect.

It has also been shown that cytochalasin inhibits the incorporation of glucosamine into glycoproteins (Holzer & Sanger 1972). This may be related to the inhibition by cytochalasin of the transport of glucose, deoxyglucose and glucosamine across plasma membranes (Kletzien *et al.* 1972; Zigmond & Hirsch 1972; Estensen & Plagemann 1972; Mizel & Wilson 1972). Amino acid transport is not inhibited by cytochalasin, so the effect is selective. A possible explanation is that phosphorylated sugars are removed from the plasma membrane by cytoplasmic protein acceptors (Kundig & Roseman 1971); if access of these proteins to the plasma membrane is reduced in cytochalasin-treated cells transport of sugars and amino sugars would be inhibited, whereas transport of amino acids and other compounds not requiring cytoplasmic protein acceptors would be unaffected.

The inhibition by cytochalasin of sugar transport has led to speculations that stopping of cellular movements also follows effects on the plasma membrane rather than on the contractile system itself. This seems unlikely, since the drug inhibits contraction of cortical microfilaments in frog eggs induced by injection of calcium ions into the cytoplasm (Ash, quoted by Wessells *et al.* 1971). Moreover, B. Elford and I (unpublished results) have found that cytochalasin B (10 μg/ml) inhibits the contraction of glycerinated smooth muscle (rabbit taenia coli) induced by ATP. Movement of isolated cytoplasm from cut cells of *Chara corralina*, displacement of organelles and rotation of chloroplasts is also inhibited by cytochalasin (Williamson 1972). In none of these situations is plasma membrane permeability a limiting factor. It therefore seems probable that cytochalasin B exerts a direct effect on the contractile system of smooth muscle and the analogous systems of other cells. Effects of cytochalasin on membranes are of interest in long-term results of treatment, such as glycoprotein synthesis, but probably do not explain the marked effects seen in cells exposed to cytochalasin for short periods. Cells treated with cytochalasin (up to 100 μM) for relatively long periods continue to synthesize protein at rates not significantly different from normal (Parkhouse & Allison 1972; P. Davies, unpublished results), so it seems unlikely that overall inhibition of metabolism is limiting (see also Estensen & Plagemann 1972). Nevertheless, it scarcely needs emphasizing that the effects of cytochalasin on contractile and other systems are still poorly defined and that further studies are needed before too much reliance is placed on experiments based solely on the effects of cytochalasin.

IDENTIFICATION OF ACTIN IN CELLS OTHER THAN MUSCLE CELLS

The presence of actin-like proteins in a variety of cell types has been demonstrated by biochemical and ultrastructural methods. One biochemical approach has depended upon the ability of crude or purified preparations to increase the ATPase activity of heavy meromyosin under appropriate conditions. Actin-like proteins have been obtained from the slime mould *Physarum* (Hatano & Oosawa 1966), *Acanthamoeba* (Weihing & Korn 1969), sea urchin eggs (Hatano *et al.* 1969), blood platelets (Bettex-Galland & Lüscher 1965), the brush border of chicken intestinal epithelial cells (Tilney & Mooseker 1971), mammalian brain (Berl & Puszkin 1970) and chick embryo fibroblasts (Yang & Perdue 1972).

A second biochemical approach has depended upon the demonstration in non-muscle cells of a protein of molecular weight about 45 000 with an actin-like amino acid composition (including 3-methylhistidine) and peptide maps indistinguishable from those of muscle actin [see Pollard & Korn (1972) for references]. A sensitive variant of this technique is the use of proteins labelled with a radioactive amino acid in cultures of chick sympathetic nerve cells to demonstrate homology with muscle actin (Fine & Bray 1971). Fine & Bray found that actin peptides from chick brain and muscle were indistinguishable and concluded that, although nothing short of the total amino acid sequences of the two proteins will establish identity, their failure to find any differences in the fingerprints of actins from brain and muscle suggests that they are the same protein. (This contrasts with the obvious difference between brain and muscle myosins discussed later.) Actin and tubulin are major bands of nerve cell cytoplasmic protein submitted to acrylamide gel electrophoresis (Fine & Bray 1971) and actin is thought to comprise some 12% of the cytoplasmic protein of many cells (Pollard & Korn 1972).

Ultrastructural identification of microfilaments as actin-like has been made possible by the demonstration that such filaments, 5–7 nm in diameter, can bind and become 'decorated' with heavy meromyosin. This procedure, originally used with skeletal muscle F-actin (Huxley 1963), has been adapted to other glycerinated cells, including chick embryo cells (Ishikawa *et al.* 1969), the slime mould *Physarum* (Nachmias *et al.* 1970), *Acanthamoeba* (Pollard *et al.* 1970), *Amoeba proteus* extracts (Pollard & Korn 1971), human blood platelets (Behnke *et al.* 1971*a*), leucocytes (Senda *et al.* 1969), newt eggs (Perry *et al.* 1971) and mouse macrophages (Allison *et al.* 1971).

Moore *et al.* (1970) have constructed three-dimensional models of F-actin, thin filaments and thin filaments decorated with heavy meromyosin. These show that the appearance of 'arrowheads' of the correct periodicity depends upon a highly specific interaction of double helical filaments of F-actin with heavy meromyosin, which is very unlikely to occur by chance association with some other material. The presence in non-muscle cells of thin filaments 5–7 nm in diameter which can be decorated with heavy meromyosin is therefore acceptable evidence that they are actin-like. The arrowheads are more readily seen in negatively stained preparations than in thin sections, but in the latter thin filaments appear much thicker, with a fuzzy outline, after decoration. Microtubules and filaments about 10 nm in diameter often seen in cells, to which no function has yet been assigned, are not decorated with heavy meromyosin.

It appears likely that in at least some cells much of the actin is present in a globular form, and that when suitably stimulated it quickly polymerizes into filaments 5–7 nm in diameter. Thus, the circulating blood platelet is disc-shaped and when rapidly fixed shows few microfilaments; in response to various stimuli, including ADP, it is transformed into an amoeboid cell in which many filaments are observed (Behnke *et al.* 1971*a*). The cytoplasmic fragments of *Amoeba proteus*, shown by Thompson & Wolpert (1963) to contract when exposed to ATP and warmed, have few thin filaments in the precursor stage; after stimulation the viscosity of the preparation increases and numerous filaments 5–7 nm in diameter appear, which can be decorated with heavy meromyosin (Pollard & Korn 1971). Later the filaments aggregate to form birefringent fibrils visible by light microscopy which interact with 16 nm filaments to constitute a contractile system. It seems likely that the actin-like filaments are formed from precursors in the groundplasm; they are unable to contract by themselves but require the myosin-like thick filaments for contraction.

Thus part of the actin in at least some cells is in a precursor form, which can polymerize when required, and part probably remains filamentous. This transition from G- to F-actin seems to represent the basis for the classical sol–gel transition, and might be important in locomotion and contact inhibition as discussed later.

Some of the actin-like filaments seem to be closely related to cell membranes. Pollard & Korn (1971) isolated purified plasma membranes from *Acanthamoeba*

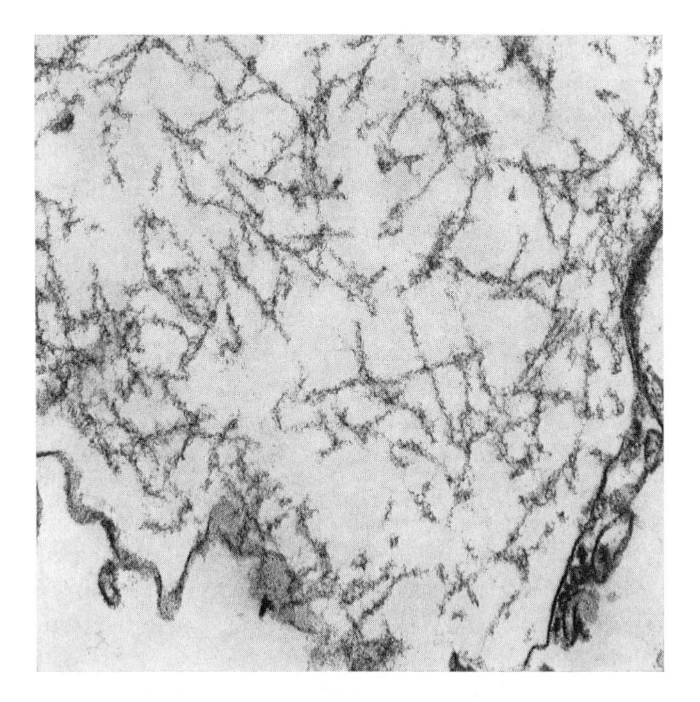

FIG. 1. Electron micrograph of the cytoplasm beneath the plasma membrane of a guinea-pig peritoneal macrophage extracted with glycerol and incubated with rabbit skeletal muscle heavy meromyosin, showing the network of actin microfilaments closely related to the plasma membrane. × 50 000 (from Allison *et al.* 1971).

and found typical actin-like filaments regularly associated with them. Crawford (1971) has obtained a fraction consisting of small vesicles and sheet membrane fragments from blood platelets. Actomyosin activity was found in this fraction. Thus membranes prepared in media that do not dissociate actin may contain these filaments, suggesting that the morphological association reflects a functional relationship.

In many cell types, such as macrophages (Allison *et al.* 1971), retinal cells (Crawford *et al.* 1972) and pancreatic β-cells (Orci *et al.* 1972), a dense network of microfilaments lies immediately beneath the plasma membrane (Fig. 1). These appear to form a cage keeping the nucleus and cytoplasmic organelles (lysosomes, mitochondria and secretory granules) away from the plasma membrane. They may also help to maintain the rounded contour of the cells by a process analogous to muscle tone in whole animals.

Whether actomyosin-like proteins are also present in organelles in which microtubules are prominent, such as the mitotic or meiotic spindle, has been much debated. Peptide mapping shows that tubulin is distinct from actin (Stephens 1971), but the concept that an actomyosin system exists alongside microtubules receives support from the observations (Gawadi 1971; Behnke *et al.* 1971*b*) on microfilaments which can be decorated with heavy meromyosin amongst the microtubules of mitotic and meiotic spindles of locust and cranefly testis cells.

It is still possible that the microtubules do not themselves contract but form a system of fairly rigid but somewhat flexible rods that can resist compression, while microfilament contraction provides the tension necessary for movement. The microtubules would thus serve a skeletal and directing function, important for polarity of movement of chromosomes and other organelles within cells. Disassembly of microtubules, especially if only possible at certain regions such as one or both ends, would allow controlled shortening of structures such as the spindle. The recent finding that calcium ions cause disassembly of microtubules (see later) as well as being the trigger for microfilament contraction implies that the two processes could readily be coordinated in cell physiology.

Kaminer & Szonyi (1972) have identified a protein closely resembling tropomyosin in the electric organs of the electric eel and *Torpedo*, so the regulatory proteins may also be found in non-muscle cells.

IDENTIFICATION OF MYOSIN-LIKE PROTEINS IN CELLS OTHER THAN MUSCLE CELLS

Three methods have been used to identify myosin-like proteins in a variety of cells: biochemical, ultrastructural and immunological. The first depends on the isolation of proteins that have a magnesium-activated ATPase activity which can be increased by addition of actin. Such materials have been isolated from human and porcine blood platelets (Bettex-Galland & Lüscher 1965; Adelstein *et al.* 1971) and slime mould (Hatano & Tazawa 1968). The proportion of myosin in non-muscle cells appears to be much less than that of actin, but further quantitative studies are required.

The ultrastructural identification of myosin filaments is at present unsatisfactory because the filaments vary widely in thickness from about 4 to 18 nm according to the material and method of preparation, and there is no convenient marker such as localizable enzyme activity or heavy meromyosin binding. However, thick filaments have been identified in a variety of preparations, such as human blood platelets (Behnke *et al.* 1971*a*) and *Amoeba* cytoplasm (Pollard & Korn 1971), and there are functional reasons for believing that these consist of myosin-like proteins.

At the light microscopical level, information about the distribution of myosin in different cell types has come from the use of fluorescent antibodies against human smooth muscle myosin (Kemp *et al.* 1971; Farrow *et al.* 1971; Allison & Brighton, unpublished results). These are either prepared by immunization of rabbits with myosin purified from fresh human uterus or are found as naturally occurring autoantibodies against smooth muscle in certain human patients with hepatitis. We have investigated the reactions of these antibodies in a variety of fixed and unfixed cells, using the 'sandwich' technique in which cells are 'stained' with antiglobulins coupled with fluorescein isothiocyanate after treatment with antibody and repeated washing. Unfixed cells, stained after exposure to 0.3M-sodium azide in the cold (to inhibit pinocytosis), show no fluorescence, from which it is clear that the outer surface of the plasma membrane, to which the antibody can gain access, does not contain demonstrable quantities of myosin. Smooth muscle from all parts of the body tested (viscera, blood vessels and bronchi) shows intense cytoplasmic fluorescence, while striated muscle (skeletal and cardiac) shows weak or no fluorescence. Fairly strong fluorescence is observed in the cytoplasm of many cells (fibroblasts, macrophages, lymphocytes, granulocytes, blood platelets, neurons and many different types of tumour cells). Staining of liver cells by this type of antibody has been described by Farrow *et al.* (1971) and localization of smooth muscle protein in myoepithelium has been reported by Archer *et al.* (1971). Most of the staining is in the peripheral cytoplasm, beneath the plasma membrane: the staining can be diffuse, as in macrophages (Fig. 2), or in linear arrays along the long axis of cells, as in fibroblasts and some tumour cells; the nucleus remains unstained.

Cells other than muscle cells are only poorly stained with antibody against skeletal muscle myosin, which stains very intensely in striated muscle cells. The antibodies against smooth muscle myosin show strong precipitation with homologous myosins in Ouchterlony gels, good reactions with smooth muscle myosins of other mammals, very weak reactions with skeletal muscle myosins and no detectable reactions with actin from either source. From these observations it seems that the non-muscle cells so far investigated contain in their peripheral cytoplasm myosin that can be shown immunologically to resemble that of smooth muscle rather than that of skeletal muscle. Striated muscle myosins are known to be heterogeneous and it would be premature to conclude that smooth muscle myosins are uniform, whether present in muscle or other cells. This variability of myosin contrasts with the apparent similarity of actin in different cells already discussed.

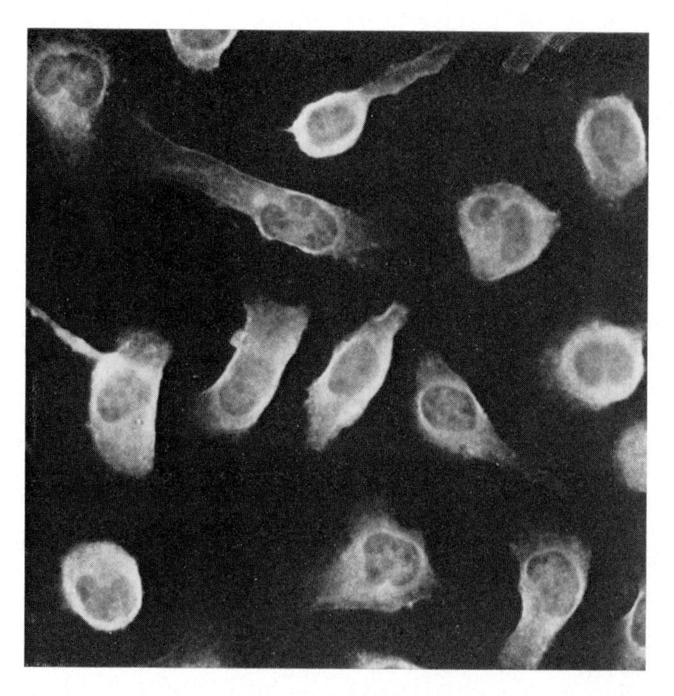

FIG. 2. Culture of mouse peritoneal macrophages fixed in acetone and stained with rabbit antibody against human smooth muscle myosin and fluorescein-conjugated porcine antibody against rabbit immunoglobulin. Myosin is shown in the cytoplasm of the cells, the highest concentration being just beneath the plasma membrane. × 600.

CAN MEMBRANES MOVE BY THEMSELVES?

A crucial question when analysing cell motility is whether membranes have any intrinsic motive power or whether they are moved by the action of extrinsic agents, such as filaments, which are attached to them. During the past few years evidence has accumulated in support of a fluid-mosaic model of membrane (see Singer & Nicolson 1972), in which the constituents are envisaged as floating in an oily fluid and not occupying a fixed position in a solid structure. Proteins are retained within a membrane because they have surface hydrophobic groups that interact with the hydrocarbon chains of lipids; nevertheless, they have lateral mobility in the plane of the membrane. Examples of such lateral movements are the mixing of membrane antigens after fusion of different cell types (Frye & Edidin 1970) and the formation of 'patches' and 'caps' in lymphocytes when surface immunoglobulins are cross-linked by bivalent anti-immunoglobulin sera (Taylor *et al.* 1971; de Petris, pp. 27-40). This reaction

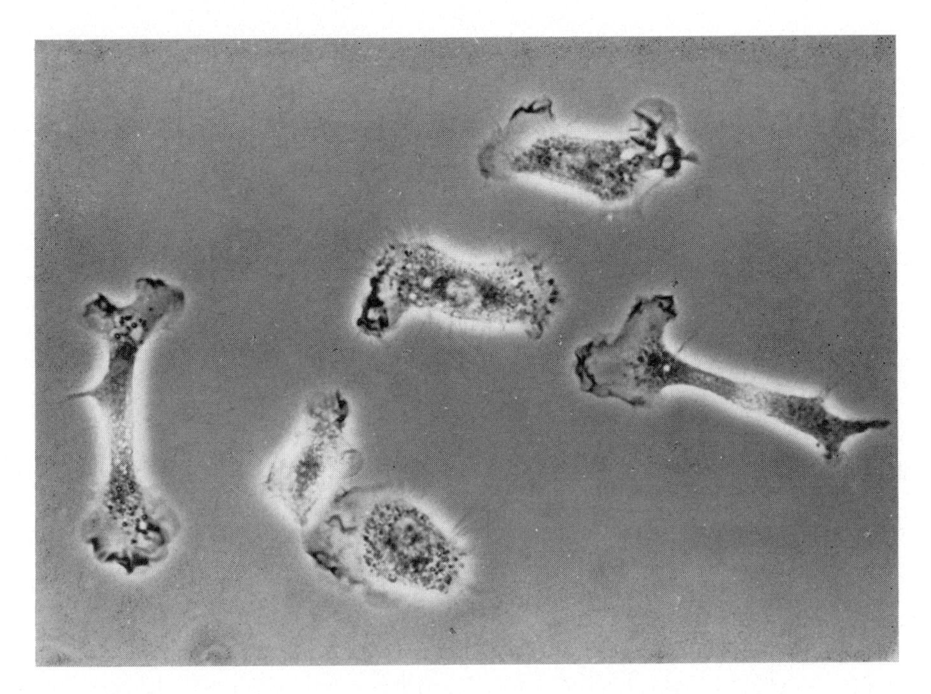

FIG. 3. Phase-contrast photomicrograph of normal mouse peritoneal macrophages in culture, showing the granule-free peripheral cytoplasm and ruffle membranes. × 700 (from Allison *et al.* 1971).

is not prevented by cytochalasin, and probably follows linkage of membrane protein units brought together by diffusion within the membrane. Formation of such protein 'patches' could, however, result in a distortion of the membrane structure to generate small spherical infoldings or outfoldings analogous to the budding of enveloped viruses (see Allison 1971*a*).

An example is possibly the invagination of membranes of erythrocytes, treated with relatively high concentrations of aminoquinolines, to form vesicles within the cell (Ginn *et al.* 1969). Analogous processes might result in the formation of small pinocytic vacuoles, such as those described later, which is unaffected by cytochalasin.

It has also been postulated that energy-driven movement of membrane proteins is involved in specific transport of ions. In liposomes, the concentric lipid bilayers developed by Bangham *et al.* (1965), alteration of the ionic composition of the medium leads to wavy motion of the lamellae in the particles. However, it seems unlikely that any of these limited types of movement known within membranes can account for the marked displacement of membranes

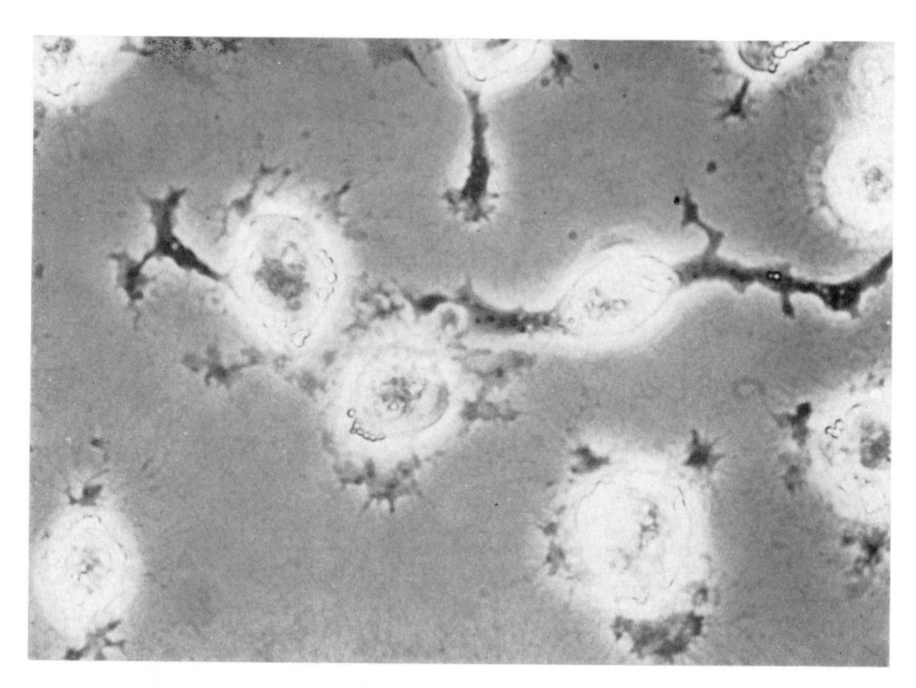

FIG. 4. Phase-contrast photomicrograph of mouse peritoneal macrophages in culture for 15 min with cytochalasin B (10 μg/ml). The cells are immobile and show thin, irregular remnants of pseudopodia and very flattened areas of peripheral cytoplasm. × 850 (from Allison *et al.* 1971).

which is necessary for ruffling, cell movement, phagocytosis, macropinocytosis and the discharge of packaged secretions. These appear to require the movement of membrane by some external force.

EFFECTS OF CYTOCHALASIN ON RUFFLE MEMBRANE ACTIVITY AND CELL LOCOMOTION

It is generally agreed that when cytochalasin B in low concentrations (0.5–5 μg/ml, i.e. 1–11 μM) is added to cells, ruffle membrane activity and cell movements cease within a few minutes and that this arrest is rapidly reversed when cytochalasin is removed and replaced by fresh medium (Fig. 3). Such changes have been described in L cells, a permanent line of mouse fibroblasts (Carter 1967), elongating nerve cells (Wessells *et al.* 1971), mouse peritoneal macrophages (Allison *et al.* 1971), BHK-21 cells (Goldman 1972) and the slime

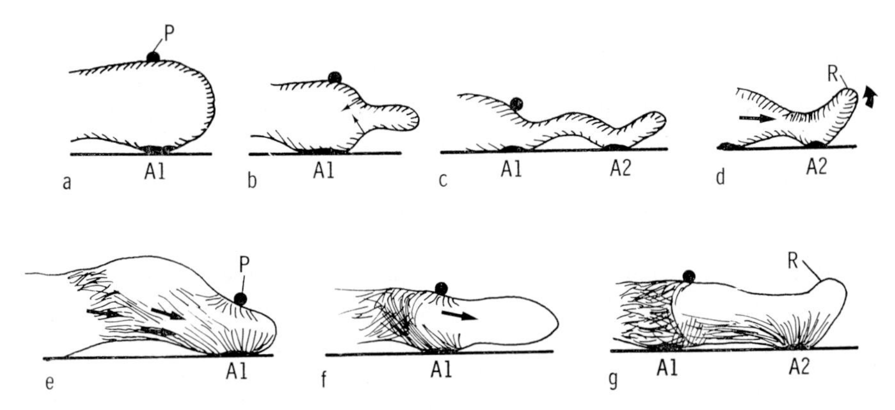

FIG. 5. Diagrammatic illustrations of two models of forward movement of cytoplasm in locomotion, assuming that the motive force is produced by contraction of a microfilament system beneath and attached to the plasma membrane, but not exerting tension in one direction only [Bray (p. 366) discusses another model in which the tension is directional]. According to the pleated-membrane model (a), the convex forward surface of the cell above a point of adhesion to the substrate, A1, is pleated by contraction of filaments which bring together the upper and lower surfaces of the plasma membrane (b). As a result the plasma membrane extends (c) and establishes a second point of adhesion to the substrate, A2. Further contraction of microfilaments leads to a movement of the cytoplasm into the pleat, thickening it from behind and pulling the membrane backwards relative to the cytoplasm. The membrane is fixed at a point of attachment below, but is pulled upwards to form a ruffle, R. The end result is influx of cytoplasm to produce a convex leading edge above the second point of adhesion, and the process is repeated.

According to the phase-transition model (e), the cytoplasm close to the leading edge of the cell is more fluid than that further away. Near a point of adhesion to the substrate, A1, actin is polymerized to form a gel-like structure which joins the remaining cytoplasmic gel. Contraction of this system squeezes the fluid cytoplasm and overlying plasma membrane forwards, a second point of adhesion, A2, is established, and the process is repeated. A contraction might initially move the leading edge upwards to form a ruffle, R. In either model a particle, P, above the initial point of adhesion will move backwards relative to the cytoplasm.

mould *Dictyostelium discoideum* (Wiklund & Allison 1972). Apart from a globule of cytoplasm (endoplasm) around the nucleus, containing nearly all the cytoplasmic organelles, the whole cell lies flat on the substratum, suggesting that whatever support there is between upper and lower layers of plasma membrane in normal cells has been lost after treatment with cytochalasin. Existing adhesions to the substrate are seldom lost, and the cells appear to establish the irregular 'spidery' appearance of the cells after treatment with the drug (Fig. 4).

The simplest interpretation of these findings is that microfilament-mediated contraction is required for ruffling and for the forward movement of the cell membrane and enclosed cytoplasm in locomotion. There is no evidence that microtubules are present in the neighbourhood of the ruffles or leading edge, nor that microfilaments can provide a microtubule-like extensible system that can resist compression. At least two models can be considered by which co-ordinated microfilament-mediated contractions can lead to cytoplasmic extension. According to one (Fig. 5), a rounded leading edge could be converted into a thin, discoid forward extending edge by contractions bringing the upper and lower parts of the plasma membrane together. This would be analogous to pleating, and if the lower surface were to adhere to the substratum, further contractions could lead to movement of the pleat upwards and backwards (ruffling). The pleating of the membrane of frog eggs after contraction of cortical microfilaments (Gingell 1970) illustrates the principle. A second model depends on the observed fluidity of cytoplasm near the site of elongation as compared with other regions. If a high proportion of actin is unpolymerized near the site of elongation, any contractions will tend to squeeze it forwards, so displacing the plasma membrane. After adhesion to the substratum, a plasma membrane change (perhaps due to apposition of adhesive units) might rapidly trigger polymerization of actin in the immediate vicinity. The cytoplasm near the point of contact would then gel, and subsequent contractions from this site would pull the cell body towards the new site of adhesion. These models are not mutually exclusive. The pattern of birefringence observed supports the frontal contraction theory of amoeboid movement (Allen 1972). The geometry of the initial forward movement would presumably depend on purely local factors such as prior constraints from adhesion sites, distribution of polymerized actin and the control of the activity. Thus, spreading of ruffling activity, with less polymerized actin filaments around the margin, might convert 'ruffling' into 'blebbing'. This happens at low cytochalasin B concentrations, as well as under unfavourable conditions, and might be due to partial inhibition of the microfilament system.

EFFECTS OF CYTOCHALASIN ON ENDOCYTOSIS

It is generally agreed that phagocytosis and macropinocytosis—the intake of fluid into vacuoles visible by light microscopy (i.e. those over 300 nm in diameter)—are markedly inhibited by cytochalasin B in the concentrations described above (Table 1). In contrast, micropinocytosis—the invagination of plasma membrane to form small vesicles (visible only by electron microscopy), which may contain ferritin, colloidal gold, horseradish peroxidase or other markers added to the extracellular fluid—is not demonstrably inhibited by cytochalasin B (Wills *et al.* 1972).

Attachment of opsonized bacteria to polymorphonuclear leucocytes or

TABLE 1

Inhibition of endocytosis by cytochalasin B

Cell type	*Endocytized material*	*Cytochalasin concentration (µM)*	*Reference*
Mouse peritoneal macrophages	*E. coli*	2–11	Allison *et al.* (1971)
Mouse peritoneal macrophages	Medium (macropinocytosis)	2–11	Allison *et al.* (1971)
Mouse peritoneal macrophages	[^{3}H]*E. coli*	4–21	S. Rivett, P. Davies, A. C. Allison & A. D. Haswell, unpublished results
Human peripheral blood polymorphs	*E. coli*	1–10	Malawista *et al.* (1971)
Rabbit alveolar macrophages	*E. coli*	1–10	Malawista *et al.* (1971)
Human peripheral leucocytes	*E. coli*	21	Davis *et al.* (1971)
Chang	[^{3}H]Sucrose	21	Wagner *et al.* (1971)
Thyroid epithelial	Colloid	6	Williams & Wolff (1971*a*, *b*)

macrophages is not inhibited by cytochalasin, but electron micrographs show that the bacteria are never completely engulfed but remain within invaginations of the plasma membrane. Lysosomal granules are seen in close apposition to the plasma membrane and are discharged into the folds containing the bacteria. Biochemical measurements show that during phagocytosis some lysosomal hydrolases are discharged into the surrounding medium; in cells treated with cytochalasin there is a similar loss, but when cells are exposed to both cytochalasin and bacteria the loss of hydrolases is much greater (Davies *et al.* 1973*a*). The loss is selective for lysosomal enzymes; the cells remain viable and the cytoplasmic enzyme lactate dehydrogenase is not found in the medium. We conclude that in the intact leucocyte a network of actin-like microfilaments prevents lysosomes from gaining access to the plasma membrane. During phagocytosis this network might function in interiorization of the membrane. As the phagocytic vacuole is formed, the filament network ceases to be continuous and lysosomes can fuse with the vacuole to form large secondary lysosomes.

Colchicine does not inhibit ruffle membrane activity, amoeboid movement of cells, phagocytosis or macropinocytosis, although the movement of endocytic vacuoles within cells may be affected (see later). Hence it is unlikely that the motive power for these activities comes from microtubules.

It has been claimed (Rabinovitch 1967) that macrophages require calcium ions in the medium for phagocytosis. This claim is based on experiments in which the cells were incubated in medium containing ethylenediaminetetraacetic acid (EDTA) which chelates magnesium and other ions as well as calcium. We found that EDTA soon produces irreversible changes in macrophages. In contrast, ethylenebis(oxyethylenenitrilo)tetraacetic acid (EGTA), which has a much greater affinity for calcium than for other ions, does not detectably inhibit phagocytosis of labelled bacteria (P. Davies & Allison, unpublished results), so that an exogenous source of calcium does not seem to be required for phagocytosis.

A POSTULATED MECHANISM FOR PHAGOCYTOSIS

The foregoing observations suggest a mechanism for phagocytosis. The initial stage is attachment of a particle to the plasma membrane. Mammalian phagocytic cells have on their plasma membranes receptors for antibody and complement (Lay & Nussenzweig 1968), and it is likely that these are involved in the initial attachment of opsonized microorganisms. Latex particles and formalin-treated erythrocytes are attached in the absence of antibody, probably owing to their hydrophobic surface (Rabinovitch & de Stefano 1970). In amoebae, the ready attachment of positively charged particles suggests that electrostatic interaction is important (see Jacques 1969).

We propose that attachment of a particle traps and immobilizes membrane proteins in the vicinity and allows them to form a cluster thereby increasing the permeability to sodium ions and depolarizing the membrane (Allison 1971*a*). Such depolarization has been reported after attachment of particles to peritoneal exudate cells (J. Kouri, personal communication).

It seems reasonable to postulate that, as in muscle, membrane depolarization is followed by release of calcium ions from an intracellular site of sequestration, probably smooth membrane vesicles analogous to sarcoplasmic reticulum. Release of calcium ions could trigger contraction of an actomyosin system of microfilaments attached to the plasma membrane, thereby inducing phagocytosis (Fig. 6). The mechanism of pinocytosis (visible by light microscopy) appears to be similar. In macrophages pinocytosis can be induced by an anticellular antibody present in calf serum (Cohn 1970); in tumour cells it can be induced by basic proteins (Ryser 1967). In either case these agents could immobilize membrane proteins as postulated above. Since the cytoplasm is often everted to form a rim over the phagocytized particle, activity at the periphery of the particle as well as below it is required (Fig. 6); this would be analogous to the 'pleating' already discussed.

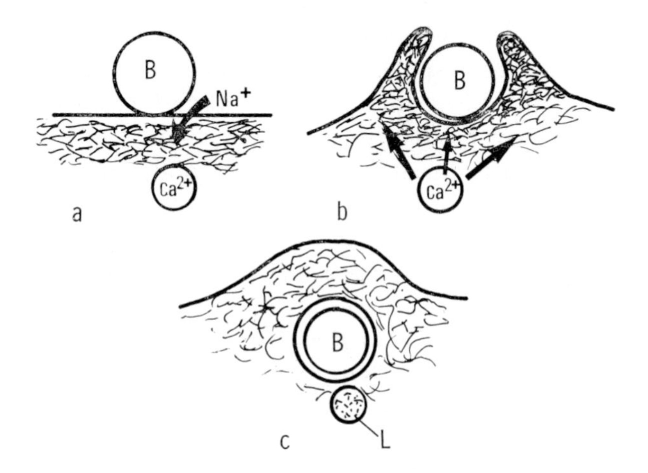

FIG. 6. Proposed mechanism of endocytosis. A bacterium (B) becomes attached to the plasma membrane and increases entry of sodium ions (a). The depolarization of the plasma membrane is followed by release of calcium from an intracellular vesicle (b), which triggers contraction of microfilaments beneath and around the particle. Consequent movement of the membrane leads to engulfment of the particle. When the phagocytic vacuole containing the bacterium enters beneath the dense ectoplasmic microfilament network (c) lysosomes (L) can fuse with it and discharge their hydrolases.

DISCHARGE OF PACKAGED SECRETIONS

Some secretory products are released from cells soon after their formation; such release is more or less continuous and is not triggered by any particular environmental stimulus. This appears to be true, for example, for the release of immunoglobulin from plasma cells. When a human or experimental animal is bearing a plasma cell tumour, for example, the immunoglobulin product accumulates in the serum as a characteristic monoclonal product, or spike. This type of secretion can be termed *continuous* and *unstimulated.*

In contrast, many secretory products are not continuously released but accumulate within cells in secretory granules. These are discharged in response to some specific stimulus, which varies from cell to cell. In the β-cells of the pancreas, for example, insulin-containing granules are released when the concentration of glucose in the environment rises. Appropriate agonists likewise trigger the release of catecholamine-containing granules from the adrenal medulla and other hormones, as well as amylase from salivary glands and other secretions. Similarly, presynaptic nerve impulses bring about release of neurotransmitters from nerve endings, and cell-bound antibody and antigen stimulate the release of histamine from mast cells. Such modes of secretion can be termed *discontinuous* and *controlled*. The underlying mechanisms are not yet fully understood, but some general properties of such systems are beginning to emerge. One is that depolarization of the plasma membrane follows the appropriate stimulation, not only in nerve cells but also in cells such as the β-cells of the pancreas and adrenal medullary cells (see Rubin 1970).

The second property is the widespread requirement of calcium ions in the extracellular medium for the release of packaged secretions. It seems likely that a rise in the concentration of calcium in the cytoplasm is the trigger for granule discharge. Thus, in the stimulated presynaptic nerve, studies with the luminescent protein aequorin have shown an increase in cytoplasmic calcium concentration and discharge of neurotransmitters (Llinas *et al* 1972). The amount of calcium accumulated in β-cells exposed to glucose is related to the rate of insulin secretion (Malaisse-Lagae & Malaisse 1971). Barium and strontium are the only ions that can substitute for calcium; increasing extracellular magnesium inhibits release. The membrane depolarization can be related to increased cytoplasmic calcium concentration by increasing the influx of this ion or release from an intracellular sequestered form after plasma membrane depolarization as in muscle.

It is therefore of interest that calcium is the universal trigger for initiating the contraction of actomyosin systems (Ebashi & Endo 1968), including smooth muscle (Hurwitz & Suria 1971) and the cortical cytoplasm of non-muscle cells (Gingell 1970). Only barium and strontium will substitute for calcium in triggering the contraction of actomyosin. Actin and myosin are now known to be present in cells from which granules are discharged, so that their role in the release mechanism deserves careful analysis. Studies with an inhibitor of actomyosin contractility might therefore provide useful information about the way in which packaged secretions are discharged.

EFFECTS OF CYTOCHALASIN ON THE RELEASE OF CELLULAR PRODUCTS

Available information on the effects of cytochalasin B on release of cellular products is summarized in Table 2. These effects vary from system to system and even within similar systems. Continuous, uncontrolled release, such as that of immunoglobulins, is unaffected by cytochalasin; presumably small vesicles containing cell products diffuse to the plasma membrane without participation of a contractile system. In certain cells cytochalasin increases the release of granular contents without lysing them, for example, lysosomal release from phagocytic cells, and this has also been reported for insulin from rat pancreatic β-cells (Orci *et al.* 1972) and histamine from human leucocytes (Gillespie &

TABLE 2

Effects of cytochalasin B on the release of cell products

Tissue	*Product*	*Effect on release*	*Cytochalasin concentration (μM)*	*Reference*
Mouse thyroid	Iodoprotein	Inhibition	1–6	Williams & Wolff (1971*b*)
Bovine pituitary	Growth hormone	Inhibition	11	Schofield (1971)
Rat mast cells	Histamine	Inhibition	10	Orr *et al.* (1972)
Human leucocytes	Histamine	Increase	11	Gillespie & Lichtenstein (1972)
Leucocyte	Histamine	Increase	0.4–10	Colten & Gabbay (1972)
Lymphocyte	Lymphotoxin	Inhibition	0.2–10	Yoshinaga *et al.* (1972)
Mouse peritoneal macrophages	Lysosomal hydrolases	Increase	0.4–21	Davies *et al.* (1973*a*)
Rabbit leucocytes	Lysosomal hydrolases	Increase	2–11	Davies *et al.* (1972, 1973*b*)
Rat pancreatic islets	Insulin	Increase	21	Orci *et al.* (1972)
Opsanus islets	Insulin	Inhibition	21	Watkins & Allison unpublished data
Mouse plasmacytoma	Immunoglobulin	None	100	Parkhouse & Allison (1972)
Sympathetic nerve endings (vas deferens)	Noradrenalin	Inhibition	6	Thoa *et al.* (1972)
Sympathetic ganglion	Acetylcholine	Inhibition		Pumplin & McClure (1972)
Rat salivary gland	Amylase	Inhibition	13	Butcher & Goldman (1972)
Adrenal	Catecholamines	Inhibition	100	Douglas & Sorimachi (1972)
Posterior pituitary		Inhibition	80	Douglas & Sorimachi (1972)

Lichtenstein 1972). However, the latter effects are highly variable; Gillespie & Lichtenstein found that with leucocytes from certain human individuals cytochalasin *inhibited* histamine release, as we found for rat mast cells (Orr *et al.* 1972). In contrast to the increased secretion of insulin by β-cells of the isolated rat pancreatic islets (which may have been altered by the enzymic digestion necessary to obtain islets), D. Watkins and I (unpublished data, 1972) found that cytochalasin *inhibited* insulin release stimulated by glucose from the islets of the toadfish, *Opsanus tau*, which do not require enzymic digestion in the course of preparation for analysis.

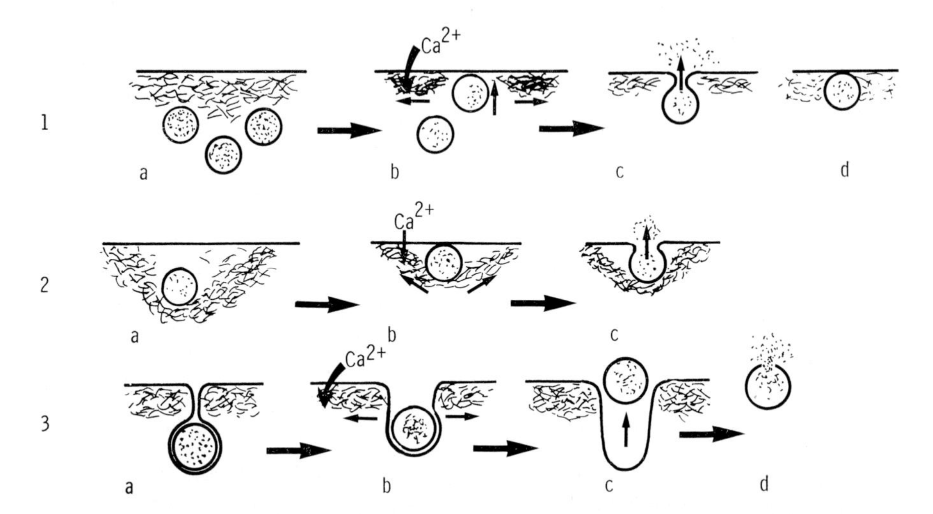

FIG. 7. Possible mechanisms by which microfilaments might release packaged secretions:

1(a), The presence of a microfilament network keeps the secretory granules away from the plasma membrane. When calcium ions trigger contraction of the filament system (b) the network ceases to form a continuous layer and allows granules to diffuse to the plasma membrane. The two membranes fuse to allow escape of granule contents (c). If the filament network is paralysed (d), granules can likewise make contact with the plasma membrane, so that unstimulated release is increased.

2, A sling-like mechanism, in which relative concentration of filaments attached to or around the granule contracts after appropriate stimulation, pulling the granule onto the plasma membrane (b), before release of the contents (c).

3, The granule is held in a flask-like invagination of the plasma membrane, the walls of which are pulled open after appropriate stimulation (b). The granule is released (c) and later the contents are discharged into the medium (d).

In the majority of tissues which have been studied, cytochalasin exerts a reversible inhibition on the discharge of packaged secretions. Cytochalasin has had no detectable effect on the entry of calcium into pancreatic islet cells in the presence of glucose (Orci *et al.* 1972) or sympathetic nerve ganglion (Pumplin & McClure, personal communication). Hence it may be paralysing the secretory mechanism itself.

The question then arises, why are basal and stimulated release sometimes *increased* by cytochalasin? As described before, the microfilament system appears to form a network beneath the plasma membrane, keeping the nucleus and granules away from the membrane. In a number of cell types treated with cytochalasin, granules (and in some cases the nucleus) are seen in close apposition to the cell membrane, so that the granule might fuse to the plasma membrane and be discharged because the restraining 'cage' function of the microfilament network is lost. Some models by which the contraction of a microfilament system might effect release of packaged secretions are illustrated in Fig. 7. The increase or decrease of secretion by cytochalasin might then be the outcome of a delicately balanced series of effects, depending on the nature and state of the cells and precise experimental conditions.

EFFECTS OF COLCHICINE ON RELEASE OF CELL PRODUCTS

Two biological activities of colchicine have long attracted attention: its capacity to arrest mitosis in metaphase and its therapeutic effect in acute gout. When the former was shown to be related to the capacity of colchicine to bind tubulin and break down labile microtubules, the possible relationship of microtubules to gout was investigated. Gout is due to ingestion of small crystals of monosodium urate by phagocytic cells in joints. Malawista & Bodel (1967) reported that colchicine has no demonstrable effect on phagocytosis in human peripheral blood leucocytes but inhibits the discharge of lysosomes into phagocytic vacuoles. Inhibition by colchicine of the release of lysosomal enzymes from human leucocytes into the medium in the presence of zymosan particles has been described by Weissmann *et al.* (1971).

Colchicine has also been reported to inhibit the release of other cell products (Table 3). In some cases, such as release of thyroid [^{125}I]iodoprotein (Williams & Wolff 1970), inhibition is observed with low concentrations of colchicine. Analogues of the drug are active in proportion to their activity on microtubular systems, and vinblastine and 2-methyl-2,4-pentanediol also inhibit secretion. In such situations it seems probable that the inhibition is mediated by microtubules, rather than some non-specific effect, and that these organelles function at some stage in stimulated thyroid hormone secretion. This is a complex process involving endocytosis of colloid at one pole, digestion in secondary lysosomes, and transport of the products to the other pole of the cell from which they are released.

In other examples the effects of colchicine are less readily reproducible in different systems or require high concentrations of the drug. Thus Lacy *et al.* (1968) have reported inhibition of insulin release from islets recovered from the rat pancreas, whereas Dr D. Hawkins and I (unpublished results) have found no inhibition of glucose-stimulated release of insulin from islets of *Opsanus*. Butcher & Goldman (1972) have reported inhibition by colchicine of amylase secretion from rat pancreas whereas Temple *et al.* (1972) were unable to find any effect of colchicine in this system. Kraicer & Milligan (1971) state that colchicine inhibits release of adrenocorticotropic hormone (ACTH) by

TABLE 3

Effects of colchicine on the release of cell products

Tissue	*Product*	*Effect on release*	*Colchicine concentration (μM)*	*Reference*
Human leucocytes	Lysosomes (into phagosomes)	Inhibition	25	Malawista & Bodel (1967)
Human leucocytes	Lysosomal hydrolases	Inhibition	1	Weissmann *et al.* (1971)
Human leucocytes	Histamine	Inhibition	300	Levy & Carlton (1969)
Rat mast cells	Histamine	Inhibition	500	Gillespie *et al.* (1968)
Rat pancreatic islets	Insulin	Inhibition	1	Lacy *et al.* (1968)
Mouse thyroid	Iodoprotein	Inhibition	0.5	Williams & Wolff (1970)
Bovine adrenal	Catecholamines	Inhibition	500	Poisner & Bernstein (1971)
Bovine pituitary	ACTH (K^+)	Inhibition	5	Kraicer & Milligan (1971)
Bovine pituitary	Growth hormone	None	1000	Schofield & Cole (1971)
Rat pituitary	ACTH (hypothalamic releasing factor)	None	5	Kraicer & Milligan (1971)
Rat pituitary	TSH, ACTH	None	50	Temple *et al.* (1972)
Rat parotid	Amylase	None	50	Temple *et al.* (1972)
Rat parotid	Amylase	Inhibition	10	Butcher & Goldman (1972)
Mouse adrenal cell culture	Corticosteroids	Increase	1	Temple *et al.* (1972)
Rat adipose	Free fatty acid	Inhibition	?	Schimmel (1972)
Mouse plasmacytoma	Immunoglobulin	None	50	Parkhouse & Allison (1972)

high concentrations of potassium from bovine pituitary slices, but not the release mediated by the hypothalamic releasing factor. Release of other pituitary hormones and of immunoglobulin from plasma cells is highly resistant to colchicine.

The only example of *stimulation* by colchicine of hormone release is that of corticosteroids from a line of mouse adrenal cells in culture. The stimulation is greater than that produced by ACTH. However, the colchicine was added for 15–16 h, and it is possible that steroid synthesis was enhanced. Other effects of colchicine must also be borne in mind. Gabbay & Tze (1972) suggest that colchicine inhibition of aldose reductase interferes with glucose-induced insulin release.

In general, it seems likely that colchicine and vinblastine can interfere with the release of some packaged secretions, apparently by effecting the microtubules. In some cases, such as thyroid secretion, release of the same product is also markedly affected by cytochalasin. Hence a colchicine-sensitive system might collaborate with a cytochalasin-sensitive system. In tissues such as the thyroid, disruption of cellular polarity might interfere with the orderly movement of material from the pole of the cell nearest the colloid to that related to blood vessels.

MICROTUBULES AND CELLULAR POLARITY

Many cells show polarity of structure, of movement over a substratum and of movement of endocytic vesicles within their cytoplasm. Polarity is most obvious in cells such as those in epithelia, which have luminal and basal borders as well as lateral contacts with other epithelial cells. Even fibroblasts and leucocytes in culture show polarity, with ruffle membrane activity often confined to one or two poles. Movement is directed towards the area of major ruffle membrane activity (Abercrombie *et al.* 1970) and pinocytosis visible by light microscopy is usually confined to this region.

Persistence of movement in one direction is readily demonstrable in fibroblasts (Vasiliev *et al.* 1970) and in macrophages (Allison 1971*b*). This appears to be a property of individual cells, and might be referred to as intrinsic polarity of movement, in contrast to extrinsic polarity of movement which is imposed by the environment, for example by having a substratum of nearly parallel fibres or in the presence of a chemotactic influence.

Miszurski (1969) reported that chick heart fibroblasts 'assume a rounded or polygonal shape' when treated with Colcemid. Vasiliev *et al.* (1970) found that Colcemid retards the migration of confluent human diploid cells into spaces created by scraping away part of the confluent layer, and affects the directional movement of fibroblasts. In fibroblasts (Gail & Boone 1971) and macrophages (Allison 1971*b*) treated with low concentrations of Colcemid or colchicine, ruffle membrane activity, instead of being confined to certain areas, spreads over the entire cell surface. Directional gliding movements are replaced by random amoeboid movements. Fibroblasts (BHK-13) treated with colchicine after cell division assume a rounded, epithelial-like appearance (Goldman 1971). After treatment with colchicine HeLa cells showed randomization of the distribution of lysosomal and Golgi elements (Robbins & Gonatas 1964), and in fibroblasts there was loss of radial organization of acid phosphatase-reactive granules around the centrosome and of directed saltatory movements of endocytic vacuoles towards the centrosomal region (Freed & Lebowitz 1970;

Bhishey & Freed 1971). Loss of polarity might participate in the inhibition of leucocyte chemotaxis by colchicine (Caner 1965), which possibly limits the number of inflammatory cells entering gouty joints.

An important function of microtubules in the maintenance of membrane structure has been recently shown. Normally phagocytosis by polymorphonuclear leucocytes (PMN) is not accompanied by internalization of membrane sites involved in specific transport of amino acids (Tsan & Berlin 1971). However treatment of PMN with 1 μM-colchicine or vinblastine leads to the internalization of large areas of membrane involved in active transport (Ukena & Berlin 1972). This suggests a loss of the mosaic-type structure normally existing in the PMN, with separation of phagocytic and specific transport sites, mediated by intact microtubules.

All these observations suggest that microtubules are prerequisites for the maintenance of the shape of cells, the disposition of organelles within them, and polarity of movement of cells in relation to the substratum as well as of endocytic vacuoles and other organelles within cells. When microtubules are laid down after cell division, they tend to be parallel, and this determines the subsequent polarity of the cell during interphase.

CALCIUM AND POLYMERIZATION OF TUBULIN

Little is known about factors that control the assembly and breakdown of microtubules in cells. An interesting advance has been made recently by Weisenberg (1972*a*), who has found that concentrated solutions of tubulin isolated from rat brain or fertilized eggs of *Spisula* can be repolymerized *in vitro* in solutions containing GTP and magnesium ions, provided that the concentration of calcium ions is very much decreased by a chelating agent such as EGTA. The high sensitivity to calcium suggests a possible mechanism for regulation of microtubule polymerizaton *in vivo*.

Evidence is also accumulating (Weisenberg 1927*b*) that there are two forms of polymerized microtubules which are present at different stages in the cycle of cell division: interphase tubulin polymers (cytoplasmic structural microtubules) and metaphase tubulin polymers (spindle microtubules). In the intervening periods (early prophase and anaphase) the polymers are broken down, as shown by the disappearance of morphologically recognizable microtubules and by biochemical experiments. The latter are based on the fact that if cells are homogenized in a microtubule-stabilizing medium, such as the spindle isolation solution of Kane (1962), and subjected to high-speed centrifugation, microtubular structures are sedimented but tubulin dimers are not. Comparison of the quantity of protein binding radioactive colchicine in the supernatant liquid with that in the sediment provides a measure of the proportion of tubulin polymerized at any stage of the cell cycle. Weisenberg (1972*b*) has found that in fertilized *Spisula* eggs the proportion of polymerized tubulin is high at interphase and metaphase, but is significantly decreased during early prophase and anaphase. Thus it appears that some factor is allowing microtubules to break down at these stages while favouring polymerization into microtubules during late prophase and telophase. In the former polymerization gives the asters and spindle, probably because of the availability of specific nucleating sites (Weisenberg 1972*a*). At anaphase, nucleation is cytoplasmic, as described by Borisy & Olmstead (1972) for brain extracts, so that cytoplasmic structural microtubules are formed in the appropriate orientation.

These observations forward the possibility that the concentration of calcium ions in the cytoplasm controls polymerization and depolymerization of tubulin. It is therefore of interest that marked variations in the activity of calcium-stimulated ATPases during the cell cycle have been found. Morill *et al.* (1971) described the existence in *Rana pipiens* of ATPase which is activated by sodium, potassium and magnesium ions during the meiotic prophase but calcium-activated during metaphase, when the activity is very high. During the first divisions of the eggs of the sea urchin, *Strongylocentrotus purpuratus*, a calcium-activated ATPase has been found by Petzelt (1972) which shows one activity peak in interphase and another during metaphase in every cell cycle. Mazia *et al.* (1972) reported that the activity of the calcium-activated ATPase is much higher in the spindle than in the surrounding cytoplasm. The mitotic apparatus has long been known to contain vesicles lined by unit membranes and it seems likely that these have calcium-activated ATPase activity and can function as calcium-sequestering systems analogous to the sarcoplasmic reticulum of muscle cells. Thus fluctuations in the activity of these vesicles could lead to variations in the concentration of calcium ions in the cytoplasm, with increases in early prophase and anaphase coincident with, and perhaps causally related to, microtubular breakdown. Investigations with the luminescent protein aequorin as an indicator of the concentration of calcium in the cytoplasm suggest that this rises during the prophase of cell division in *Xenopus* (Baker & Warner 1972), but other observations are required to define this type of change further.

In anaphase a rise in cytoplasmic calcium could trigger not only spindle microtubule breakdown but contraction of filaments, giving rise to the threshing movements so obvious in time lapse cinematography, as well as ring contraction for cytokinesis.

EFFECTOR SYSTEMS FOR HORMONES AND DRUGS

Evidence is accumulating that hormones and drugs bind first to specific receptors on the plasma membranes of target cells. Indirect, and more recently direct, measurements of agonist binding show that it is reversible, with association constants in the range 10^8–10^{11} l/mol (see Pastan & Perlman 1971). Elsewhere (Allison 1972*a*, *b*) I have suggested that attachment of agonists to receptors on plasma membranes has one of two consequences. First, it can depolarize the membrane, thereby activating an actomyosin system and triggering contraction of smooth muscle, endocytosis or granule discharge. This can be termed the *contractile* or *A effector system* (A because it depends on actomyosin filaments and also because it was the first to be discovered).

The alternative consequence involves changes in the *metabolism* of target cells; attachment of an agonist to the target cell plasma membrane is followed by activation of adenyl cyclase or an analogous enzyme, increase in the concentration of cyclic AMP or another nucleotide and stimulation of protein kinase activity. As a result substrates such as the enzyme phosphorylase are phosphorylated. This is termed the *B* or *metabolic effector system*, the activity of which has been extensively discussed (Robinson *et al.* 1968; Pastan & Perlman 1971). Hormones shown to raise intracellular concentrations of cyclic AMP in target tissues include catecholamines (through β-receptors), glucagon, ACTH, MSH, LH, vasopressin, parathyroid hormone, prostaglandins and thyrocalcitonin.

The A and B systems may work synergistically or antagonistically in different cells. In the thyroid cyclic AMP stimulates endocytosis, but this effect is blocked by cytochalasin (Williams & Wolff 1971*b*); in this organ the two systems act together. In smooth muscle, catecholamine α-receptor stimulation activates the A effector system whereas stimulation of catecholamine β-receptors activates the B system, increases cyclic AMP concentrations, and antagonizes histamine-

TABLE 4

Comparison of effector systems

	A system	*B system*
Target:	Ca^{2+} transport site	Adenyl cyclase
Timing:	Seconds, short duration	Minutes or hours, long duration
Effector:	Actomyosin	Cyclic AMP, protein kinases
Result:	Contraction	Altered metabolism
Effect:	Smooth muscle contraction, secretory granule extrusion, endocytosis	Glycogen breakdown, gluconeogenesis, hormone synthesis, etc.

mediated contraction of bronchial smooth muscle (Axelsson 1971) and uterine smooth muscle (Mitznegg *et al.* 1971). The release reaction triggered by ADP in blood platelets is probably an A system reaction mediated by microfilament contraction, and is antagonized by cyclic AMP (Salzman 1972). Higher concentrations of cyclic AMP in blood mast cells inhibit histamine release (Bourne *et al.* 1972). The distinctive features of the two effector systems are summarized in Table 4 and examples of their effects are given in Tables 5 and 6.

The repertoire which cells have at their disposal is limited, and often a system is adapted for various purposes in the course of evolution. It seems that the primitive contractile system has been modified to trigger the release of packaged secretions, while the primitive control system based on cyclic AMP has been used to antagonize and so modulate this activity. Such antagonism

TABLE 5

Examples of type A or contractile effector systems

Tissue	*Agonist*	*Effect*	*Antagonist*
Vascular muscle, nictitating membrane, vas deferens, dilator pupillae, rabbit uterus	Adrenergic compounds	Contraction	Adrenergic α-blockers
Bronchial smooth muscle	Histamine	Contraction	Catecholamines (β)
Intestinal smooth muscle	Acetylcholine	Contraction	Catecholamines (β)
Adrenal cortex	ACTH	Acute release of corticosteroids	
Thyroid	TSH	Endocytosis of colloid; release of thyroxin	
Pancreatic β-cells	Glucose, Tolbutamide	Acute release of insulin	Catecholamines (β)
Adrenal medulla	Acetylcholine	Release of catecholamines	
Mast cells	Reaginic antibody and antigen	Release of histamine, 5-hydroxytryptamine, SRS-A*	Catecholamines (β)
Macrophages	Antibody, antigen	Endocytosis	
Polymorphonuclear leucocytes	Leucocidin	Exocytosis of granules	
Blood platelets	ADP, thrombin, antibody with antigen	Exocytosis of granules	

* SRS-A, slow reaching substance A.

TABLE 6

Examples of type B or metabolic effector systems

Tissue	*Agonist*	*Effect*	*Enzyme*
Liver	Catecholamines, glucagon	Glycogenolysis, gluconeogenesis	Phosphorylase, phosphopyruvate carboxylase
Skeletal muscle	Catecholamines	Glycogenolysis	Phosphorylase
Cardiac muscle	Catecholamines, glucagon	Glycogenolysis	Phosphorylase
Adipose tissue	Catecholamines, glucagon, ACTH	Lipolysis	Lipase
Kidney cortex	Parathyroid hormone	Gluconeogenesis, phosphaturia	
Kidney medulla	Vasopressin	Water reabsorption	
Bone	Parathyroid hormone	Calcium resorption	
Skin	Melatonin, noradrenalin, melanocyte-stimulating hormone	Melanocyte dispersion	
Thyroid	Thyroid-stimulating hormone	Iodination, glucose oxidation	
Adrenal cortex	ACTH	Increased steroid synthesis	
Corpus luteum	Luteinizing hormone	Increased steroid synthesis	

recurs not only in the hormonal and drug effects already described but possibly also in inhibition of macrophage (Pick 1972) and fibroblast motility (Johnson *et al.* 1972) by cyclic AMP, and restoration of contact inhibition of malignant cells by dibutyryl cyclic AMP (see Pastan & Perlman 1971).

PARALYSIS OF THE CONTRACTILE SYSTEM

Inhibition of the A effector system can be envisaged as taking place at two levels. First, inhibition could interfere with the mechanism of activation, for example, antagonism of α-adrenergic effects by appropriate drugs or prevention by disodium cromoglycate of increased permeability to cations induced by cell-bound antibody and antigen. Alternatively, the contractile system itself could be inhibited. This can be termed a C (cytochalasin-like) effect. General anaesthetics administered by inhalation (such as halothane, chloroform or cyclopropane) can inhibit the movement of the slime mould *Dictyostelium discoideum* in concentrations required for surgical anaesthesia (Wiklund & Allison 1972). In view of the fact that microfilaments might mediate neurotransmitter release (Table 2) and the close apposition of cells at tight junctions, inhibition of microfilament-mediated functions might induce anaesthesia.

GENERAL COMMENT

Many aspects of membrane movement appear to require participation of a microfilament-related contractile system. The activity of this system is closely coordinated with events in the cell membrane itself, particularly in the induction of endocytosis and release of packaged secretions. The activity of microfilaments is also correlated with the disposition and behaviour of microtubules. These three cellular components function together in the orderly movements of and within cells involved in locomotion, endocytosis and exocytosis. The use of inhibitors of microfilament and microtubule-related activities provides useful preliminary information about their role in cell physiology, but further studies with other methods are required to resolve their complex interactions.

References

ABERCROMBIE, M., HEAYSMAN, J. E. M. & PEGRUM, S. M. (1970) The locomotion of fibroblasts in culture. I. Movements of the leading edge. *Exp. Cell Res.* **59**, 393-398

ADELSTEIN, R. S., POLLARD, T. D. & KUEHL, W. M. (1971) Isolation and characterization of myosin and two myosin fragments from human blood platelets. *Proc. Natl. Acad. Sci. U.S.A.* **68**, 2703-2707

ALLEN, R. D. (1972) Pattern of birefringence in the giant amoeba, *Chaos carolinensis*. *Exp. Cell Res.* **72**, 34-45

ALLISON, A. C. (1971*a*) The role of membranes in the replication of animal viruses. *Int. Rev. Exp. Pathol.* **10**, 181-142

ALLISON, A. C. (1971*b*) Discussion in *Immunologic Intervention* (Uhr, J. W. & Landy, M., eds.), pp. 294-298, Academic Press, New York

ALLISON, A. C. (1972*a*) Role of membranes in effector systems for hormones and drug action. *Chem. Phys. Lipids* **7**, 118-129

ALLISON, A. C. (1972*b*) Analogies between triggering mechanisms in immune and other cellular reactions, *Cell Interactions: Third Lepetit Colloquium* (L. G. Silvestri, ed.), North Holland Press, Amsterdam, pp. 156-161

ALLISON, A. C., DAVIES, P. & DE PETRIS, S. (1971) Role of contractile microfilaments in movement and endocytosis. *Nat. New Biol.* **232**, 153-155

ARCHER, F. L., BECK, J. S. & MELVIN, J. M. O. (1971) Localisation of smooth muscle protein in myoepithelium by immunofluorescence. *Am. J. Path.* **63**, 109-118

AXELSSON, J. (1971) Catecholamine functions. *Ann. Rev. Physiol.* **33**, 1-30

BAKER, P. F. & WARNER, A. E. (1972) Intracellular calcium and cell cleavage in early embryos of *Xenopus laevis*. *J. Cell Biol.* **53**, 579-581

BANGHAM, A. D., STANDISH, M. M. & WATKINS, J. C. (1965) Diffusion of univalent ions across the lamellae of swollen phospholipids. *J. Mol. Biol.* **13**, 238-252

BEHNKE, O., KRISTENSEN, B. I. & NIELSEN, L. E. (1971*a*) Electron microscopical observations on actinoid and myosinoid filaments in blood platelets. *J. Ultrastruct. Res.* **37**, 351-369

BEHNKE, O., FORER, A. & EMMERSEN, J. (1971*b*) Actin in sperm tails and meiotic spindles. *Nature (Lond.)* **234**, 408-410

BERL, S. & PUSZKIN, S. (1970) Mg^{2+}–Ca^{2+}-activated adenosine triphosphatase system isolated from mammalian brain. *Biochemistry* **9**, 2058-2067

BETTEX-GALLAND, M. & LÜSCHER, E. F. (1965) Thrombosthenin—the contractile protein from blood platelets and its relation to other contractile proteins. *Adv. Protein Chem.* **20**, 1-35

BHISHEY, A. N. & FREED, J. J. (1971) Altered movement of endosomes in colchicine-treated cultured macrophages *Exp. Cell Res.* **64**, 430-438

BORISY, G. G. & OLMSTEAD, J. B. (1972) Nucleated assembly of microtubules in porcine brain extracts. *Science (Wash. D.C.)* **177**, 1196-1197

BOURNE, H. R., LICHTENSTEIN, L. M. & MELMON, K. L. (1972) Pharmacologic control of allergic histamine release *in vitro:* evidence for an inhibitory role of 3′, 5′-adenosine monophosphate in human leukocytes. *J. Immunol.* **108**, 695-705

BUTCHER, F. R. & GOLDMAN, R. H. (1972) Effect of cytochalasin B and colchicine on the stimulation of α-amylase release from rat parotid tissue slices. *Biochem. Biophys. Res. Commun.* **48**, 23-29

CANER, J. E. Z. (1965) Colchicine inhibition of chemotaxis. *Arthritis Rheum.* **8**, 757-764

CARTER, S. (1967) Effects of cytochalasins on mammalian cells. *Nature (Lond.)* **213**, 261-264

CLONEY, R. A. (1966) Cytoplasmic filaments and cell movements: epidermal cells during ascidian metamorphosis. *J. Ultrastruct. Res.* **14**, 300-328

COHN, Z. (1970) Endocytosis and intracellular digestion in *Mononuclear Phagocytes* (van Furth, R., ed.), pp. 121-132, Blackwell, Oxford

COLTEN, H. & GABBAY, K. H. (1972) Histamine release from human leukocytes: modulation by a cytochalasin B sensitive barrier. *J. Clin. Invest.* **51**, 1927-1931

CRAWFORD, N. (1971) The presence of contractile proteins in platelet microparticles isolated from human and animal platelet-free plasma. *Br. J. Haematol.* **21**, 53-69

CRAWFORD, B., CLONEY, R. A. & CAHN, R. D. (1972) Cloned pigmented retinal cells: the effects of cytochalasin B on ultrastructure and behaviour. *Z. Zellforsch. Mikrosk. Anat.* **130**, 135-151

CREASEY, W. A., BENSCH, K. G. & MALAWISTA, S. E. (1971) Colchicine, vinblastine and griseofulvin. Pharmacological studies with human leukocytes. *Biochem. Pharmacol.* **20**, 1579-1588

DAVIES, P., ALLISON, A. C., FOX, R. I., POLYZONIS, M. & HASWELL, A. D. (1972) The exocytosis of polymorphonuclear leucocyte lysosomal enzymes induced by cytochalasin B. *Biochem. J.* **128**, 78P-79P

DAVIES, P., ALLISON, A. C. & HASWELL, A. D. (1973*a*) Selective release of lysosomal hydrolases from phagocytic cells by cytochalasin B. *Biochem. J.*, in press

DAVIES, P., FOX, R. I., POLYZONIS, M., ALLISON, A. C. & HASWELL, A. D. (1973*b*) The inhibition of phagocytosis and facilitation of exocytosis in rabbit PMN by cytochalasin B. *Lab. Invest.* **28**, 16-22

DAVIS, A. T., ESTENSEN, R. & QUIE, P. G. (1971) Cytochalasin B. 3. Inhibition of human polymorphonuclear leukocyte phagocytosis. *Proc. Soc. Exp. Biol. Med.* **137**, 161-164

DOUGLAS, W. W. & SORIMACHI, M. (1972) Effects of cytochalasin B and colchicine on secretion of posterior pituitary and adrenal medullary hormones. *Br. J. Pharmacol.* **45**, 143P-144P

EBASHI, S. & ENDO, M. (1968) Calcium ion and muscle contraction. *Prog. Biophys. Mol. Biol.* **18**, 123-183

ESTENSEN, R. D. & PLAGEMANN, P. G. W. (1972) Cytochalasin B: inhibition of glucose and glucosamine transport. *Proc. Natl. Acad. Sci. U.S.A.* **69**, 1430-1434

FARROW, L. J., HOLBOROW, E. J. & BRIGHTON, W. D. (1971) Reaction of human smooth muscle antibody with liver cells. *Nat. New Biol.* **232**, 186-187

FINE, R. E. & BRAY, D. (1971) Actin in growing nerve cells. *Nat. New Biol.* **234**, 115-118

FITZGERALD, T. J., WILLIAMS, B. & UYEKI, E. M. (1971) Colchicine on sodium urate-induced paw swelling in mice: structure-activity relationship of colchicine derivatives. *Proc. Soc. Exp. Biol. Med.* **136**, 115-120

FORER, A., EMMERSEN, J. & BEHNKE, O. (1972) Cytochalasin B: does it affect actin-like filaments? *Science (Wash. D.C.)* **175**, 774-776

FREED, J. J. & LEBOWITZ, M. M. (1970) The association of a class of saltatory movements with microtubules in cultured cells. *J. Cell Biol.* **45**, 334-354

FRYE, L. D. & EDIDIN, M. (1970) The rapid intermixing of cell surface antigens after formation of mouse-human heterokaryons. *J. Cell Sci.* **7**, 319-335

GABBAY, K. H. & TZE, W. J. (1972) Cytochalasin B sensitive emigate in the beta-cell. *Diabetes* **21** (Suppl.) 327

GAIL, M. H. & BOONE, C. W. (1971) Effect of Colcemid on fibroblast motility. *Exp. Cell Res.* **65**, 221-227

GAWADI, N. (1971) Actin in the mitotic spindle. *Nature (Lond.)* **234**, 410

GIBBONS, I. R. & FRONK, E. (1972) Some properties of bound and soluble dynein from sea urchin sperm flagella. *J. Cell Biol.* **54**, 365-381

GILLESPIE, E. & LICHTENSTEIN, L. M. (1972) Histamine release from human leukocytes: studies with deuterium oxide, colchicine and cytochalasin B. *J. Clin. Invest.* **51**, 2941-2947

GILLESPIE, E., LEVINE, R. J. & MALAWISTA, S. E. (1968) Histamine release from rat peritoneal mast cells: inhibition by colchicine and potentiation by deuterium oxide. *J. Pharmacol. Exp. Ther.* **164**, 158-165

GINGELL, D. (1970) Contractile responses at the surface of an amphibian egg. *J. Embryol. Exp. Morphol.* **23**, 583-609

GINN, F. L., HOCHSTEIN, P. & TRUMP, B. (1969) Membrane alterations in hemolysis: internalization of plasmalemma induced by primaquine. *Science (Wash. D.C.)* **164**, 843-845

GOLDMAN, R. H. (1971) The role of three cytoplasmic fibres in cell motility. 1. Microtubules and the effects of colchicine. *J. Cell Biol.* **51**, 752-762

GOLDMAN, R. D. (1972) The effects of cytochalasin B on the microfilaments of baby hamster kidney (BHK-21) cells. *J. Cell Biol.* **52**, 246-254

HATANO, S. & OOSAWA, F. (1966) Isolation and characterization of plasmodium actin. *Biochim. Biophys. Acta* **127**, 488-498

HATANO, S. & TAZAWA, M. (1968) Isolation, purification and characterization of myosin B from myxomycete plasmodium. *Biochim. Biophys. Acta* **154**, 507-519

HATANO, S., KONDO, H. & MIKI-NOUMURA, T. (1969) Purification of sea urchin egg actin. *Exp. Cell Res.* **55**, 275-277

HEATH, D. A., PALMER, J. S. & AURBACH, G. D. (1972) The hypocalcemic action of colchicine. *Endocrinology* **90**, 1589-1593

HOLZER, H. & SANGER, J. W. (1972) Cytochalasin B: microfilaments, cell movement and what else? *Dev. Biol.* **21**, 444-446

HURWITZ, L. & SURIA, A. (1971) The link between agonist action and response in smooth muscle. *Annu. Rev. Pharmacol.* **11**, 303-326

HUXLEY, H. E. (1963) Electron microscope studies on the structure of natural and synthetic protein filaments from striated muscle. *J. Mol. Biol.* **7**, 281-308

ISHIKAWA, H., BISCHOFF, R. & HOLTZER, H. (1969) Formation of arrowhead complexes with heavy meromyosin in a variety of cell types. *J. Cell Biol.* **43**, 312-328

JACQUES, P. (1969) Endocytosis in *Lysosomes in Biology and Pathology* (Dingle, J. T. & Fell, H. B., eds.), pp. 395-420, vol. 2, North Holland Press, Amsterdam

JOHNSON, G. S., MORGAN, W. D. & PASTAN, I. (1972) Regulation of cell motility by cyclic AMP. *Nature (Lond.)* **235**, 54-56

KAMINER, B. & SZONYI, E. (1972) Tropomyosin in electric organs of eel and *Torpedo*. *J. Cell Biol.* **55**, Abstract 257

KANE, R. E. (1962) The motitic apparatus: isolation by controlled pH. *J. Cell Biol.* **12**, 47-55

KEMP, R. B., JONES, B. M. & GRÖSCHEL-STEWART, U. (1971) Aggregative behaviour of embryonic chick cells in the presence of antibodies directed against actomyosins. *J. Cell Sci.* **9**, 103-122

KLETZIEN, R. F., PERDUE, J. F. & SPRINGER, A. (1972) Cytochalasin A and B. Inhibition of sugar uptake in cultured cells. *J. Biol. Chem.* **247**, 2964-2966

KRAICER, J. & MILLIGAN, J. V. (1971) Effect of colchicine on *in vitro* ACTH release induced by high K^+ and by hypothalamus-stalk-median eminence extract. *Endocrinology* **89**, 408-412

KUNDIG, W. & ROSEMAN, S. (1971) Sugar transport. II. Characterization of constitutive membrane-bound enzymes II of the *Escherichia coli* phosphotransferase sytem. *J. Biol. Chem.* **246**, 1407-1418

LACY, P. E., HOWELL, S. L., YOUNG, D. A. & FINK, C. J. (1968) New hypothesis of insulin secretion. *Nature (Lond.)* **219**, 1177-1179

LASH, J., CLONEY, R. A. & MINOR, R. R. (1970) Tail adsorption in Ascidians: effects of cytochalasin B. *Biol. Bull.* **139**, 427-428

LAY, W. H. & NUSSENZWEIG, V. (1968) Receptors for complement on leukocytes. *J. Exp. Med.* **128**, 991-1007

LEVY, D. A. & CARLTON, J. A. (1969) Influence of temperature on the inhibition by colchicine of allergic histamine release. *Proc. Soc. Exp. Biol. Med.* **130**, 1333-1336

LLINAS, R., BLINKS, J. R. & NICHOLSON, C. (1972) Calcium transport in presynaptic terminal of squid giant synapse: detection with aequorin. *Science (Wash. D.C.)* **176**, 1127-1129

MALAISSE-LAGAE, F. & MALAISSE, W. J. (1971) Stimulus-secretion coupling of glucose-induced insulin release. III. Uptake of ^{45}calcium by isolated islets of Langerhans. *Endocrinology* **88**, 72-80

MALAWISTA, S. E. & BODEL, P. T. (1967) The dissociation by colchicine of phagocytosis from increased oxygen consumption in human leukocytes. *J. Clin. Invest.* **46**, 786-796

MALAWISTA, S. E., GEE, J. B. & BENSCH, K. G. (1971) Cytochalasin B reversibly inhibits phagocytosis: functional, metabolic and ultrastructural effects in human blood leukocytes and rabbit alveolar macrophages. *Yale J. Biol. Med.* **44**, 286-300

MARGULIS, L. (1973) Microtubules. *Int. Rev. Cytol.*, in press

MAZIA, D., PETZELT, C., WILLIAMS, R. O. & MEZA, I. (1972) A Ca-activated ATPase in the mitotic apparatus of the sea-urchin egg (isolated by a new method). *Exp. Cell Res.* **70**, 325-332

MISZURSKI, B. (1969) Effects of colchicine on resting cells in tissue cultures. *Exp. Cell Res.* suppl. 1, 450-451

MITZNEGG, P., HACH, B. & HEIM, F. (1971) The influence of guanosine 3′, 5′-monophosphate and other cyclic nucleotides on contractile responses induced by oxytocin in isolated rat uterus. *Life Sci.* **10**, 1285-1289

MIZEL, S. B. & WILSON, L. (1972) Nucleotide transport in mammalian cells. Inhibition by colchicine. *Biochemistry* **11**, 2573-2578

MOORE, P. B., HUXLEY, H. E. & DEROSIER, D. J. (1970) Three dimensional reconstruction of F-actin, thin filaments and decorated thin filaments. *J. Mol. Biol.* **50**, 279-295

MORILL, G. A., KOSTELLOW, A. B. & MURPHY, J. B. (1971) Sequential forms of ATPase activity correlated with changes in cation binding and membrane potential from meiosis to first cleavage in *R. pipiens*. *Exp. Cell Res.* **66**, 289-298

NACHMIAS, V. T., HUXLEY, H. E. & KESSLER, D. (1970) Electron microscope observations on actomyosin and actin preparations from *Physarum polycephalum* and on their interaction with heavy meromyosin subfragment I from muscle myosin. *J. Mol. Biol.* **50**, 83-90

ORCI, L., GABBAY, K. H. & MALAISSE, W. J. (1972) Pancreatic β-cell web: its possible role in insulin secretion. *Science (Wash. D.C.)* **175**, 1128-1130

ORR, T. S. C., HALL, D. E. & ALLISON, A. C. (1972) A role of contractile microfilaments in the release of histamine from mast cells. *Nature (Lond.)* **236**, 350-351

OWELLEN, R. J., OWENS, A. H., JR. & DONIGIAN, C. (1972) The binding of vincristine, vinblastine and colchicine to tubulin. *Biochem. Biophys. Res. Commun.* **47**, 685-691

PARKHOUSE, R. M. E. & ALLISON, A. C. (1972) Failure of cytochalasin or colchicine to inhibit secretion of immunoglobulins. *Nat. New Biol.* **235**, 220-222
PASTAN, I. & PERLMAN, R. L. (1971) Cyclic AMP in metabolism. *Nat. New Biol.* **229**, 5-7
PERRY, M. M., JOHN, H. A. & THOMAS, N. S. T. (1971) Actin-like filaments in the cleavage furrow of newt egg. *Exp. Cell Res.* **65**, 249-253
PETZELT, C. (1972) Ca^{2+}-activated ATPase during the cell cycle of the sea urchin *Strongylocentrotus purpuratus*. *Exp. Cell Res.* **70**, 333-339
PICK, E. (1972) Cyclic AMP affects macrophage migration. *Nat. New Biol.* **238**, 176-177
POISNER, A. M. & BERNSTEIN, J. (1971) A possible role of microtubules in catecholamine release from the adrenal medulla: effect of colchicine, vinca alkaloids and deuterium oxide. *J. Pharmacol. Exp. Ther.* **177**, 102-108
POLLARD, T. D. & KORN, E. D. (1971) Filaments of *Amoeba proteus*, II. Binding of heavy meromyosin by thin filaments in motile cytoplasmic extracts. *J. Cell Biol.* **48**, 216-219
POLLARD, T. D. & KORN, E. D. (1972) The 'contractile' proteins of *Acanthamoeba castellani. Cold Spring Harbor Symp. Quant. Biol.* **37**, 573-584
POLLARD, T. D., SHELTON, E., WEIHING, R. R. & KORN, E. D. (1970) Ultrastructural characterization of F-actin isolated from *Acanthamoeba castellanii* and identification of cytoplasmic filaments as F-actin by reaction with rabbit heavy meromyosin. *J. Mol. Biol.* **50**, 91-97
PORTER, K. R. (1966) Cytoplasmic microtubules and their functions in *Principles of Biomolecular Organization (Ciba Found. Symp.)*, p. 308, Churchill, London
PUMPLIN, D. W. & MCCLURE, W. O. (1972) Effects of cytochalasin B and vinblastine on the release of acetylcholine from a sympathetic ganglion. *Proc. Natl. Acad. Sci. U.S.A.*, in press
RABINOVITCH, M. (1967) The dissociation of the attachment and ingestion phases of phagocytosis by macrophages. *Exp. Cell Res.* **46**, 19-28
RABINOVITCH, M. & DE STEFANO, M. (1970) Interactions of red cells with phagocytes and the wax-moth *(Galleria mellonella*, L) and mouse. *Exp. Cell Res.* **59**, 272-282
ROBBINS, E. & GONATAS, N. K. (1964) Histochemical and ultrastructural studies on HeLa cell cultures exposed to spindle inhibitors with special reference to the interphase cell. *J. Histochem. Cytochem.* **12**, 704-711
ROBINSON, G. A., BUTCHER, R. W. & SUTHERLAND, E. W. (1968) Cyclic AMP. *Annu. Rev. Biochem.* **37**, 149-174
ROSENBAUM, J. L., MOULDER, J. E. & RINGO, D. L. (1969) Flaggellar elongation and shortening in *Chlamydomonas*. The use of cycloheximide and colchicine to study the synthesis and assembly of flagellar proteins. *J. Cell Biol.* **41**, 601-619
RUBIN, R. P. (1970) The role of calcium in the release of neurotransmitter substances and hormones. *Pharmacol. Rev.* **22**, 389-428
RYSER, H. J. P. (1967) Studies on protein uptake by isolated tumor cells: 3. Apparent stimulations due to pH, hypertonicity, polycations, or dehydration and their relation to the enhanced penetration of infectious nucleic acids. *J. Cell Biol.* **32**, 737-750
SALZMAN, E. W. (1972) Cyclic AMP and platelet function. *N. Engl. J. Med.* **286**, 358-363
SANGER, J. W., HOLZER, S. & HOLZER, H. (1971) Effects of cytochalasin B on muscle cells in tissue culture. *Nat. New Biol.* **229**, 121-123
SCHIMMEL, R. J. (1972) Inhibition of free fatty acid mobilization by colchicine. *Fed. Proc.* **31**, 351 abs
SCHOFIELD, J. G. (1971) Cytochalasin B and release of growth hormone. *Nat. New Biol.* **234**, 215-216
SCHOFIELD, J. G. & COLE, E. N. (1971) Behaviour of systems releasing growth hormone *in vitro*. *Mem. Soc. Endocrinol.* **19**, 185-201
SCHROEDER, T. E. (1970) The contractile ring. I. Fine structure of dividing mammalian (HeLa) cells and the effects of cytochalasin B. *Z. Zellforsch. Mikrosk. Anat.* **109**, 431-449
SCHROEDER, T. E. (1972) The contractile ring. II. Determining its brief existence, volumetric changes and vital role in cleaving *Arbacia* eggs. *J. Cell Biol.* **53**, 419-434
SENDA, N., SHIBATA, N., TATSUMI, N., KONDO, K. & HAMADA, K. (1969) A contractile protein

from leucocytes: its extraction and some of its properties. *Biochim. Biophys. Acta* **181**, 191-200

SINGER, S. J. & NICOLSON, G. L. (1972) The fluid mosaic model of the structure of cell membranes. *Science (Wash. D.C.)* **175**, 720-731

SPUDICH, J. A. & LIN, S. (1972) Cytochalasin B, its interaction with actin and actomyosin from muscle. *Proc. Natl. Acad. Sci. U.S.A.* **69**, 442-446

STEPHENS, R. E. (1971) Microtubules in *Biological Macromolecules* (Timasheff, S. N. & Fasman, G. D., eds.), pp. 355-391, chap. 8, vol. 5, Marcel Dekker, New York

TAYLOR, R. B., DUFFUS, W. P. H., RAFF, M. C. & DE PETRIS, S. (1971) Redistribution and pinocytosis of lymphocyte surface immunoglobulin molecules induced by anti-immunoglobulin antibodies. *Nature (Lond.)* **233**, 225-229

TEMPLE, R., WILLIAMS, J. A., WILBER, J. F. & WOLFF, J. (1972) Colchicine and hormone secretion. *Biochem. Biophys. Res. Commun.* **46**, 1454-1461

THOA, N. B., WOOTEN, G. F., AXELROD, J. & KOPIN, I. J. (1972) Inhibition of release of dopamine-β-hydroxylase and norepinephrine from sympathetic nerves by colchicine, vinblastine or cytochalasin B. *Proc. Natl. Acad. Sci. U.S.A.* **69**, 520-522

THOMPSON, C. M. & WOLPERT, L. (1963) The isolation of motile cytoplasm from *Amoeba proteus*. *Exp. Cell Res.* **32**, 156-160

TILNEY, L. G. & MOOSEKER, M. (1971) Actin in the brush-border of epithelial cells of the chicken intestine. *Proc. Natl. Acad. Sci. U.S.A.* **68**, 2611-2615

TSAN, M. F. & BERLIN, R. D. (1971) Effect of phagocytosis on membrane transport of electrolytes. *J. Exp. Med.* **134**, 1016-1035

UKENA, T. E. & BERLIN, R. D. (1972) Effect of colchicine and vinblastine on the topographical separation of membrane functions. *J. Exp. Med.* **136**, 1-7

VASILIEV, JU. M., GELFAND, I. M., DOMNINA, L. V., IVANOVA, O. Y., KOMM, S. G., & OLSHEVSKAJA, L. V. (1970) Effect of Colcemid on the locomotory behaviour of fibroblasts. *J. Embryol. Exp. Morphol.* **24**, 625-640

WAGNER, R., ROSENBERG, M. & ESTENSEN, R. (1971) Endocytosis in Chang liver cells. Quantitation by sucrose-^{3}H uptake and inhibition by cytochalasin B. *J. Cell Biol.* **50**, 804-817

WEIHING, R. R. & KORN, E. D. (1969) Amoeba actin: the presence of 3-methylhistidine. *Biochem. Biophys. Res. Commun.* **35**, 906-912

WEISENBERG, R. C. (1972*a*) Changes in the organization of tubulin during meiosis in the eggs of the surf clam, *Spisula solidissima*. *J. Cell Biol.* **54**, 266-278

WEISENBERG, R. C. (1972*b*) Microtubule formation *in vitro* in solutions containing low calcium concentrations. *Science (Wash. D.C.)* **177**, 1104-1105

WEISENBERG, R. C. & TIMASHEFF, S. N. (1970) Aggregation of microtubule subunit protein: the effects of divalent cations, colchicine and vinblastine. *Biochemistry* **9**, 4110-4116

WEISENBERG, R. C., BORISY, G. G. & TAYLOR, E. W. (1968) The colchicine-binding properties of mammalian brain and its relation to microtubules. *Biochemistry* **7**, 4466-4479

WEISSMANN, G., DUKOR, P. & ZURIER, R. B. (1971) Effect of cyclic AMP on release of lysosomal enzymes from phagocytes. *Nat. New Biol.* **231**, 131-135

WESSELLS, N. K., SPOONER, D. S., ASH, J. F., BRADLEY, M. O., LUDUEÑA, M. A., TAYLOR, E. L., WRENN, J. T. & YAMADA, K. M. (1971) Microfilaments in cellular and developmental processes. *Science (Wash. D.C.)* **171**, 135-143

WIKLUND, R. & ALLISON, A. C. (1972) The effects of anaesthetics on the motility of *Dictyostelium discoideum:* evidence for a possible mechanism of anaesthesia. *Nature (Lond.)* **239**, 221-222

WILLIAMS, J. A. & WOLFF, J. (1970) Possible role of microtubules in thyroid secretion. *Proc. Natl. Acad. Sci. U.S.A.* **67**, 1901-1908

WILLIAMS, J. A. & WOLFF, J. (1971*a*) Thyroid secretion *in vitro:* multiple actions of agent affecting secretion. *Endocrinology* **88**, 206-217

WILLIAMS, J. A. & WOLFF, J. (1971*b*) Cytochalasin B inhibits thyroid secretion. *Biochem. Biophys. Res. Commun.* **44**, 422-425

WILLIAMSON, R. E. (1972) A light microscope study of the action of cytochalasin B on the cells and isolated cytoplasm of the *Characeae*. *J. Cell Sci.* **10**, 811-819

WILLS, E. J., DAVIES, P., ALLISON, A. C. & HASWELL, A. D. (1972) The failure of cytochalasin B to inhibit pinocytosis by macrophages. *Nat. New Biol.* **240**, 58-60

WILSON, L. (1970) Properties of colchicine binding protein from chick embryo brain: interactions with vinca alkaloids and podophyllotoxin. *Biochemistry* **6**, 4999-5007

YANG, Y.-Y. & PERDUE, J. F. (1972) Contractile proteins of cultured cells. 1. The isolation and characterization of an actin-like protein from cultured chick embryo fibroblasts. *J. Biol. Chem.* **247**, 4503-4509

YOSHINAGA, M., WAKSMAN, B. H. & MALAWISTA, S. E. (1972) Cytochalasin B inhibits lymphotoxin production by antigen-stimulated lymphocytes. *Science (Wash. D.C.)* **176**, 1147-1149

ZIGMOND, S. H. & HIRSCH, J. G. (1972) Cytochalasin B. Inhibition of D-2-deoxyglucose transport into leukocytes and fibroblasts. *Science (Wash. D.C.)* **176**, 1432-1434

Discussion

Abercrombie: How does this contractile system produce endocytosis?

Allison: In our opinion, the microfilament system forms a network beneath the plasma membrane, exerting a certain amount of tone all the time. In other words it is a little contracted; this contributes to the rounded contour of the cell. One of the reasons why cells treated with cytochalasin lie flat on glass is that such muscle-like tone is lost. When you attach something to the membrane you produce an alteration, probably by cross-linking certain membrane constituents such as glycoproteins (cf. p. 23). Not only attachment of particles but also anti-cellular antibody initiates endocytosis. The latter is the standard way in which Cohn (1970) induces endocytosis by placing macrophages in bovine serum which contains an anti-cellular antibody. One can also initiate endocytosis with basic proteins. This is particularly effective for amoebae and tumour cells, as Chapman-Andresen (1965) and Ryser (1967) have shown. All these agents can cross-link membrane constituents, probably glycoproteins. There is some evidence for this, but it is not yet compelling. The next event, we propose, is depolarization of the membrane, followed by local contraction of the subjacent microfilament system. This would result in the pinching together of the plasma membrane at the edge of the attachment site, forming a continuous fold around the endocytic vacuole. This is drawn together in a purse-string fashion.

Abercrombie: The contraction of an annulus?

Allison: Yes, exactly, followed by fusion of the folds to separate the vacuole from the membrane (see Fig. 6).

In relation to cell movement an important question is whether extension is the result of contraction alone or whether you need some kind of resistance to compression. This is a basic distinction in the science of materials. It seems conceivable that microfilaments can operate only through exerting tension whereas microtubules are able to resist compression.

Wessells: Spudich (1973*b*) has shown that if highly purified actin is combined with tropinin-tropomyosin the addition of cytochalasin has no effect whatsoever on the morphology of the system; conversely, without troponin–tropomyosin, the F-actin breaks up in cytochalasin into short segments, which are amazingly similar to the masses of material seen in the cytoplasm of a number of cell types treated with cytochalasin.

Allison: That is probably like skeletal muscle.

Wessells: That's right. This seems to resolve the controversy with Forer *et al.* (1972) on that point, since they probably did not have highly purified actin to treat with cytochalasin.

To investigate the inhibition of glucose transport by cytochalasin, we worked on the salivary gland (Taylor & Wessells 1973) and on locomotion of glial cells and elongation of axons (Yamada & Wessells 1973). We used a glucose-free medium, and found no inhibition of cell locomotion, axon elongation or salivary morphogenesis, nor any effect upon the morphology of microfilaments.

Gail: Adelstein & Conti (1972) recently reported the chemical isolation of both actin and myosin from fibroblasts, which confirms your observations. It might be relevant that Johnson *et al.* (1972) claim that L-cell motility is diminished after treatment with theophylline and dibutyryl cyclic AMP.

Allison: It is well known in the pharmacological field that smooth muscle contraction in a variety of systems is inhibited by increasing the concentration of cyclic AMP (Mitznegg *et al.* 1971). Possibly this is a general phenomenon.

Goldman: From your fluorescent studies, it appears as if there is an enormous amount of myosin in the same location as microfilaments. If this were so, wouldn't you expect to be able to see the myosin with the electron microscope?

Allison: But our results do not necessarily represent a large amount of material. The technique is sensitive. Our attempts at quantitation are crude, but we have little doubt that there is less myosin than actin.

Goldman: Spudich (1973*a*) has shown that at low concentrations of cytochalasin, the intrinsic viscosity of muscle and platelet actin changes, but this does not necessarily mean that the gross fibrous structure of actin is altered. This might help to explain why we find microfilaments with normal appearance at the low effective doses of cytochalasin B. In other words, it is not essential to see simultaneous functional and structural changes with the electron microscope after treatment with cytochalasin B.

Allison: One idea that appeals to me, and I would like to know Dr Wessells' opinion, is that the polymerized filaments are hardly affected by cytochalasin, whereas the drug markedly inhibits their rapid formation from monomers.

Wessells: This is an excellent idea. A number of contradictions in the literature could be explained if sensitivity to cytochalasin is a function of the state of polymerization of microfilaments and of the presence of subsidiary molecules (as myosinoid elements; see Behnke *et al.* 1971).

Goldman: The electron microscopy of BHK-21 cells in suspension (not attached to a substrate) demonstrates that they possess very few microfilaments but, upon attachment to a substrate, microfilaments are rapidly assembled (Goldman & Knipe 1973). If cells are suspended in the presence of concentrations of cytochalasin ranging from 1 to 10 μg/ml and then allowed to attach and spread on a substrate, microfilament bundles still form. There are, of course, fewer microfilament bundles formed at the higher doses which produce spiky cells (Goldman & Knipe 1973). On this evidence, we feel that cytochalasin does not act by binding and preventing the assembly of microfilament subunits, and therefore is not analogous to colchicine, which does bind to microtubule subunits and prevents their assembly. This conclusion is also supported by Forer *et al.* (1972) and Spudich (1973) who demonstrate that the transformation of G- into F-actin is not inhibited by cytochalasin.

Allison: In that sense, your results are different from those of Spudich which you have just reported.

Goldman: No, I feel that they are consistent, especially at the minimum effective concentrations of cytochalasin. However, it is extremely difficult, especially with the information available, to make comparisons between *in vitro* and *in vivo* or *in situ* studies.

Allison: Treatment of skeletal muscle cells from early embryos in culture with cytochalasin results in marked disorganization of their sarcomeric structure, whereas contractility and structure are almost unaffected in cells from older embryos.

Trinkaus: How do you explain that?

Wessells: It can be explained in terms of the degree of differentiation of muscle cells. The development of the eight-day-old chick ventricles, which we used, was highly asynchronous: some cells have highly organized sarcomeres, and others have sarcomeres just in the process of forming. Dr D. Kelly, who has studied the development of amphibian muscle, has examined our electron micrographs and suggests that, until a means is found to synchronize differentiation of muscle cells, it will be difficult to interpret the effects of cytochalasin upon muscle. There is no doubt, however, that cytochalasin does produce dramatic effects on the filamental systems of heart muscle cells that are in the earlier stages of development.

Allison: What about the validity of the other evidence I mentioned, such as the inhibition of cortical contraction in frog eggs induced by calcium injection and the work on glycerinated smooth muscle? The results suggest to me that cytochalasin exerts an effect directly on the microfilament system, and not just on the plasma membrane.

Bray: Williamson (1972) has recently shown that the streaming of isolated drops of cytoplasm extruded from *Chara* cells is arrested by cytochalasin. Since these droplets probably do not have a membrane this seems to support your ideas.

Allison: It would be comforting to be sure that cytochalasin affects microfilaments directly, because it is such a useful tool. As far as overall inhibition of metabolism is concerned, we found that when cytochalasin inhibits endocytosis, macrophages release their lysosomes into the medium and continue to synthesize large quantities of lysosomal enzymes. Also, the specific synthesis of immunoglobulin in plasma cells is not inhibited by cytochalasin. If there had been a marked interference with the overall metabolism of the cells it is inconceivable that they could have synthesized protein so well. So I do not think cytochalasin in the concentrations used (about 5 μg/ml) is so generally toxic or inhibitory to metabolism as some people have tried to make out.

Goldman: You also imply that phagocytosis is inhibited by cytochalasin B but that micropinocytosis is not. I agree; we obtain similar results (see Fig. 10 on p. 99).

Allison: Yes. That is why I suggested that micropinocytosis is a pure membrane phenomenon. You can obtain a micropinocytosis-like response in a red blood cell exposed to aminoquinolines and other agents (Ginn *et al.* 1969).

Goldman: We found that BHK-21 cells can take up horseradish peroxidase and colloidal gold even at doses of cytochalasin B as high as 50 μg/ml (see Fig. 10).

Allison: This is just what we have found with macrophages (Wills *et al.* 1972).

Wessells: With regard to your comments on cytochalasin and the Gingell effect, Mizel & Wilson (1972), working on the inhibition of glucose and deoxyglucose transport by cytochalasin, noted that the drug had no effect on calcium transport across the membrane. This is important, since calcium is likely to be an essential ingredient in the control of microfilament activity.

Allison: Yes. Further, D. W. Pumplin & W. O. McClure (personal communication) inform me that they have found the same lack of effect of cytochalasin when they measured calcium influx into sympathetic ganglion cells.

Wessells: More research on calcium effects would be worthwhile. We have been working with papaverine (Ash *et al.* 1973), which is thought to inhibit movement of calcium into smooth muscle so that relaxation results. When applied to the salivary gland, where microfilaments might be active, papaverine inhibits morphogenesis and causes loss of clefts which are believed to be due to contractions of microfilaments. However, the morphology of microfilaments is not altered by papaverine. This would be expected if papaverine acts on calcium availability and not on filaments *per se*. Analogous effects are produced by lanthanum (E. L. Taylor & Wessells, unpublished results), which inhibits utilization of extracellular calcium by cells. All this means that calcium might be pivotal in cell locomotion and tissue morphogenesis.

Trinkaus: What is known about the diffusion of cytochalasin in the cell?

Allison: Nothing.

Trinkaus: We have had a puzzling result with cytochalasin B. During normal *Fundulus* epiboly the egg is constricted by the margin of the syncytial periblast and by the margin of the enveloping layer adhering to it (see Fig. 2, Trinkaus 1971). You can therefore imagine our excitement when we discovered that these marginal regions are packed with circumferentially oriented microfilaments about 6 nm in diameter (Betchaku & Trinkaus 1973). Our first thought, obviously, was that these are contractile and responsible for the constriction. So we removed blastoderms and placed them in cytochalasin B. The intense marginal constriction normally observed in such isolates was reversibly inhibited by the drug—as expected (Trinkaus & Ramsey 1973). However, when an egg deprived of its blastoderm was placed in cytochalasin B, the constriction of the marginal periblast was unaffected. This was very curious, especially in view of the effect of cytochalasin B on the rest of the periblast. Within seconds, the periblast starts expanding and flattening on the substratum; it becomes flaccid and no longer resists gravity. (Incidentally, microfilaments are present too.) This suggests that cytochalasin B is penetrating the periblasts. But nevertheless the marginal constriction remains. Could you suggest an explanation for this?

Allison: Apart from the impermeable yolk cytoplasmic layer, is there not lipid in yolk cells? Cytochalasin, which is highly lipid-soluble, might be concentrated in such granules, reducing the concentration elsewhere.

Goldman: Is there any published evidence which demonstrates that cytochalasin B does enter cells?

Allison: I am not aware of any evidence one way or the other. It would be feasible to test this if we had labelled cytochalasin.

References

ADELSTEIN, R. & CONTI, M. A. (1972) The characterization of contractile proteins from platelets and fibroblasts. *Cold Spring Harbor Symp. Quant. Biol.* **37**, 599-606

ASH, J. F., SPOONER, B. S. & WESSELLS, N. K. (1973) Effects of papaverine and calcium-free medium on salivary gland morphogenesis. *Dev. Biol.*, in press

BEHNKE, O., KRISTENSEN, B. I. & NIELSEN, L. E. (1971) Electron microscopy of actinoid and myosinoid filaments in blood platelets. *J. Ultrastruct. Res.* **37**, 351-369

BETCHAKU, T. & TRINKAUS, J. P. (1973) Marginal contacts of the *Fundulus* enveloping layer with the periblast before and during epiboly, in press

CHAPMAN-ANDRESEN, C. (1965) The induction of pinocytosis in amoebae. *Arch. Biol.* **76**, 189-207

COHN, Z. (1970) Endocytosis and intracellular digestion in *Mononuclear Phagocytes* (van Furth, R., ed.), pp. 121-132, Blackwell, Oxford

FORER, A., EMMERSEN, J. & BEHNKE, O. (1972) Cytochalasin B: does it affect actin-like filaments? *Science (Wash. D.C.)* **175**, 774-776

GINN, F. L., HOCHSTEIN, P. & TRUMP, B. (1969) Membrane alterations in hemolysis: internalization of plasmalemma induced by primaquine. *Science (Wash. D.C.)* **164**, 843-845

GOLDMAN, R. D. & KNIPE, D. M. (1973) The functions of cytoplasmic fibers in non-muscle cell motility. *Cold Spring Harbor Symp. Quant. Biol.* **37**, 523-534

JOHNSON, G. S., MORGAN, W. D. & PASTAN, I. (1972) Regulation of cell motility by cyclic AMP. *Nature (Lond.)* **235**, 54-56

MITZNEGG, P., HACH, B. & HEIM, F. (1971) Influence of guanosine 3′, 5′-monophosphate and other cyclic nucleotides on contractile responses induced by oxytocin in isolated rat uterus. *Life Sci.* **10**, 1285-1289

MIZEL, S. B. & WILSON, L. (1972) Inhibition of hexose transport in mammalian cells by cytochalasin B. *J. Biol. Chem.* **247**, 4102-4105

RYSER, H. J. P. (1967) A membrane effect of basic polymers dependent on molecular size. *Nature (Lond.)* **215**, 934-936

SPUDICH, J. A. (1973*a*) On the effects of cytochalasin B on actin filaments. *Proc. Natl. Acad. Sci. U.S.A.*, in press

SPUDICH, J. A. (1973*b*) On the effects of cytochalasin B on actin filaments. *Cold Spring Harbor Symp. Quant. Biol.* **37**, 585-594

TAYLOR, E. L. & WESSELLS, N. K. (1973) Cytochalasin B: alterations in salivary gland morphogenesis not due to glucose depletion. *Dev. Biol.*, in press

TRINKAUS, J. P. (1971) Role of the periblast in *Fundulus* epiboly. *Ontogenesis* **2**, 401-405

TRINKAUS, J. P. & RAMSEY, W. S. (1973) Contractility in the *Fundulus* gastrula, in press

WILLIAMSON, R. E. (1972) A light microscope study of the action of cytochalasin B on the cells and isolated cytoplasm of the *Characeae*. *J. Cell Sci.* **10**, 811-819

WILLS, E. D., DAVIES, P., ALLISON, A. C. & HASWELL, D. (1972) The failure of cytochalasin B to inhibit pinocytosis in macrophages. *Nature (Lond.)* **240**, 58-60

YAMADA, K. M. & WESSELLS, N. K. (1973) Cytochalasin B: effects on membrane ruffling, growth cone and microspike activity, and microfilament structure not due to altered glucose transport. *Dev. Biol.*, in press

Microtubules in intracellular locomotion

KEITH R. PORTER

Department of Molecular, Cellular and Developmental Biology, University of Colorado

Abstract An array of microtubules parallel to an axis of a cell is generally agreed to lead to the development of an anisometric cell form. It is clear that the tubules do not only provide a frame but must help in the restructuring of the inside of the cell. Experiments with colchicine have shown that microtubules are essential to elongation.

Studies of the arms of the melanophores of the killifish, *Fundulus heteroclitus*, and more extensively on the erythrophore of *Holocentrus ascensionis* show populations of microtubules parallel to the long axis of the arms and also to the direction of pigment flow. One population appears to be in the central apparatus and the other in the cell cortex. The motion of the pigment outwards along the arms is distinguishable from that inwards by being slower and irresolute.

The ultrastructure of the erythrophore when the pigment is both dispersed and aggregated has been examined. Neither in melanophores nor in erythrophores are there any bridges between the microtubules and the pigment granules, but in the erythrophores the microtubules are found associated with a randomly-dispersed flocculent component of the matrix. It is proposed that when the microtubules disassemble, the gel with the pigment granules undergoes rapid syneresis and that during reassembly it helps to define the channels of motion of the pigment. A reasonable conclusion is that cell locomotion depends on the integrity of the microtubules, their orientation and their assembly and disassembly. Thus, a frame of microtubules emanating from the centrosphere is oriented predominantly in one direction. As the frame extends in that direction, cytoplasm is carried forward with it, the whole providing a channel for the movement of other materials, such as vesicles going to the plasma membrane. As new microtubules are laid down in the direction of motion, those trailing disassemble and the central apparatus must follow the bulk of the cytoplasm.

There is now general agreement that the assembly of microtubules parallel to one axis of a cell or cell process is associated with the development of anisometric cell form. One interpretation is that the tubules provide a frame of cytoskeleton along which other cytoplasmic components are transported. But there is more to the phenomenon than that in the sense that the whole interior of the cell is restructured. The protoplasm (including the nucleus) is redistributed or translocated relative to an axis coincident with the long axis of the cell and generally (in animal cells) passing through the centrosphere with its centrioles. During this translocation, microtubules display a non-random distribution, frequently emanating from the centrosphere or some defined zone in the cell cortex. They are usually oriented in arrays parallel to the long axis being established. Their assembly is programmed, in the sense that some factor (or factors) of the intracellular milieu initiates and promotes their assembly from the pool of dimer tubulin units at a particular time (Porter 1966).

These phenomena were observed initially in cells of the presumptive lens ectoderm during cell elongation in lens placode formation (Byers & Porter 1964), and have also been observed in the posterior epithelium of the lens (Pearce & Zwaan 1970) and in cells of the neural plate (Burnside 1971). The implication that microtubules are essential to the elongation is strengthened by experiments with the anti-mitotic drug colchicine and low temperatures. Applied at the appropriate time they prevent tubule assembly (or induce disassembly) and the associated cell elongation (Pearce & Zwaan 1970; Handel & Roth 1971).

Such studies leave certain basic questions unanswered. Do the tubules provide an elastic (but relatively rigid) frame by which a motive force can exert its influence on the surrounding solated cytoplasm or does the existence of the microtubules as an anisometric component of the gelated cytoplasmic continuum serve to distort or induce a shape change in the entire cytoplasmic matrix? Conceivably both these possibilities are operative and essential for shape changes as well as for the translocation of cytoplasmic particulates.

These questions encouraged us early on to look for a system in which the phenomena associated with tubules could be monitored with both light and electron microscopes. The chromatophore was an obvious choice (Bikle *et al.* 1966). Electron microscopy readily showed that the long slender arms of melanophores of the killifish, *Fundulus heteroclitus*, contain populations of microtubules which are oriented parallel to both the long axis of the arms and the direction of pigment flow, and which appear to define channels wherein the pigment moves. We presented evidence that there are two populations: one 'based' in the central apparatus and the other in the cell cortex. Since microtubules are evident in the arms of the melanophore after the pigment has moved to the centre of the cell in aggregation it has been assumed that both populations persist (unlike the homologous situation in erythrophores, see later) when it is possible that only the cortical population survives aggregation. This problem remains to be investigated.

The inward and outward motions of the pigment are recognizably different, as Green (1968) has also described. The velocity of pigment granules during aggregation is uniform, once motion has been established. The distance traversed may be as much as 60 μm and aggregation can be completed in two minutes. The dispersion of pigment takes 2–3 times longer than aggregation and the motion is far less uniform in rate. Individual granules move back and forth, or show saltation, during their outward dispersion. This suggests that different mechanisms underlie the two motions. The inward motion reminds one of syneresis, essentially a contraction of a continuum in which the granules are suspended and which is, in the dispersed state, distorted by the organized array of microtubules. Take away the tubules and the asymmetric system collapses (with a release of energy). In dispersion possibly the system has to be reconstructed through the new assembly of microtubules (with energy stored). We have found little evidence of this in melanophores but then there is no evidence against it. (The streaming of pigment granules centripetally in melanophores does not include mitochondria which suggests that the granules are incorporated in a system that is selective for melanosomes.)

Green (1968) interprets the pigment motion similarly, further suggesting that the granules are part of a continuum which has elastic properties and that the continuum 'contracts' during aggregation and 'expands' during pigment dispersion.

Subsequently, we have examined the effects of several factors, including metabolic inhibitors, colchicine and vinblastine, and have concluded that the microtubules are essential to pigment movement (Bennett *et al.* 1970; Junquiera & Porter 1969). When the tubules are disassembled or prevented from forming by colchicine, pigment migration is also blocked.

THE ERYTHROPHORE

For several reasons, some technical, the melanophore offered fewer experimental possibilities than another chromatophore, the erythrophore, which is found in a number of fishes including *Holocentrus ascensionis* (Osbeck). These cells contain a red pigment in the form of small droplets which, like fat droplets, lack a limiting membrane. The responses of the cells to changing concentrations of Na^+ and K^+ ions as well as other factors resemble the responses of melanophores (Smith & Smith 1935).

Erythrophores are commonly found just under the epidermis on the undersides of the scales, and they vary in number with the size of the scales. The size and form also varies with the location of the scale on the fish. For the most part the erythrophores are disc-shaped cells about 70–80 μm in diameter.

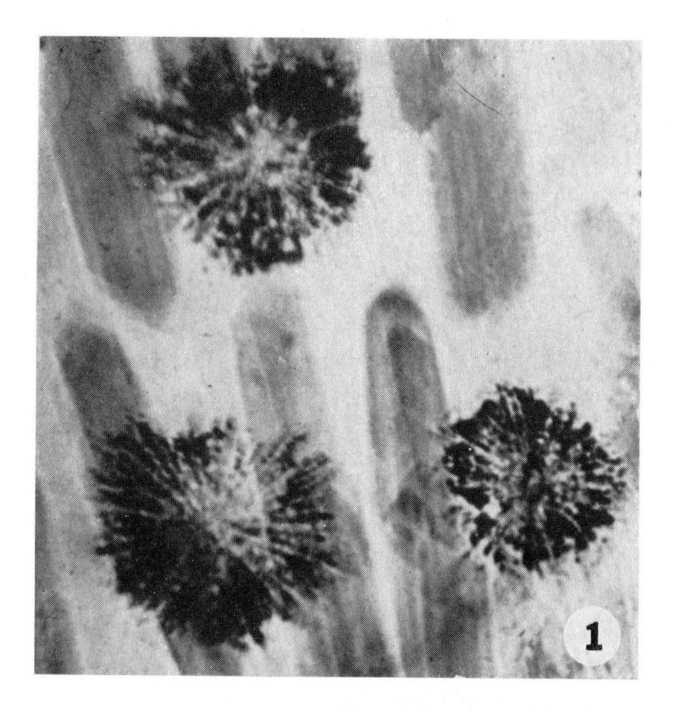

FIG. 1. Photomicrograph of three erythrophores with their pigment in the dispersed or disaggregated state. × 450.

The nucleus and the mitochondria of a typical erythrophore are located peripherally in the cell, as though pushed there by the very large and prominent centrosphere (aster) which is the dominant feature of these cells. All the pigment granules are arranged in radial rows within this centrosphere with the larger droplets of pigment toward the outside (Fig. 1).

Another remarkable feature of the erythrophore is its capacity to aggregate its pigment and disperse it again in a few seconds. The inward motion, which can be initiated with 0.1M-KCl, is completed in about 3–4 s. Since the distance traversed is about 30–40 μm the velocity is of the order of 10 μm/s. The aggregating velocity is uniform over the entire distance just as in the melanosomes, and also in chromosomes during anaphase of mitosis. The word 'resolute' seems to describe the motion. The outward motion, or dispersion, which can be observed in cells that have begun a rhythmic movement of their pigment, is slower and the velocity of the individual particles is less uniform. Again, as in the melanophores, there is some saltation in the motion of the particles during dispersion and 'irresolute' seems an apt description. Furthermore, one gains the impression that the mechanism providing the motive force is different for the two motions; in dispersion the mechanism seems to be being constructed as the motion proceeds. It is important to note that the prominent radial organization of the pigment in the dispersed state is lost when the pigment completes its aggregation, only reappearing as the pigment begins its redispersion.

The control of these pigment cells involves an extensive innervation which might be both cholinergic and adrenergic on the same cell. Both kinds of endings have been identified but the exact distribution of the synapses has not been worked out.

THE ERYTHROPHORE WITH PIGMENT DISPERSED

As expected, the erythrophores with pigment dispersed were found to contain large numbers of microtubules (Figs. 2 and 3). Apart from the pigment granules, which show no limiting membrane, the microtubules (24 nm diameter) dominate the picture; hundreds of them (many more than in melanophores) are embedded in a matrix or continuum which seems to possess few other fibrous components. There are a few profiles of the endoplasmic reticulum and the only other membrane-limited structure represents irregular tubular invaginations of the plasma membrane. These appear to be analogous to the transverse tubular system of striated muscle (Fig. 4).

For the most part the microtubules are oriented radially relative to the centrosphere of the cell (Fig. 2 and more clearly in Fig. 3). The tubules are not perfectly straight and appear as if fixation had caught them in an undulating configuration (Fig. 3). Though not well shown in these pictures, there is a population of tubules associated with the cell cortex, possibly with their proximal ends anchored in the cortex and their distal ends free in the cytoplasm. By counting tubules, we have determined that the population of microtubules around the outside is about two-times greater than in the centre of a picture such as Fig. 3. The conclusion is that not all tubules reach the central area or are inserted in the dense components of the central apparatus.

The central apparatus is represented in part after staining by strands of electron-dense material (Figs. 5 and 6). Their texture as well as their density is reminiscent of the wall of the centriole and associated material which is finely punctate. For the most part these strands are sickle-shaped. They cluster around the centrioles with their long axes parallel to the plane of the cell (Fig. 5). Microtubules seem to insert into them in large numbers and clearly extend from them in an array which matches the disc shape of the cell (Fig. 6). If the strands of the central apparatus are arranged in any repeating or distin-

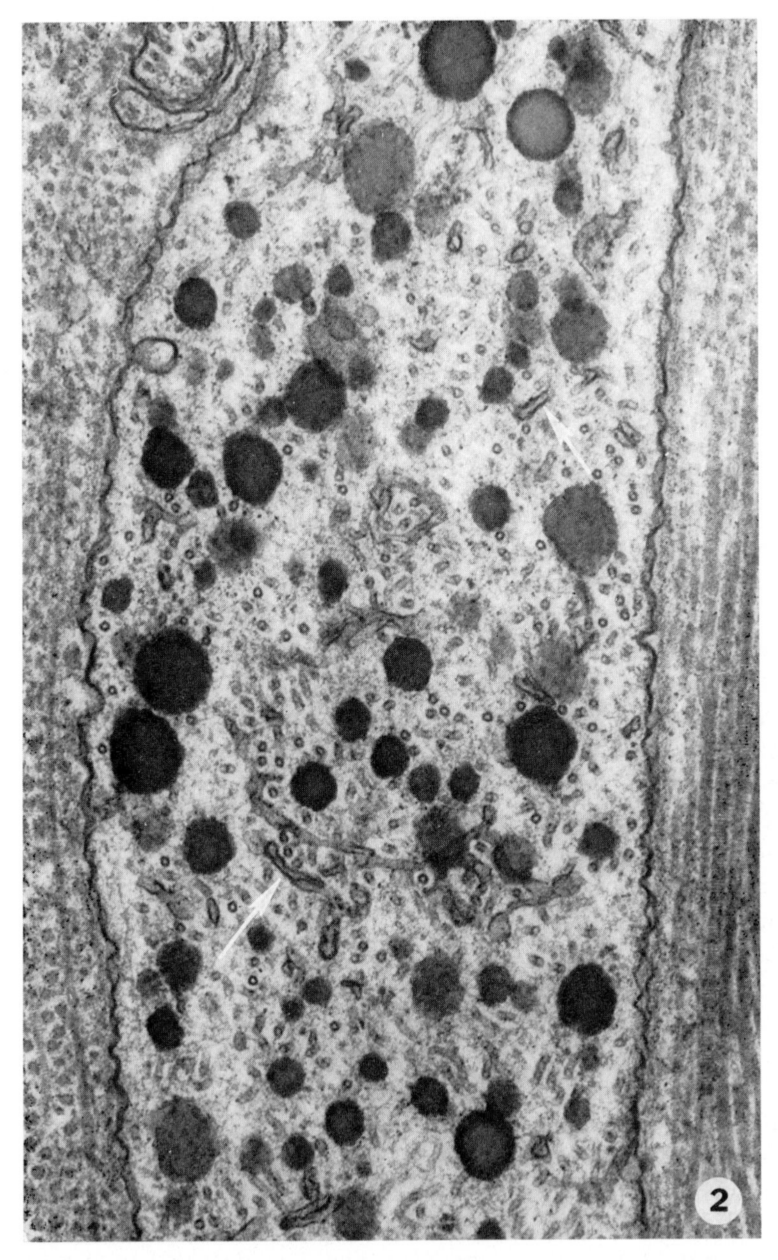

FIG. 2. Electron micrograph of a vertical section through an erythrophore cut somewhat off centre; the vertical axis of the cell runs from left to right. The dense spherical bodies are pigment droplets. Microtubules appear as small circles in cross section near the centre of the image, and as short rods in oblique section toward the margins. Collagen fibres occupy the extracellular space. The vesicular, membrane-limited elements mingling with the microtubules are slender invaginations of the cell surface (arrows). $\times$ 52 000.

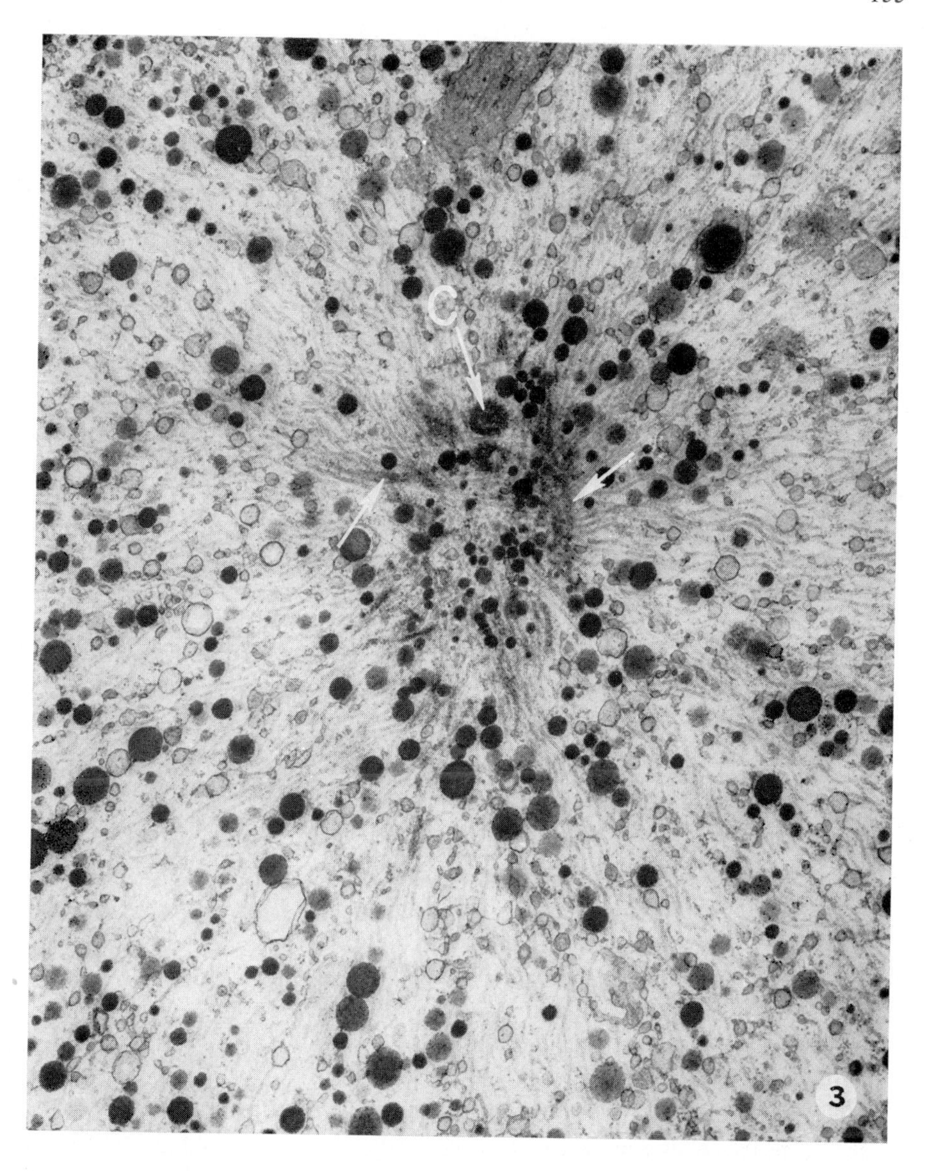

FIG. 3. A horizontal section of an erythrophore. The pigment granules appear as dense spheres arranged generally in radial rows. Large numbers of microtubules radiate from the central part of the cell. They appear to focus on parts of the central apparatus represented by centrioles (c) and thin strands of dense material (arrows) which resemble in their density and texture the wall of the centrioles (c). Mitochondria are notably absent from this part of the cell. The membrane-limited 'vesicles' are interpreted as parts of the endoplasmic reticulum and invaginations of the plasma membrane. × 15 400.

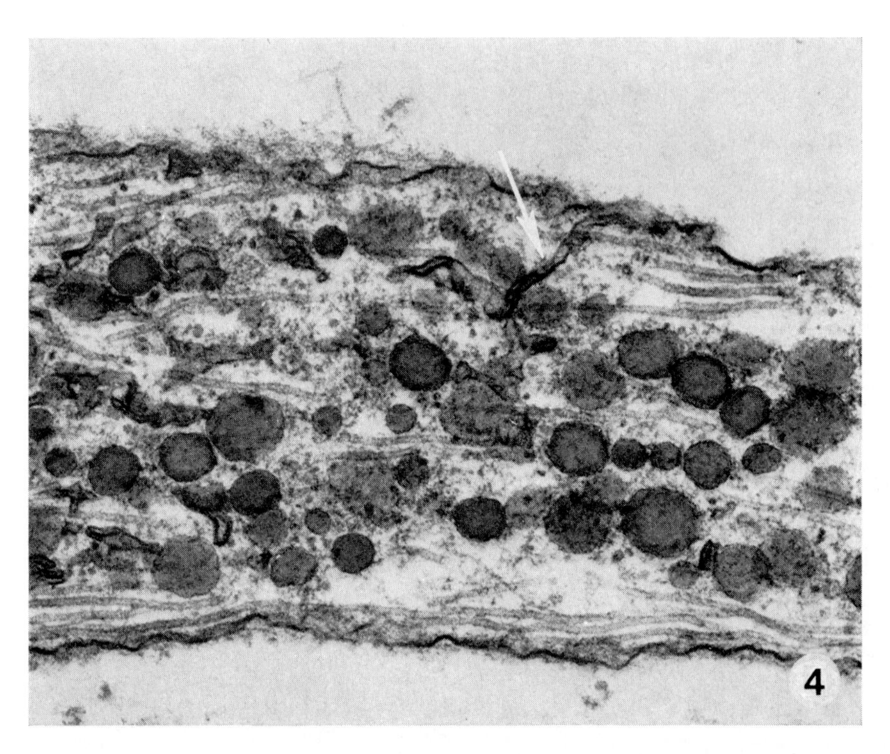

FIG. 4. A vertical section through the margin of an erythrophore in which the pigment is dispersed. The microtubules are easily identified as lying mostly in the plane of the section. A flocculent component of the cytoplasmic continuum (ground substance or matrix) associates with the surfaces of the tubules and pigment granules. A slender invagination of the surface membrane is depicted at the arrow. × 50 400.

guishable pattern, we have not discovered it. We hope, however, that stereomicroscopy of thick (0.5–1.0 μm) sections with a high-voltage electron microscope will provide a better image of the overall construction of the tubule initiating and orienting apparatus.

This large central, disc-shaped structure with radiating microtubules (the aster) appears to dominate the central region of the cell and displaces the nucleus from the more central position it normally occupies to the cell margins. It also displaces the mitochondria and during aggregation of pigment they remain in the margins as though, unlike the pigment, they are not an integral part of the central astral continuum.

The response of the erythrophore microtubules to physiologically acceptable concentrations of colchicine and vinblastine was essentially as expected; they disassembled, or failed to reassemble whichever is appropriate. After treatment

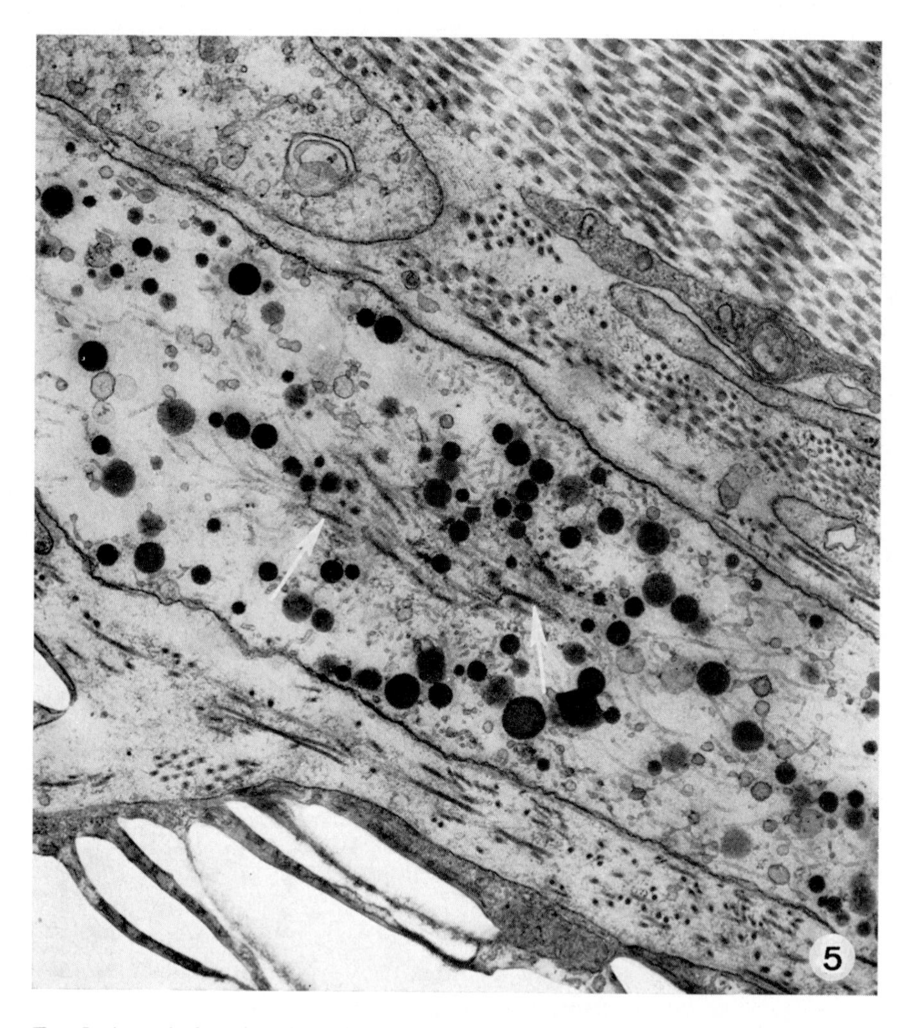

FIG. 5. A vertical section through the central region of the cell. This micrograph includes images of the dense strands (arrows) which make up the central apparatus. Many microtubules appear to arise from this structure. Others populate the cortical zone of the cell. Collagen fibrils are numerous in the surrounding connective tissue. × 17 600.

with vinblastine, typical crystals (of tubulin and vinblastine) appeared throughout the cell. The normal radial arrangement of pigment disappeared and the cells ceased to respond to K^+ or to pulsate in Ringer's solution. It is logical to conclude that microtubules are an essential part of the mechanism for aggregation and disaggregation of pigment.

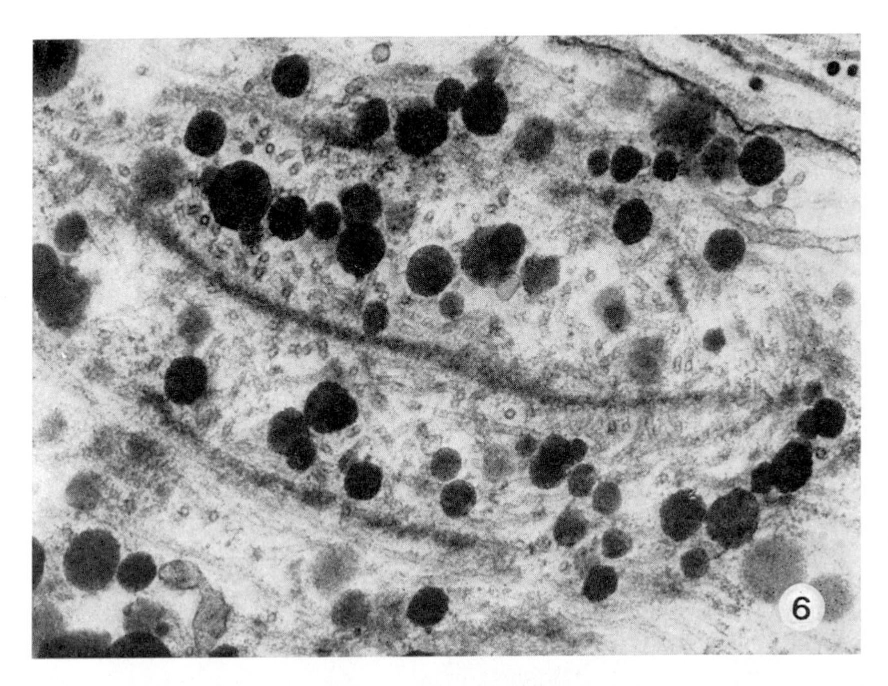

FIG. 6. A higher magnification of a few dense components of the central apparatus. It is evident that large numbers of microtubules impinge on and become part of these dense strands. The tubules may be said to insert into or arise from these structures. They probably represent initiating sites for tubule assembly. × 44 800.

THE ERYTHROPHORE WITH PIGMENT AGGREGATED

Ordinarily cells in a single scale may be kept under observation in Ringer's solution for many hours. During this time they may spontaneously begin a rhythmic aggregation and dispersion of pigment. If kept in a perfusion chamber arranged for light microscopy, the cells (and scale) may be fixed while under observation. In this way one can record whether dispersion or aggregation was under way when the fixation stopped the motion. Alternatively the chamber may be perfused with 0.1M-KCl or adrenalin to induce aggregation or with 0.1M-NaCl or melanocyte-stimulating hormone to induce dispersion and the cells fixed in any state desired.

A cell in which aggregation was induced with KCl and immediately fixed is shown in Figs. 7 and 8. The pigment is compactly aggregated around the cell centre. Some of the radial organization of pigment particles persists, but to a large extent the linear or channelled distribution vanishes. Vertical sections through cells in which the pigment is aggregated show the pigment mass to be

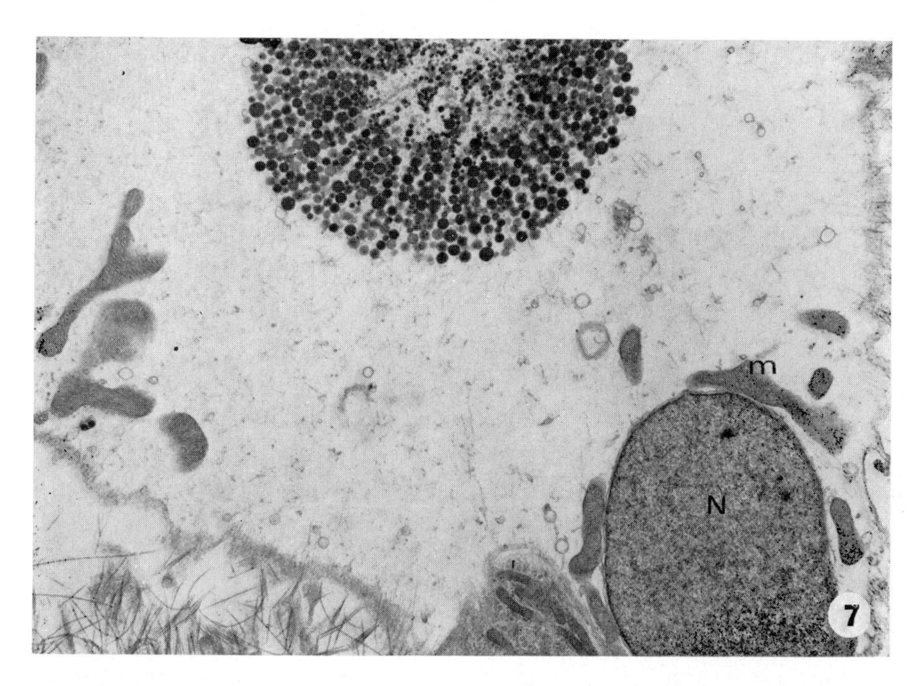

FIG. 7. A horizontal section through an erythrophore that has been stimulated to aggregate its pigment by exposing the scale to 0.1M-KCl. The pigment is clustered around the central apparatus at the top of the image. Mitochondria (m) and the nucleus (N) remain in the marginal regions of the cytoplast. The cell membrane is evident at the lower left and along the right hand margin of the micrograph. Except for a few short segments, microtubules are no longer evident in the cytoplasm. × 6 375.

spherical and the peripheral margins of the cell to have thinned. Thus during aggregation the total astral complex (continuum plus tubules plus pigment) changes from a discoidal to a spherical form.

It is evident in Fig. 7 that during pigment aggregation the mitochondria and nucleus stay in the cell margin. Profiles of vesicles, presumably part of the endoplasmic reticulum, also remain dispersed. But the striking change is that the microtubules have essentially vanished from the cytoplasmic matrix beyond the pigment mass (compare Figs. 3 and 7). Thus in the aggregation process the bulk of the microtubules comprising the central population has disassembled and only a flocculent residue is apparent. The cortical population of tubules, arranged closely adjacent to the plasma membrane, appears to survive the aggregation process (Fig. 8) and this makes it appear that the observed behaviour of the central complex of tubules is genuine and not an artifact of observation conditions or of fixation.

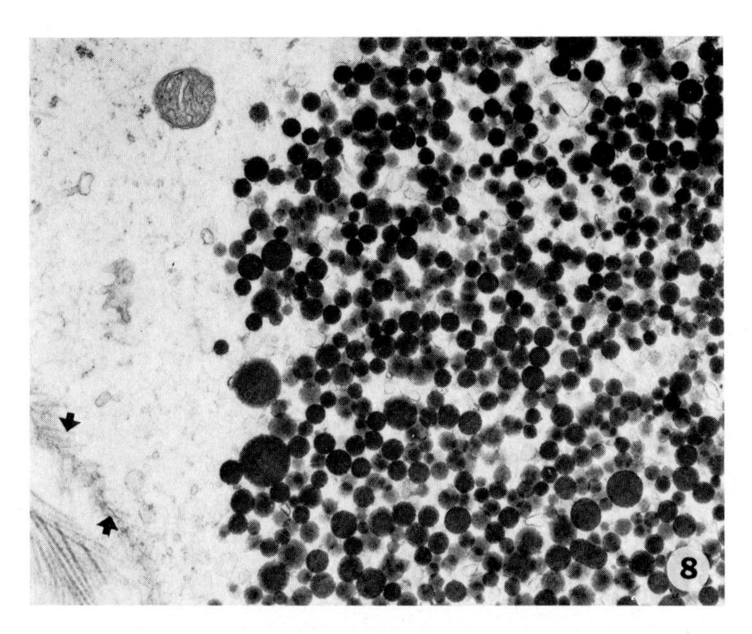

FIG. 8. An aggregated mass of pigment at a higher magnification. The pigment granules have lost their usual radial arrangement and only a very few microtubules persist among them. Beyond the margins of the mass no tubules are evident except in the cell cortex of which a small part is shown at the lower left (arrows). × 13 500.

THE ERYTHROPHORE DURING PIGMENT DISPERSAL

The disaggregation of pigment can be followed in pulsating cells or induced by replacing 0.1M-KCl with cold-blooded Ringer's solution. The earliest identifiable stage in this phenomenon is marked by a realignment of pigment granules into radially arranged rows (Fig. 9). These it appears are separated by small arrays of microtubules (apparently bundles) of which a few extend beyond the still aggregated mass of pigment. Evidently the pigment disperses as the microtubules reassemble. This reassembly continues distally as the pigment moves out toward the cell periphery (Figs. 10 and 11). Occasionally microtubules are found extending a few microns beyond the limits of the pigment mass but in general the limits of tubule assembly and the front of pigment migration coincide. Thus the motive force, whatever its nature, is effective only where tubules have reassembled. In other words, restructuring of the disc-shaped aster continuum (centrosphere) is apparently dependent on the tubules and essential to the redistribution (dispersion) of the pigment.

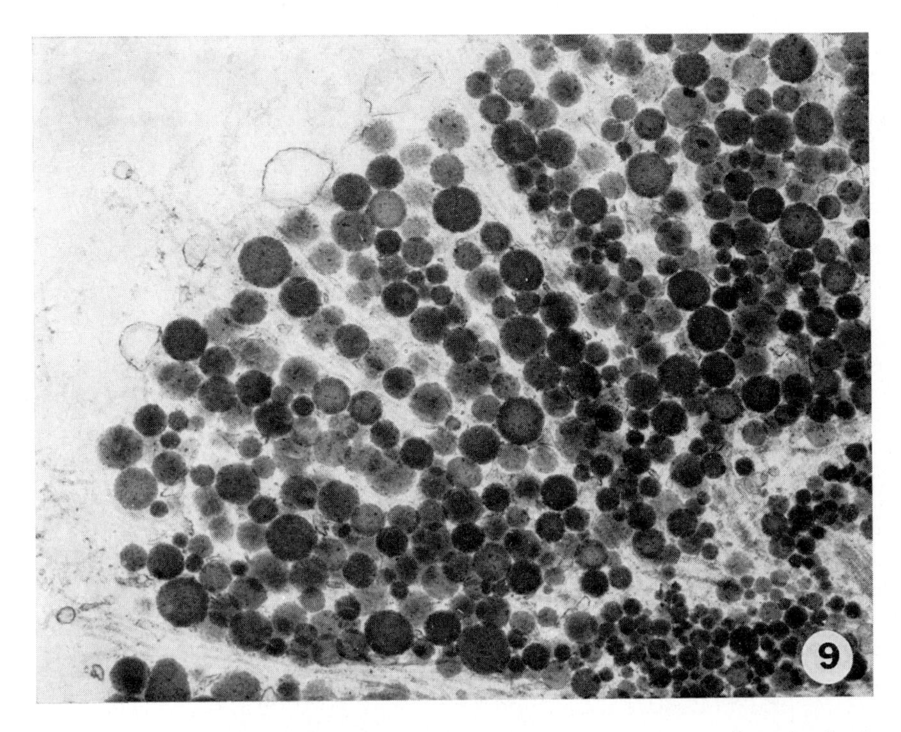

FIG. 9. A section of the pigment mass in an erythrophore where pigment dispersion has just been initiated. Microtubules have begun to reassemble and occur as small bundles between rows of pigment granules. In a few places they extend for a short distance only beyond the margin of the pigment mass. $\times$ 21 600.

THE MOTIVE FORCE

The concept of filament sliding over filament has been transferred from muscle to microtubules; arms or bridges that might develop the motive force for the sliding are not the exclusive property of thick myofilaments (McIntosh *et al.* 1969; McIntosh & Porter 1967). They are found in the (9 + 2) complex and also between the microtubules of the axoneme in *Actinosphaerium* and the axostyle, for example of *Saccinobacullus*. In both instances the arms (dynein) have been identified with ATPase activity (Gibbons 1963; Mooseker & Tilney 1972) and in the cilium, even though they do not bridge the gap between adjacent peripheral doublets, they probably participate in the sliding movement which the doublets execute with respect to one another (Summers & Gibbons 1971).

In erythrophores and melanophores there are no discernible arms or bridges between microtubules and granules (Green 1968; Bikle *et al.* 1966). Instead the

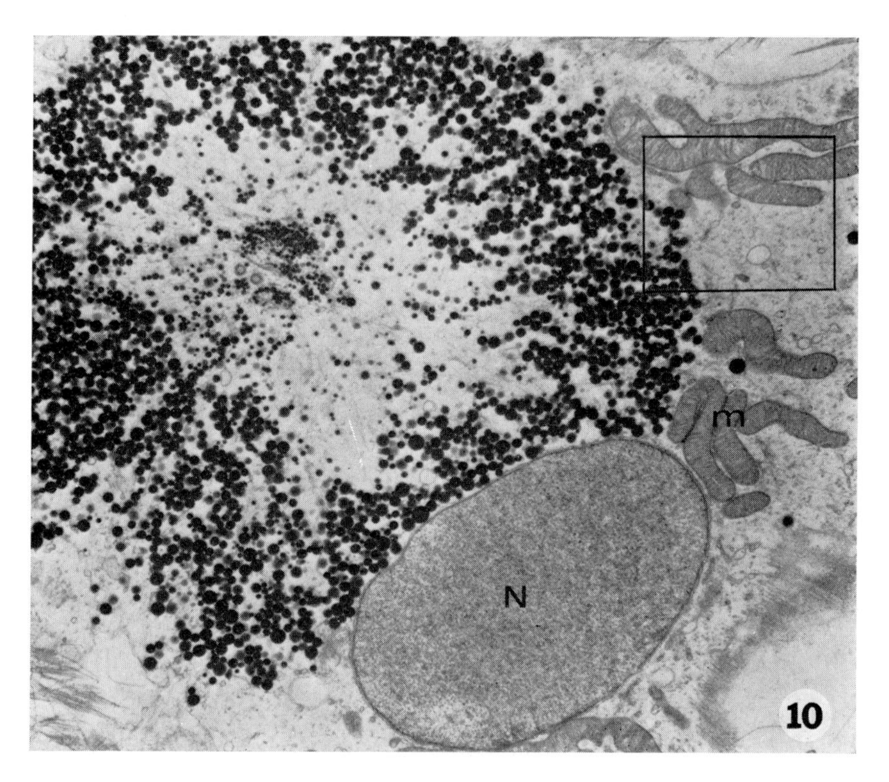

FIG. 10. A horizontal section of a cell where the pigment had moved part way to the cell margins before being fixed. Mitochondria (m) and the cell nucleus (N) are in their usual position at the periphery. The pigment moves out as a wave with only a relatively small number of granules remaining in the cell centre. A large number of microtubules radiate from the central apparatus which includes two centrioles. × 8 100.

tubules (in the fixed and stained preparations) are associated structurally with a flocculent component of the matrix or continuum. A similar material appears to coat the pigment droplets and to be continuous as well with the dense bodies of the central apparatus (Fig. 4). It is not distributed along the tubules in regularly dispersed blebs or bridges but rather seems to be randomly dispersed. A conservative judgement would classify it as a precipitated component of the cytoplasmic matrix, but representative nevertheless of a material existing on and around the tubules. In the living cell it might be part of a gel in which the tubules exist as skeletal elements and in which the pigment might at times be embedded. Disassemble the microtubules, and the gel (matrix) with associated pigment granules undergoes a rapid syneresis. It is further postulated that in their reassembly the tubules and associated matrix (more rigidly gelled

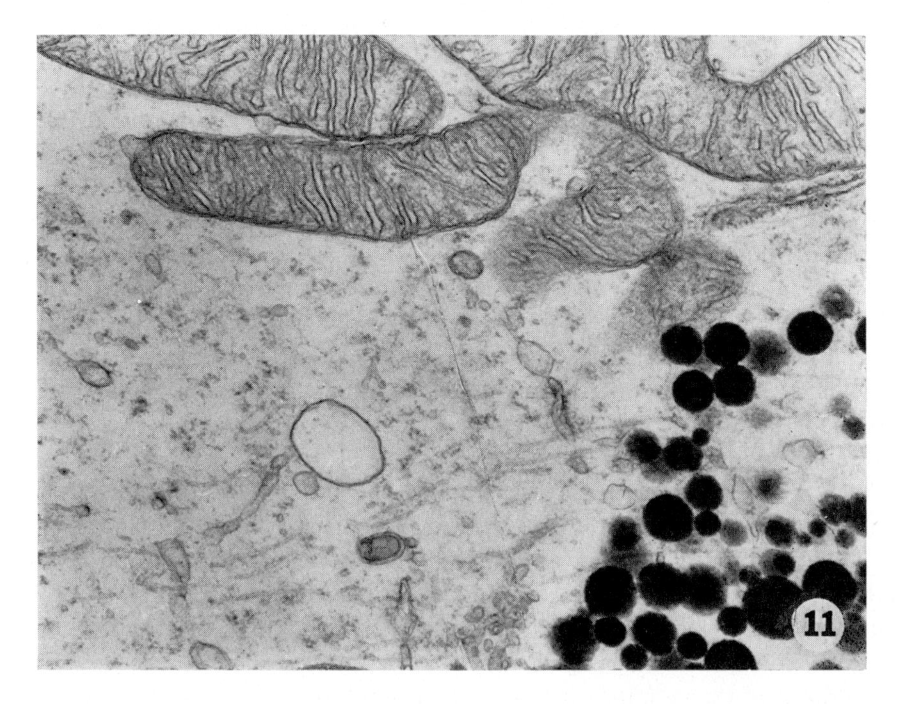

FIG. 11. An enlargement of the area inside rectangle in Figure 10. Microtubules extend beyond the margins of the pigment but only a short distance and in relatively small numbers. × 22 500.

regions) define channels along which the pigment in the more solated matrix is propelled. The nature of the motive force involved in this outward motion of pigment over the tubules or the tubule-supported gel remains obscure.

TUBULE ASSEMBLY IN ERYTHROPHORES AND THE DEVELOPMENT OF CELL FORM

It seems improbable of course that these mechanisms for moving pigment or achieving a translocation of cytoplasm have evolved solely for the erythrophore. On the basis of our observations it will be important to look again at the melanophore to see whether there also the tubule behaviour of erythrophores is repeated. Actually there are other cells in which parallel events seem to occur.

In an earlier study of lens formation in the chick embryo it was observed that a frame of microtubules appears as the cell elongates (Byers & Porter 1964). These tubules emanate from the apical diplosomes (centrioles) and the surrounding dense cytoplasm which is the centrosphere equivalent of these cells. The major difference between these cells and the erythrophores is that the microtubules are *all* initiated and oriented from one side of the centrosphere. As they assemble and elongate the cell elongates and the bulk of the cytoplasmic matrix (with organelles and particulates) is redistributed along the tubule frame.

Very similar phenomena are seen in mammalian spermiogenesis when the spermatid elongates and the bulk of the cytoplasm is translocated caudally before being discarded (Fawcett *et al.* 1971). Initially a ring of density appears in the cell cortex as a shoulder around the posterior part of the nucleus. The assembly of a cylindrical array of microtubules (the caudal sheath or Manchette) is initiated in the ring and oriented posteriorly in the direction of subsequent spermatid elongation. Again a portion of the cytoplasmic matrix with ribosomes and other components is translocated as tubules are assembled.

That the restructuring of the aster (centrosphere) in erythrophores involves only the central matrix of the cytoplasm plus pigment granules does not seem to set it apart from the cases cited above and others of a similar nature. In other words, even though the shape of the erythrophore is not dramatically altered by the events of pigment aggregation it does not follow that the internal translocations are not analogous to those in cell elongations. The lack of withdrawal of the peripheral margins of the erythrophore with each pigment aggregation is probably a reflection of the persistence and supportive action of the close cortical array of microtubules and also the existence of a basement lamina which is closely associated structurally with the plasma membrane on one side and collagen fibres on the other side. To the extent that the shape of the astral mass changes from a disc to a sphere upon pigment aggregation, the cell profile in vertical section thickens in the centre and thins at the margins.

THE MICROTUBULE FRAME AND CELL LOCOMOTION

Offhand one would probably not assign microtubules any role in cell locomotion and yet the idea should not be dismissed without examination.

A fibroblast or other migrating tissue cell usually moves over a substrate by extending a leading lamella in the direction of motion [strictly repetitive extension and withdrawal (Abercrombie *et al.* 1970)]. As a rule, the nucleus and central apparatus are on an axis which coincides with the direction of motion, and the nucleus trails behind. Freed & Lebowitz (1970) have shown that in an extended cell process the bulk of microtubules are oriented parallel to the axis of extension (or motion). They also demonstrated that acid phosphatase-rich particles (lysosomes) and other particulates are similar to pigment granules in moving, sometimes rapidly, along channels defined by microtubules. When cells showing a radial and ordered distribution of lysosomes are treated with colchicine the distribution becomes entirely random. According to Vasiliev *et al.* (1970) colchicine blocks the directional motion of a fibroblast as though the mechanism for translocation of cytoplasm had been destroyed.

With these observations in mind it is reasonable to suggest that locomotion of cells over a substrate depends on the integrity of microtubules and also on their orientation and active assembly and, when appropriate, disassembly. On this scheme, a cell will advance in a direction defined by the orientation of the majority of the longer microtubules emanating from the centrosphere. As a frame these would involve a portion of the cytoplasmic continuum in their further growth (and thus translocate) and also define extending channels for the directed motion of (among other things) Golgi-derived membrane-rich vesicles to the leading and expanding edge of the cell for insertion into the plasma membrane. Presumably new populations of microtubules would continue to assemble always further in the direction of motion while others in the trailing part of the cytoplasm might disassemble. The central apparatus would be obliged to follow gradually in the direction taken by the bulk of the cytoplasm.

ACKNOWLEDGEMENTS

These studies were made with the collaboration of Gudrun S. Bennett and Lewis C. Junqueira and supported by Grant Number GM-00707, from the National Institutes of Health.

References

Abercrombie, M., Heasman, J. E. M. & Pegrum, S. M. (1970) The locomotion of fibroblasts in culture. I. Movements of the leading edge. *Exp. Cell Res.* **50**, 393-398

Bennett, G. S., Junquiera, L. C. & Porter, K. R. (1970) Microtubules in intracellular pigment migration. Seventh Intern. Cong. of Electron Microscopy, Grenoble, France, pp. 945-946. (Abstract)

Bikle, D., Tilney, L. G. & Porter, K. R. (1966) Microtubules and pigment migration in the melanophores of *Fundulus heteroclitus* L. *Protoplasma* **61**, 322-345

Burnside, B. (1971) Microtubules and microfilaments in newt neuralation. *Dev. Biol.* **26**, 416-441

Byers, B. & Porter, K. R. (1964) Oriented microtubules in elongating cells of the developing lens rudiment after induction. *Proc. Natl. Acad. Sci. U.S.A.* **52**, 1091-1099

Fawcett, D. W., Anderson, W. A. & Phillips, D. M. (1971) Morphogenetic factors influencing the shape of the sperm head. *Dev. Biol.* **26**, 220-251

Freed, J. J. & Lebowitz, M. M. (1970) The association of a class of saltatory movements with microtubules in cultured cells. *J. Cell Biol.* **45**, 334-354

GIBBONS, I. R. (1963) Studies on the protein components of cilia from *Tetrahyonema pyriformis*. *Proc. Natl. Acad. Sci. U.S.A.* **50**, 1002-1010

GREEN, L. (1968) Mechanism of movements of granules in melanocytes of *Fundulus heteroclitus*. *Proc. Natl. Acad. Sci. U.S.A.* **59**, 1179-1186

HANDEL, M. A. & ROTH, L. E. (1971) Cell shape and morphology of the neural tube: implications for microtubule function. *Dev. Biol.* **25**, 78-95

JUNQUEIRA, L. C. & PORTER, K. R. (1969) Mechanisms for pigment migration in melanophores and erythrophores. *J. Cell Biol.* **43**, 62a (Abstract)

MCINTOSH, J. R. & PORTER, K. R. (1967) Microtubules in the spermatids of the domestic fowl. *J. Cell Biol.* **35**, 153-173

MCINTOSH, J. R., HEPLER, P. K. & VAN WIE, D. G. (1969) Model for mitosis. *Nature (Lond.)* **224**, 659-663

MOOSEKER, M. S. & TILNEY, L. G. (1972). Isolation and reactivation of the axostyle. Evidence for a dynein-like ATPase in the axostyle. *J. Cell Biol.* **56**, 13-26

PEARCE, T. L. & ZWAAN, J. (1970) A light and electron microscopic study of cell behavior and microtubules in the embryonic chicken lens using Colcemid. *J. Embryol. Exp. Morphol.* **23**, 491-507

PORTER, K. R. (1966) Cytoplasmic microtubules and their functions in *Principles of Biomolecular Organizations (Ciba Found. Symp.)* (Wolstenholme, G. E. W. & O'Connor, M., eds.) pp. 308-345, J. & A. Churchill, London

SMITH, D. C. & SMITH, M. T. (1935) Observations on the color changes and isolated scale erythrophores of the squirrel fish, *Holocentrus ascensionis* (Osbeck). *Biol. Bull.* **68**, 131-139

SUMMERS, K. E. & GIBBONS, I. R. (1971) Adenosine triphosphate-induced sliding of tubules in trypsin treated flagella of sea urchin sperm. *Proc. Natl. Acad. Sci. U.S.A.* **68**, 3092-3096

VASILIEV, JU. M., GELFAND, I. M., DOMNINA, L. V., IVANOVA, O. Y., KOMM, S. G. & OLSHEVSKAJA, L. V. (1970) Effect of Colcemid on the locomotory behaviour of fibroblasts. *J. Embryol. Exp. Morphol.* **24**, 625-640

Discussion

Trinkaus: Is it possible that the microtubules are oriented by the stress imposed on the fibroblast by its locomotory mechanism?

Porter: All one can say is that the cell seems to depend in some way on the microtubule for its shape. If the tubules are disassembled, the cell loses its normal asymmetry.

Huxley: In the melanophores is it only the pigment granules that move or do other cell organelles move also?

Porter: Mitochondria and elements of the endoplasmic reticulum do not move. The pigment seems to be locked selectively into a continuum which the microtubules shape or distort. One can think of this continuum as being elastic: if the frame of microtubules is taken away, the continuum contracts into a smaller sphere in the cell centre.

Wolpert: Were there no microfilaments in the cytoplasm at all?

Porter: Very few; not enough to be significant.

Allison: In melanocytes there is apparently a complicated interplay between tubules and filaments in movement of pigment granules (Malawista 1971). If the layers were of intact cells, there might have been a problem of penetration of fixative.

Porter: Yes; possibly the filaments were not preserved.

Allison: Presumably this postulated lack of penetration would also make experiments with cytochalasin difficult.

Porter: It is not easy to bring such compounds into contact with the cell because of the overlying epidermis and the layers of collagen in the scale.

Allison: Why do the pigmented cells contract synchronously?

Porter: Probably because they share a common innervation. One should bear in mind that the response to potassium and sodium may be by way of this innervation.

Steinberg: Did cytochalasin have any effect?

Porter: We have not tried cytochalasin.

Gail: If the cells you described are chilled or somehow deprived of energy, do they rest in the contracted or the expanded position?

Porter: At first the pigment aggregates on chilling, but after 30–60 min it diffuses throughout the cell and loses its radial organization, and in general behaves as it does after colchicine treatment.

Gail: Could the energy be required to re-establish the highly organized microtubular structure found in the expanded state, with contraction being a thermodynamically spontaneous reversion from a low entropy (expanded) to a higher entropy (contracted) state?

Porter: That is right. The matrix in which the pigment resides (as in gel) seems to contract when the tubules disassemble. But, since the pigment diffuses from the central mass after colchicine or vinblastine the matrix as well must be sensitive to these drugs.

Albrecht-Bühler: Must disassembly of the microtubules be the reason that they are not visible in the electron micrograph? Couldn't the explanation be simply that they are not stained any more?

Porter: They would probably stain if they were there. It is more reasonable to suggest that they are not fixed or are so labile they disassemble before the fixative reaches them.

Wohlfarth-Bottermann: It is probable that they are fixed, because it is known from the work on microtubules from flagellae that they can be depolymerized, even *in vitro*.

Porter: Systems of microtubules that are constantly polymerizing and depolymerizing as in dividing cells are just as labile as those in erythrophores and are readily fixed. Inoué has described the changes in birefringence, evident in the interzone during cytokinesis, as resembling 'Northern Lights'. What he is seeing probably represents a rapid assembly and disassembly of microtubules.

Goldman: Inoué & Sato (1967) have also suggested that the controlled disassembly of spindle microtubules at the poles is responsible for chromosome movement. Have you studied the intermediate stages in the movement of granules and can you observe disassembly of microtubules behind the granules as they move in?

Porter: No, we have not, but I realize their importance.

Huxley: In the excitatory mechanism do you think some internal membrane system is connected with the surface membrane of these cells?

Porter: These phenomena of aggregation and disaggregation are influenced by calcium and magnesium: if these are removed from the environment by EDTA, for example, the pigment aggregates, and when they are restored, the pigment disperses. The response is quite rapid. The innervation to the chromatophore is still intact, and might be responsible.

If the stimulation of pigment aggregation has anything in common with the stimulation of muscle contraction, some structural device might transmit the excitation from the surface into the interior of the cell. There are after all several micrometres for diffusion between the cell surface and the central masses of microtubules which show the response. We have therefore looked especially at membrane systems which do not seem to belong to the endoplasmic reticulum (see Figs. 2 and 6). Such membrane systems are common in tubule-rich portions of the cytoplasm, their membranes are thick like the plasma membrane, and they include profiles of tubules and vesicles which are dissimilar in dimension and outline to elements of the smooth endoplasmic reticulum. In a few instances we have found fairly clear evidence of continuity between these thick membranes and the membrane limiting the cell. While we have not yet succeeded in getting a marker such as ferritin or peroxidase into the system and thus demonstrating its continuity with the external environment, we feel that the system is a derivative, essentially an invagination, of the cell surface designed, like the transverse tubule system of muscle, to carry the excitation rapidly into the sensitive pigment-moving mechanism.

Wohlfarth-Bottermann: One hypothesis is that the motive force for the motion of the chromosomes results from polymerization and depolymerization of microtubules. A pool of monomeric G-tubulin in the cytoplasm delivers the material for generation of microtubules. Another idea is that of a sliding mechanism as in muscle (and in flagellae). Which do you prefer?

Porter: Neither! I suggest that the chromosomes move in much the same manner as the pigment and in response to similar events in the microtubule systems of the spindle. When, therefore, at the end of metaphase the pole-centred microtubules begin to disassemble, the chromosomes respond to some elastic component of the matrix and begin their migration. While the velocity is much less than in pigment motion it is similarly uniform over most of the distance traversed. Saltation and the response to colchicine are also similar.

Wolpert: Do you discount the recent reports of actin-like filaments in the mitotic apparatus (Behnke *et al.* 1971; Gawadi 1971)?

Porter: Yes, because I suspect contamination in experiments which use heavy meromyosin to detect actin.

References

BEHNKE, D., FORER, A. & EMMERSEN, J. (1971) Actin in sperm tails and meiotic spindles. *Nature (Lond.)* **234**, 408-409

GAWADI, N. (1971) Actin in the mitotic spindle. *Nature (Lond.)* **234**, 410

INOUÉ, S. & SATO, H. (1967) Cell motility by labile association of molecules. The nature of the mitotic spindle fibers and their role in chromosome movement in *The Contractile Process (Symp. N. York Heart Assoc.)*, pp. 259-292, Little Brown, New York

MALAWISTA, S. E. (1971) Cytochalasin B reversibly inhibits melanin granule movement in melanocytes. *Nature (Lond.)* **234**, 354-355

Cell adhesion and locomotion

A. S. G. CURTIS and T. E. J. BÜÜLTJENS

Department of Cell Biology, University of Glasgow

Abstract The hypothesis that cell adhesion controls the locomotory activity of cells has been experimentally tested. The adhesiveness of cells has been both increased and decreased from the control level by alteration of their lipid composition. These changes in adhesiveness affect neither the average speed of movement of the cells nor the rate of initiation of pseudopods. A new method of measuring contact inhibition of movement is described. This method examines the cytoplasmic overlap of the cells. Changes in cell adhesion do not affect the contact inhibition shown by cells. It is suggested that cells can 'sense' mechanical strains acting on them, and suitably adjust their motile processes to compensate for changes in cell adhesion.

It has often been suggested that cell adhesion must be important in the control of cell movement if only for the simple reason that too much adhesion would prevent the cell from starting locomotion. However, speculation has gone much further than this. It has been suggested that the relative degree of adhesion between one cell and another cell, and with their substrates, controls such phenomena as contact inhibition of movement (Abercrombie & Ambrose 1962), sorting of cells (Steinberg 1964) and the direction and speed of cell movement. Surprisingly this last assumption has apparently never been tested experimentally. Most of us consider cells to be like racing cars driven at a nearly constant engine power. Under these conditions direction and speed are considerably influenced by the adhesion of the tyres to the road. Loss of adhesion starts a skid. There is another analogy worth considering. Suppose that cells are like a piece of heavy earth-moving machinery such as a bulldozer or a scraper. The speed of such machines is nearly constant however great the adhesion and resistance of the soil, until stalling takes place. Greater engine power is used if the substrate is very sticky, less if the substrate becomes less adhesive. Engine power is matched to adhesion, etc., to produce a constant speed. If cells behave as the machines in this second analogy there will be little relation between the adhesiveness of a cell to its substrate and to fellow cells and such matters as rate of movement and rate of initiation of pseudopods. Contact inhibition of movement will be unrelated to adhesiveness. If the bulldozer analogue correctly describes cells, there then arises the problem of how cells sense the mechanical forces of resistance tending to impede their progress and how they adjust their power output to match the resistance thus producing a constant speed.

In essence the experiments we have set up to test between these two hypotheses consist in examining how various aspects of cell movement depend on cell adhesiveness. Cell adhesiveness has been altered by changing the lipid composition of the plasmalemmae. Three parameters have been examined: (i) the rate of cell movement, as measured by the rate of extension of an outgrowth from an explant; (ii) the rate of ruffling in a pseudopod, that is the frequency with which a ruffle (visible as a dark line under phase microscopy) appears at the front of a pseudopod; and (iii) the contact inhibition of movement shown by the cells.

Recent experimental work in our laboratory has shown that it is possible to alter the adhesiveness of neural retina cells from embryonic chicks by incubation in a simple medium containing ATP, CoA and a chosen fatty acid. This follows the finding by Fischer *et al.* (1967) that reacylation of lysolecithin can be stimulated in the plasmalemma by these components, and our discovery that accumulation of lysolecithin in the cell surface lowers adhesiveness. Further, we found extensive incorporation of fatty acid into the phosphatidyl compounds of the neural retina plasmalemma in both R^1 and R^2 positions in the presence of ATP and CoA. If an excess of one particular fatty acid (within the range C_{14}—C_{22}) or fatty acid-CoA compound is present in the incubation medium this is preferentially incorporated by comparison with any fatty acid released by the plasmalemma during production of lysolecithin. In this way the fatty acid composition of the phosphatidyl compounds in the cell surface can be altered. Either of the alkyl groups R^1 and R^2 can be replaced in about 30% of the phosphatidyl molecules present. Increasing the unsaturation or decreasing the chain length of the fatty acid incorporated diminishes the adhesion.

We repeated this work on EDTA-dispersed ventricle cells from embryonic chicks. The cells were incubated for 20 min at 37 °C in CoA–ATP–fatty acid media, and the adhesiveness was measured by the collision efficiency method (Curtis 1969). Results are shown in Table 1(i). (It is of interest that incubation in Hank's medium diminishes adhesion.)

However, this type of measurement does not clearly establish whether the adhesiveness of cells in culture can be changed by incubation. Consequently in a second type of experiment seven-day-old ventricle cells were grown in

TABLE 1

(i) Adhesiveness of ventricle cells from seven-day-old chicks after incubation in various media for 20 min or 3 h. Adhesiveness is expressed as collision efficiency percentage value (± S.D.)

Medium	*Time (min)*	*Adhesiveness*	*n*
Unincubated[a]	—	32.7 ± 5.1	8
Hanks[a]	20	12.3 ± 3.5	5
Linoleate[b]	20	12.8 ± 4.5	5
Linoleate[b]	180	7.4 ± 0.8	4
Stearate[b]	20	38.8 ± 6.4	11
Stearate[b]	180	38.0 ± 4.5	4

(ii) Adhesiveness of the same cells as in (i) after culture for 24 h in control medium[c] and then a further 3 h in the media shown below. Measurement in H + 199 medium.

Medium	*Adhesiveness*	*n*
Control[c]	13.3 ± 3.5	8
Linoleate[d]	3.0 ± 1.3	7
Stearate[d]	31.1 ± 6.3	4

[a] Measured in Hanks.
[b] Medium: CoA (5 μmol/dm^3), ATP (12.5 μmol/dm^3), fatty acid (10 mg/dm^3) in Hanks' saline in the experiments with linoleate and stearate.
[c] Medium: minimal Eagle's, calf serum (10%) and embryo extract (2%).
[d] Concentrations of CoA, ATP and fatty acid as in footnote *b*, but in medium *c*.

culture (Eagle's MEM, 10% calf serum and 2% embryo extract) from trypsinized suspensions, for 24 h. The medium was then replaced by new medium: in the controls fresh medium of the same type was used but in the experimental series CoA–ATP–linoleate or CoA–ATP–stearate was added to the fresh medium. The cultures were incubated for a further three hours and then the cells were harvested with EDTA to overcome their adhesion. The cell suspensions were made up in H+199 medium and their adhesiveness was measured by the collision efficiency method [see Table 1(ii)].

We have not yet completed testing whether linoleic and stearic acid are incorporated in the plasmalemmal phospholipids of the ventricle cells but what is important is that cell adhesion can be considerably altered by this treatment without affecting the viability of the cell. The difference in collision efficiency between stearate- and linoleate-treated cells corresponds to a 10 000-fold change in the energy of cell adhesion. If the cells are incubated continuously in the linoleate medium for three hours there is a further small decrease in adhesiveness [see Table 1(i)]. Continuous incubation in stearate for three hours maintains the adhesion at the high level found after incubation in this medium for only 20 min.

In order to investigate the effect of changing cell adhesiveness on cell locomotion, cultures of the seven-day-old chick ventricle cells were prepared. The cells were grown in Eagle's MEM–serum–embryo extract medium, either as explants or after being allowed to settle onto glass surfaces from trypsinized suspensions. Cells were grown in this control medium for 24–36 h. The culture medium was then replaced in our initial experiments with a fresh medium of the same nature to which linoleate, CoA and ATP had been added in the experimental series. These additions were made at the same concentrations as in the experiments on the degree of adhesion. We found, however, that treatment of the cultures with the control or experimental medium led to a total cessation of cell movement and pseudopodal activity for about three hours. So an alternative technique was devised for adding the CoA, ATP and fatty acid components to the cultures. In both the control and experimental series, a small volume of Tris–saline was injected into the cultures, but in the experimental series this addition also contained the appropriate amounts of fatty acid, CoA and ATP. There was no obvious immediate response by the cells to the injection of Tris–saline alone or of CoA–ATP–stearate, since the form of the cells showed no sudden change and pseudopodal activity continued unimpaired. However the addition of CoA–ATP–linoleate produced much rounding up and detachment of the cells during the 20 min after injection. Pseudopodal activity was also suppressed and cell movement stopped. It is possible that this is a direct consequence of the reduction in cell adhesiveness induced by incorporation of linoleate into plasmalemmal lipids. However, after incubation for three hours, the cells recovered extended fibroblast-like form and both pseudopodal activity and migration were active. Therefore measurements of locomotory behaviour of the linoleate-treated cells were made after three hours' incubation. It is of interest that the measurements of cell adhesiveness show that the cells do not recover their normal adhesion after three hours' incubation in the presence of linoleate, indeed the cells are slightly less adhesive after three hours than after treatment for 20 min. Thus the recovery of normal or near-normal morphology and motile properties is not due to a regain of adhesion but to some other process which might offset the effects of loss of adhesion.

EFFECTS OF CHANGED ADHESION ON MOTILE PROPERTIES

Cell locomotory rate

If motility is controlled by the adhesiveness of the substrate it would be expected that a change in that adhesion would affect locomotory rate. Of course we do not know whether an increase in adhesion would accelerate the cell,

TABLE 2

Speed of cell movement

	Speed (μm/h)	*(S.D.)*
Control	64.4	7.0
Linoleate treated	67.0	11.3
Stearate treated	64.5	3.8
		$n = 10$

owing to the slip being reduced so that the cell had better purchase, or whether an increase in adhesion would slow the cell because of the greater frictional forces to be overcome. The rate of cell locomotion was calculated from the mean outward rate of movement of cells at the edge of an outgrowth: measurements were made only on those cells that remained in contact with the outgrowth during the period of measurement. Abercrombie & Heaysman (1966) have shown that movement is strongly affected by the contact number of the cell under measurement. Measurements on linoleate-treated cells refer only to the second three hour period after treatment. The results (see Table 2) show that locomotory rate is unaffected by these treatments which alter adhesiveness so greatly. However, measurements of the rate of locomotion are necessarily rather inaccurate, and it is possible that changes in adhesiveness produce effects which tend to compensate for any change in locomotory rate (which is simply the consequence of a change in adhesion). For example the extent of contact inhibition of the cells might affect locomotion. A decrease in adhesion might diminish contact inhibition thus tending to slow movement outward from an explant though the drop in adhesion itself might tend to accelerate intrinsic motility. For this reason it is particularly desirable to examine other features of cell motile behaviour, such as contact inhibition of movement and pseudopodal activity.

Pseudopodal activity

Ingram (1969) and Abercrombie *et al.* (1970*b*) have shown that the protrusion of the lamellopodium at the front end of a fibroblast is primarily responsible for the locomotion of these cells. The alternate protrusion and withdrawal of the lamellopodia and in particular their dorsal bending gives the appearance of 'ruffling' which is seen in this region of the cell. Abercrombie *et al.* (1970*a*) suggest that the rate of ruffling is a measure of the locomotory activity of the cell. For this reason one of us (A.C.) observed the ruffling behaviour of ordinary fibroblasts and those treated with linoleate or stearate. The lifetime of a ruffle is extremely variable as some travel much further back on a fibroblast than others. Moreover a ruffle can become temporarily stationary. However the frequency of initiation of a ruffle at the free front end of a fibroblast appears to be fairly constant within any period of activity. Observations were made visually. In the control series a new ruffle appeared every 42 ($\pm$ 5) s, in the linoleate-treated series every 44 ($\pm$ 6) s and in the stearate-treated series every 41 ($\pm$ 5) s (S.D. in parentheses). (These figures were obtained from measurements during periods of activity but do not include initiating or terminating ruffles or periods of inactivity.) There is some evidence that the ruffling frequency is bimodal in all three series, with one period being almost twice the other. There is, however, no obvious difference between the cells under the different treatments. The times between ruffles are longer than those reported by Abercrombie *et al.* (1970*a*), but it should be noted that the locomotory rate of the cells is also considerably higher.

Contact inhibition of movement

It has been suggested (see Abercrombie & Ambrose 1962) that contact inhibition of movement is largely controlled by the relative adhesiveness of the cells and their substrate. This idea was examined more fully by Carter (1965). We have examined the question of whether changes in cellular adhesiveness would affect contact inhibition. It would be expected that cell–substrate adhesions would be much less affected by linoleate or stearate treatment than cell–cell adhesions simply because the inert substrates would not be altered by these treatments. This relative change between cell–cell and cell–substrate adhesions after linoleate or stearate treatment might affect contact inhibition.

We felt that it would be interesting to examine cytoplasmic overlap rather than the nuclear overlap normally measured in studies of contact inhibition. One reason for confining measurement to nuclear overlap has been the lack of techniques for measuring cytoplasmic overlap. However by combining a Chalkley counting method (to measure the relative areas of overlap and non-overlap) with high resolution phase contrast microscopy of the cells fixed in aqueous formaldehyde–saline it is possible to see and measure areas of cytoplasmic overlap, except at very high population densities. A great advantage of this technique is that cell interaction can be examined at the earliest stages, long before there is any appreciable probability of nuclear overlap even if there is no contact inhibition. If A is the area of one randomly distributed cell, the expected area of overlap in an area T containing n cells is given by:

$$\frac{n(n-1)A^2}{2T}\left(\frac{T}{T'}\right)^2$$

where T' is the augmented area of the field due to the fact that cells mainly outside the field can overlap cells within the field. This boundary condition appears to have been ignored in earlier treatments of expected overlap. This treatment can be derived from those treatments given by Kendall & Morgan (1963) (we thank Dr R. Elton for bringing this treatment to our notice).

If cytoplasmic overlaps can be recognized, areas of cells can be accurately measured. Curtis & Varde (1964) found that the plan area of cells (assuming no overlap) fell very rapidly in a curvilinear fashion with increasing cell population density. This suggests either that cells overlap (cytoplasmically) fairly appreciably or that they have some precise method for controlling their plan area. Preliminary work showed that the former explanation is correct and that the plan area of these cells is actually unaffected by population density, the mean size being constant. Thus we can ignore possible effects of population density on the area A in the above formula.

The mean overlap area as a proportion of cell area is given in Table 3, together with the overlap area (as a proportion) expected on random distribution, and the ratio of real overlap to expected overlap, for the two treatments and the control series. It can be seen that there is no significant difference in the ratio of real to expected overlap between the three treatments. There is a large variance in the measurements but this probably arises from the fact that each set of measurements was carried out on small numbers of cells (17–65) because of the requirement of high resolution microscopy. Thus it appears that the contact inhibition of movement, as measured by cytoplasmic overlap, is unaffected by large changes in cell–cell adhesion. It should be noted that there is less cytoplasmic overlap than would be expected on random grounds; in other words

TABLE 3

Cytoplasmic overlap: means and standard deviations

	Overlap (as proportion of cell area)			
	Measured	*Expected*	*Measured/Expected*	*n*
Control	0.066	0.165	0.400 ± 0.32	16
Linoleate-treated	0.064	0.161	0.397 ± 0.21	21
Stearate-treated	0.076	0.145	0.523 ± 0.22	12

Sampling area 55990 μm²; cell population 17–65.

contact inhibition of movement can be detected as a lack of cytoplasmic overlaps as well as a lack of nuclear overlaps.

A question yet to be resolved is whether there is a greater probability of cytoplasmic overlap than nuclear overlap, after corrections have been made for the relative areas of nucleus and cytoplasm. About 5–10% of each cell is, on average, involved in cytoplasmic overlap so that if nuclear and cytoplasmic overlaps are distributed equally we should find about 20 μm^2 of nuclear overlap in amongst 30 cells. Preliminary measurements give a value close to this which suggests that contact inhibition acts equally on all parts of a cell.

DISCUSSION

Our experimental results show that large changes in cell–cell adhesiveness do not affect locomotory features of cell behaviour. The mean rate of locomotion, the rate of pseudopodal or lamellopodal initiation and the degree of overlap of the cells are not affected by large changes in cell–cell adhesion presumably accompanied by relatively smaller changes in cell to substrate adhesion. This constancy suggests that a cell is able to compensate for changes in its adhesiveness and adjust its locomotory mechanisms accordingly. For example, though the very marked diminution in adhesiveness consequent on linoleate treatment initially changes cell shape and totally inhibits cell movement, the cell recovers its locomotory ability even though its adhesiveness is still low. Returning to the analogy given at the start of this paper we can now say that cells appear to behave like bulldozers rather than racing cars. They must presumably have a sensitive means of determining the effect adhesion has on various parts of the cell surface in restraining motion, so that they can balance these restraints by increasing contractile and locomotory activity in the required parts of their bodies. It is perhaps a little fanciful to speculate about the nature of such a control system but it is tempting to suggest that microtubule and microfilament contraction are subject to an internal feedback system which senses the external strain being placed on the tubule or filament, and adjusts the contraction to match the strain. If tubules or filaments are attached to the surface this will link adhesion to a directly compensating effector system. Vasiliev *et al.* (1970) and Berlin & Ukena (1972) have provided evidence indicating that microtubules are linked to cell surface components. Of course areas of cell surface attached to microtubules might be especially adhesive and possibly this is how the microtubular systems in groups of cells link up.

We find generally that the assumption that cell adhesiveness directs and controls cell movement is not supported by the experimental results. Other factors might override differences or changes in adhesion. Nevertheless there may still be situations, such as the strain-specific morphogenic control of cell adhesion which determines simple morphogenesis in *Ephydatia* (Curtis & van de Vyver 1971), where adhesion changes can determine the direction of cell movement.

ACKNOWLEDGEMENTS

We thank the SRC for a grant and the University of Glasgow for facilities. We are grateful to the Wellcome Trust for a research scholarship for one of us (T.E.J.B.) and we also thank Miss Rose McKinney for much invaluable technical assistance.

References

ABERCROMBIE, M. & AMBROSE, E. J. (1962) The surface properties of cancer cells: a review. *Cancer Res.* **22**, 525-548

ABERCROMBIE, M. & HEAYSMAN, J. E. M. (1966) The directional movement of fibroblasts emigrating from cultured explants. *Ann. Med. Exp. Biol. Fenn.* **44**, 161-165

ABERCROMBIE, M., HEAYSMAN, J. E. M. & PEGRUM, S. M. (1970*a*) The locomotion of fibroblasts in culture. II. 'Ruffling'. *Exp. Cell Res.* **60**, 437-444

ABERCROMBIE, M., HEAYSMAN, J. E. M. & PEGRUM, S. M. (1970*b*) The locomotion of fibroblasts in culture. III. Movements of particles on the dorsal surface of the leading lamella. *Exp. Cell Res.* **62**, 389-398

BERLIN, R. D. & UKENA, T. E. (1972) Effect of colchicine and vinblastine on the agglutination of polymorphonuclear leucocytes by concanavalin A. *Nat. New Biol.* **238**, 120-122

CARTER, S. B. (1965) Principles of cell motility: the direction of cell movement and cancer invasion. *Nature (Lond.)* **206**, 1183-1187

CURTIS, A. S. G. (1969) The measurement of cell adhesiveness by an absolute method. *J. Embryol. Exp. Morphol.* **22**, 305-325

CURTIS, A. S. G. & VARDE, M. (1964) Control of cell behavior: topological factors. *J. Nat. Cancer Inst.* **33**, 15-26

CURTIS, A. S. G. & VAN DE VYVER, G. (1971) The control of cell adhesion in a morphogenetic system. *J. Embryol. Exp. Morphol.* **26**, 295-312

FISCHER, H., FERBER, E., HAUPT, I., KOHLSCHÜTTER, A., MODOLELL, M., MUNDER, P. G. & SONAK, R. (1967) Lysophosphatides and cell membranes. *Protides Biol. Fluids Proc. Colloq. Bruges* **15**, 175-184

INGRAM, V. M. (1969) A side view of moving fibroblasts. *Nature (Lond.)* **222**, 641-644

KENDALL, M. G. & MORAN, P. A. P. (1963) *Geometrical Probability*, Griffin, London

STEINBERG, M. S. (1964) The problem of adhesive selectivity in cellular interactions in *Cellular Membranes in Development* (Locke, M., ed.), pp. 321-66, Academic Press, New York

VASILIEV, JU. M., GELFAND, I. M., DOMNINA, L. V., IVANOVA, O. Y., KOMM, S. G. & OLSHEVSKAJA, L. V. (1970) Effect of Colcemid on the locomotory behaviour of fibroblasts. *J. Embryol. Exp. Morphol.* **24**, 625-640

Discussion

Gail: We measured motility in terms of D^*, the augmented diffusion constant, and we tried to quantitate cell–substrate (not cell–cell) adhesivity. Whether we developed a useful measure is for you to consider, but we measured cell–substrate adhesivity by the distraction method (Gail & Boone 1972). We distinguished three categories of cell–substrate adhesivity—low, control and high. We found that the motility of 3T3 fibroblasts and SV3T3 transformants decreased with increasing cell–substrate adhesivity.

Curtis: The differences of interpretation will only be resolved when the appropriate experiments are done, but it is possible to reconcile easily our interpretation with yours at a theoretical level. Büültjens and I suggest that cells are able to compensate their locomotory activity for changes in adhesion over a fairly wide range. Obviously extreme changes are going to make the cells either too slippery or too sticky to move. Thus one can plot locomotory activity against adhesion as in Fig. 1. The plateau in the graph exists because of the compensation that the cell can produce in its locomotory activity. I suggest that you varied the adhesion between, say, points A and B, and thus affected motility. We might have worked at the plateau with the result that perhaps a much bigger change in adhesion, between, say, C and D, had no effect on locomotion.

Gail: I assume, as you do, that there is an optimal range of cell–substrate adhesivities below which cells move more slowly, but I have not been able to demonstrate this putative effect of decreasing cell–substrate adhesivity.

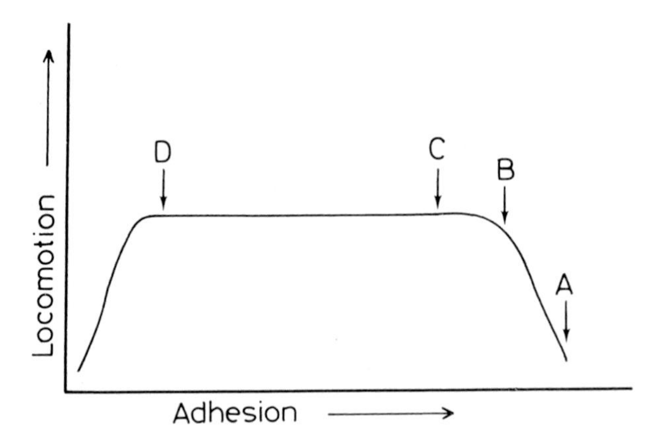

FIG. 1 (Curtis). Diagrammatic representation of the relationship between adhesion and locomotion.

Our experiments evidently fall in a supra-optimal adhesivity range where increasing cell–substrate adhesivity decreases motility.

Wolpert: Professor Curtis, are you saying that if you decrease the adhesiveness 10 000 times there is no obvious effect?

Curtis: There is no obvious effect on cell motility or contact inhibition of movement even after three hours of culture in the presence of the linoleate incorporation medium.

Huxley: What do you mean by adhesiveness? Do you mean that the amount of force exerted would be 10 000 times smaller?

Curtis: The evidence showed that the energy of adhesion had decreased 10 000-fold.

Wolpert: But then the cells do not stick.

Allison: A more serious criticism is that you have measured adhesion by the establishment of contact at any early stage in the experiment but you have measured the movement later on.

Curtis: Cell adhesion has been measured both after 20 min (cells incubated in suspension) and after three hours incubation in culture with the cells growing on a substrate. Adhesion has the same low value in both cases for linoleate-treated cells. Of course the measurements after three hours incubation form a closer control of the adhesive state of the cells actually being examined for their behavioural properties than the short-term incubations, but it could be argued that because these cells must be dissociated before adhesion measurement, the very dissociation might have produced the low stickiness found. I feel that we have some fairly good evidence against this sort of artifact because incubation in a linoleate incorporation medium lowers the stickiness of the cells whether the incorporation is carried out before or after dissociation. In addition, cells treated in other ways (e.g. stearate incorporation) show the same high stickiness whether dissociated before or after treatment.

Trinkaus: You measured the adhesiveness of the cells to each other, and then the rest of your studies were on the adhesiveness of the cells to the glass substratum. Furthermore, we do not know that the cells are adhering to each other by an overlap.

Curtis: We advance the theory that cells can compensate their motile mechanisms for very large changes in adhesion because behaviour is not affected even when large changes in adhesion occur. Of course these are measurements of cell–cell adhesion but unless one believes that cell–inert substrate adhesion is different in kind from cell–cell adhesion (see Curtis 1967), it seems reasonable to assume that a change in the stickiness of the cell would also have an effect on the cell–substrate adhesion. In addition, it seems likely that cell–cell adhesion is at least a component of the contact inhibition phenomenon. Thus we feel that cells can compensate for changes in adhesion over wide ranges. This compensation might take place by the cell altering its internal locomotory components. Obviously this compensation might take some time to effect and this is possibly the reason that the linoleate-treated cells initially round up and occasionally come off the substrate. Later these cells respread even though their adhesiveness (per unit area) does not appear to have changed.

Steinberg: Two terms seem to be in current use. I do not think the word adhesiveness is made any more scientific by having -ity as the suffix instead of -ness. Adhesivity and adhesiveness are both informal terms, hospitable to many meanings. We do not *measure* adhesiveness, we *define* it by a given operation.

Professor Curtis, I don't think you have shown that collision efficiencies, as measured by your procedures, are determinants of the phenomena that you are trying to use these measurements to explain. I cannot see any logically rigorous connection between the collision efficiencies of cells as they briefly collide in suspension and their behaviour as they creep and interact on the substratum.

Trinkaus: I agree. Also, wouldn't it be proper to conclude that the results obtained with one measure of adhesiveness do not allow one to predict what the results would be with another measure of adhesiveness?

Curtis: Certainly these problems are a source of difficulty. There seem to me to be three separate questions into which these objections can be analysed. First, are the various methods of measuring adhesion (formation of adhesion, 'collision' and 'distraction') in principle comprehensible in terms of one common cell property? Second, do cells change this property as they adhere or soon afterwards, so that all measurements made by a collision method will never agree with those made by a distraction method? Third, will a measurement of the change in the adhesion between two cells on altering, say, the culture medium give a good prediction for the change of the adhesion between one of these cells and a substrate, when their medium is similarly altered?

The first question I feel can be answered. Certain forces will be responsible for maintaining a cell in adhesion. The magnitudes of these forces define the adhesiveness. If you have a method of measuring these forces, or their sum, ideally per unit of cell surface, uncomplicated by extraneous matters, this should give the same result for a given cell contact as another similarly absolute method, provided of course that neither technique alters the adhesion. I think that the collision efficiency method (Curtis 1969) goes far, perhaps the whole way, to removing extraneous factors from the measurement. Unfortunately other methods [e.g. distraction (see p. 183)] do not yet do this. Problems owing to cell shape, size, viscosity of the medium, contact area, hydrodynamic forces etc., are absent with the collision efficiency technique. It should be remembered that this technique can be used to interpret results on a wide variety of theories of adhesive mechanism even though I preferred to use a lyophobic colloid interpretation (Curtis 1969).

The second question of changes in adhesion after adhesions form cannot yet be answered completely. I have argued that these changes are absent at least in some situations (Curtis 1967).

The third question raises the problem of whether you feel that the same mechanism is used in cell–cell adhesion as in cell–substrate contacts. Coman (1961) has argued in favour of a difference, but I have produced contrary evidence (Curtis 1967). If the mechanisms are common to both situations it seems likely that raising cell–cell adhesion will raise cell–substrate adhesion (and *vice versa*) because at least one component is more adhesive. Moreover, it seems likely that one of the phenomena we have studied, contact inhibition of movement, involves cell–cell adhesion (Abercrombie & Ambrose 1962) so that measurements of cell–cell adhesion should certainly be directly relevant in this case.

Wolpert: But a 10 000-fold change in adhesion ought to be measurable even with the crudest technique.

Curtis: We can observe—not measure—that the linoleate treatment apparently decreases the adhesiveness of the cells to the substrate.

Allison: Distraction experiments at different times could be instructive.

Curtis: I don't agree. Distraction experiments are potentially very inaccurate, because the separation of the cell from the substrate is a peeling process, so that besides the adhesive forces one must take into account the cell profile, the areas of contact at varying separations, the internal rheology of the cell, and the energy involved in pulling medium into the gap between the separating cell and substrate as well as some fairly complicated hydrodynamic problems about the application of the shear forces. All these parameters may vary widely from one experimental situation to another. As Weiss (1967) pointed out, such measurements might lead to the cell being torn in half. When this happens the measurement has nothing at all to do with adhesion.

Albrecht-Bühler: If cell–cell adhesiveness is reduced, cells should be released from contact inhibition of mitosis. Does incubation of a monolayer in culture with linoleate stimulate overgrowth?

Curtis: That is a very interesting experiment which we have not done.

Ambrose: The topology of the cell surface might also be an important factor here; the contours of the surface, also the shapes and dimensions of pseudopodia, will affect the extent to which areas of cell–substrate or cell–cell contact can be established.

Curtis: You are quite right. This is one of the great difficulties with the distraction type of experiment. Now, although the morphologies of the cells, treated in various ways, in our experiments are apparently the same when the cells are spread in culture, there may be small differences. We hope to carry out a careful quantitative examination of morphology for such differences.

In the collision efficiency experiments, the difficulty of unlike morphology and cell profiles is avoided because the cells are spheres or at least subspherical.

Gingell: Returning to this huge divergence in adhesiveness, was this a collision efficiency?

Curtis: The collision efficiency difference, as a percentage, was 3–38%. I have described the relation of the collision efficiency to the force constant of adhesion previously (Curtis 1969).

Gingell: The greatest differences that Parsegian and I could derive for adhesive strengths in model interactions (Parsegian & Gingell 1973) were for the interaction of the cell with metals of high density, for example palladium, by comparison with materials such as PTFE, yet these only differ in attractive energy by about a factor of ten.

Curtis: I am surprised that you did not get an infinite difference, since that is simply a comparison between a non-adhesive and an adhesive situation. Areas of initial contact are constant in our measuring system.

Gingell: We are talking about attractive energies solely in adhesive situations. You have not included repulsive energies in your calculations.

Curtis: No; they can be allowed for. They might have an energy of some $50kT$/contact for an almost completely dispersed (non-adherent) state of the cells. It would not be impossible to have values of 10^4kT or even 10^5kT: such figures have been obtained in certain systems (Curtis 1967).

Gingell: You would only get such big energies in their primary minimum where the surfaces are effectively in 'molecular contact'.

Curtis: Yes, the higher values only apply to adhesions of the primary minimum type. I have said nothing about the type of adhesion in our experiments. Possibly the 'stearate-adhesion' is of a different type from the weak 'linoleate-adhesion'.

Gingell: Com ng back to the adhesive behaviour question, would you not expect a difference in behaviour between cells in molecular contact with the substratum and those that were floating above it in a secondary minimum adhesion?

Curtis: For once I agree with the electron microscopists that cells with seemingly different types of contact often behave similarly.

Gingell: The behaviour of cells on palladium and cellulose acetate (Carter 1965, 1967) as well as on PTFE and glass (Weiss & Blumenson 1967) are apparently instances where cell motility reflects the substratum.

Curtis: We don't know that there is any difference in the adhesiveness. Changes in cell behaviour might be due to the mild toxicity of palladium.

Harris: I have found that cells will accumulate on palladium in preference to cellulose acetate but on glass in preference to palladium, which cannot be explained by toxicity; palladium does not seem to be toxic (Harris 1973).

Steinberg: In his model, Curtis (1960, 1966, 1967) proposes a particular nature for the attractive forces, whereas the actual attractive forces between cells might involve a variety of kinds of bonding, including antigen–antibody-like reactions. H. M. Phillips and I have been trying to measure the energies of intercellular adhesion for many years. The parameters of adhesiveness we are measuring are ones that can be rigorously shown to be capable of determining the configurations adopted by populations of mobile, adhesive cells. These parameters are the *specific interfacial free energies* (σ) of cell populations. These are defined as the reversible work that must be done to expand a cell aggregate's interfacial areas by a unit amount. Our preliminary estimates of the cohesive energies of aggregates of four chick embryonic tissues fall in the range 6–30 erg/cm^2, that is, well above the maximum values expected from Curtis' and similar models. These measurements seem to rank these tissues in the same sequence we had previously deduced from our studies of their mutual spreading and sorting behaviour. A more cohesive cell population evidently does sort out internally to a less cohesive one (H. M. Phillips & Steinberg, unpublished observations).

Curtis: This energy of adhesion is comparable with moderately good epoxy resin bonding of, say, metal to metal.

Steinberg: It is in the same range of interfacial energies as the water–benzene interface (35 erg/cm^2 at 20 °C) or the water–diethyl ether interface (10.7 erg/cm^2 at 20 °C). The bonds between such liquids are easily disrupted. The different measurements of 'adhesiveness' by no means all represent the same thing. At some stage one has to justify the relevance of a given measurement to the phenomenon one wants to explain through its use.

Curtis: I think you have confused the lateral cohesion of the benzene surface, which you can hardly break because of the high surface free energy and the low viscosity of the benzene (so that new surface is formed as you try to break it), and the internal cohesion of the benzene which is low.

Middleton: Using a modification of their collecting aggregate technique, Weston & Roth (1969) demonstrated that it was possible to alter cell–cell adhesion without altering cell–substrate adhesion. They found that treatment of embryonic chick heart fibroblasts with 300mM-urea reduced their mutual adhesion but had no effect on their adhesion to glass. Additionally they noted that monolayer cultures of these fibroblasts showed a significant increase in overlap index in the presence of 160mM-urea and they attributed this increase to the decreased adhesion of the cells to one another.

Curtis: But they measured nuclear overlap, whereas we measured total overlap. Possibly nuclear overlap is affected differently.

References

ABERCROMBIE, M. & AMBROSE, E. J. (1962) The surface properties of cancer cells: a review. *Cancer Res.* **22**, 525-548

CARTER, S. B. (1965) Principles of cell motility: the direction of cell movement and cancer invasion. *Nature (Lond.)* **208**, 1183-1187

CARTER, S. B. (1967) Haptotaxis and the mechanism of cell motility. *Nature (Lond.)* **213**, 256-260

COMAN, D. R. (1961) Adhesiveness and stickiness: two independent properties of the cell surface. *Cancer Res.* **21**, 1436-1438

CURTIS, A. S. G. (1960) Cell contacts: some physical considerations. *Am. Nat.* **94**, 37-56

CURTIS, A. S. G. (1966) Cell adhesion. *Sci. Prog.* **54**, 61-86

CURTIS, A. S. G. (1967) *The Cell Surface: its molecular role in morphogenesis*, Logos Press, Academic Press, London & New York

CURTIS, A. S. G. (1969) The measurement of cell adhesiveness by an absolute method. *J. Embryol. Exp. Morphol.* **22**, 305-325

GAIL, M. H. & BOONE, C. W. (1972) Cell–substrate adhesivity: a determinant of cell motility. *Exp. Cell Res.* **70**, 33-40

HARRIS, A. K. (1973) Behavior of cultured cells on substrata of variable adhesiveness. *Exp. Cell Res.* **77**, 285-297

PARSEGIAN, V. A. & GINGELL, D. (1972) *J. Adhes.*, in press

WEISS, L. (1967) *The Cell Periphery, Metastasis, and other Contact Phenomena* p. 380, North Holland, Amsterdam

WEISS, L. & BLUMENSON, L. E. (1967) Dynamic adhesion and separation of cells *in vitro.* II. Interactions of cells with hydrophilic and hydrophobic surfaces. *J. Cell Physiol.* **70**, 23-32

WESTON, J. A. & ROTH, S. A. (1969) Contact inhibition: behavioural manifestations of cellular adhesive properties *in vitro* in *Cellular Recognition* (Smith, R. T. & Good, R. A., eds.), pp. 29-34, Appleton-Century-Crofts, New York

General discussion I

CELL ADHESION

Heaysman: We are using a technique to study cell–substrate adhesion and cell–cell adhesion with the minimum of disturbance to the spatial arrangement of the cells (Abercrombie *et al.* 1971). We grow the cells on small pieces of Araldite and then embed them in more Araldite. When we section, we do not get any fracture along the Araldite–Araldite plane; clearly we are not disturbing the cells and are retaining as true a cell–substrate relationship as possible. We photograph the cells before and after fixation and can thus select a given cell for sectioning. In this way we have been able to section through a ruffle (Fig. 1). When filming cells, we view them from above and the ruffles appear as dark lines which pass back over the surface of the leading lamella. We refer to the long thin processes of constant width which extend from the front of the lamella as lamellipodia (Abercrombie *et al.* 1970). They are protruded at the front end of the lamella parallel with the substrate and then they may lift up and pass back as ruffles until they are slowly pulled back into the surface. Often more than one can be seen at a time.

Porter: How do you interpret the material between the Araldite and the cell?

Heaysman: This is material that is laid down in nearly all situations where cells are growing or medium containing serum is used.

Porter: So there is not a direct contact between the cell and the Araldite?

Heaysman: No; not only is the contact not directly with the Araldite, but it is possible that there is a surface coat on top of the unit membrane which does not show up in our electron micrographs. Associated with the leading lamella we often observe a structure which we believe is connected with cell–substrate adhesion. Along the ventral surface of the cell the cortical structure is reasonably uniform except at certain regions where the membrane comes close to the substrate. Wherever we have this region of close approach we see an electron-

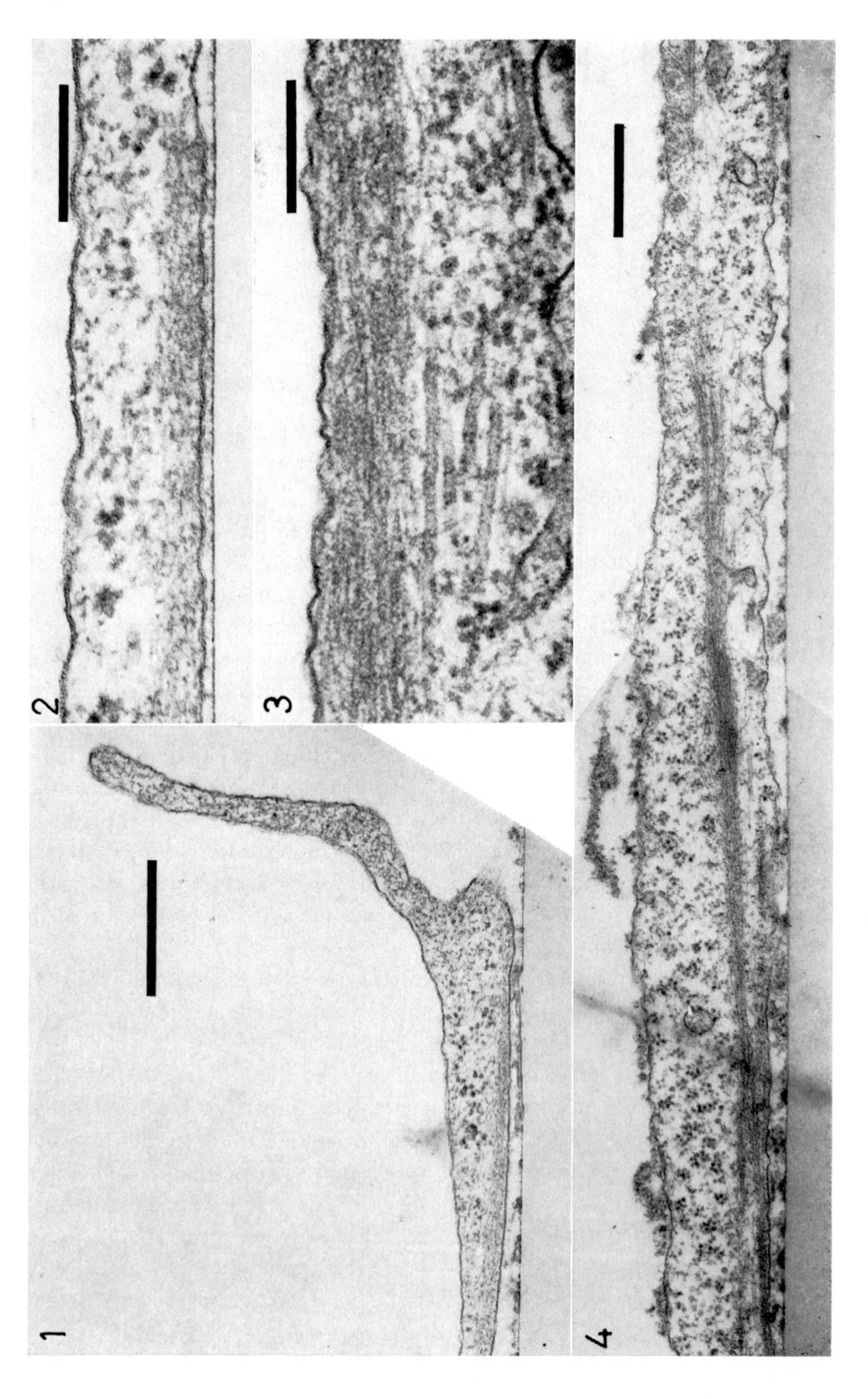
1
2
3
4

dense area in the cell, which we refer to as a plaque (Fig. 2). We most frequently find plaques close to the tip of the lamella but never beneath a lamellipodium. In Fig. 3 the cortical filaments at the dorsal surface of the cell can be seen. We often find transverse fibres apparently passing from these down to the region of the plaque (Fig. 4). Study of cells where we would expect contact inhibition reveals a similar situation (Heaysman & Pegrum 1973). Usually when two cells approach and collide, they retract after anything up to seven minutes—a situation which is typical of contact inhibition. We watched two cells approaching each other and directly they collided we fixed them. This is the closest we can get to the point of collision and we claim that it is within 20 s of collision. In Fig. 5a–d one can see different stages of development of the rather plaque-like structure between the cells rather than between the cells and the substrate. Over most of the area of overlap there is a gap of 15–20 nm between the unit membranes of the two cells but at certain points this narrows to less than 10 nm. Wherever this happens, one can see the beginning of the development of electron-dense areas. Fig. 5 is not a time series but various examples of the development of these areas within about 20 s of collision. In Fig. 6a the great resemblance of the cell–substrate plaque to the cell–cell plaque, one minute after collision, can be seen. Fig. 6b shows the next stage: a sudden lining up of filaments in each cell towards this area of closer apposition. Two minutes after collision, the lining up of fibres between the cells is very marked (Fig. 7) and the surfaces of the two cells have become distorted. We believe that in contact inhibition there is always this sort of contraction; the cells pull towards each other after collision. The lining up of the filaments shown here appears to give a certain basis for the origin of the tension seen in these cells. Fig. 8 is an ordinary light micrograph of two cells in contact and then pulling apart whilst Fig. 9a and b show once again the apparent tension; not only are fibres lined up but the cells, which usually have a fairly constant gap between their ventral surfaces and the substrate, have been pulled up away from the substrate. Fig. 10 shows two cells which have separated within the last 20 s before fixation. We suggest that the differentiated areas have already disappeared.

Trinkaus: Have you looked at these under higher magnification? It would be

---

←

FIG. 1 (Heaysman). Vertical longitudinal section of a ruffle (lamellipodium). The bar represents 0.5 μm.

FIG. 2 (Heaysman). Vertical longitudinal section of plaque. The bar represents 0.2 μm.

FIG. 3 (Heaysman). Vertical longitudinal section. Microfilaments form a dorsal cortical layer. The bar represents 0.2 μm.

FIG. 4 (Heaysman). Vertical longitudinal section showing oblique strand of microfilaments running anteriorly towards the plaque and posteriorly towards the dorsal cortical filaments. The bar represents 0.5 μm.

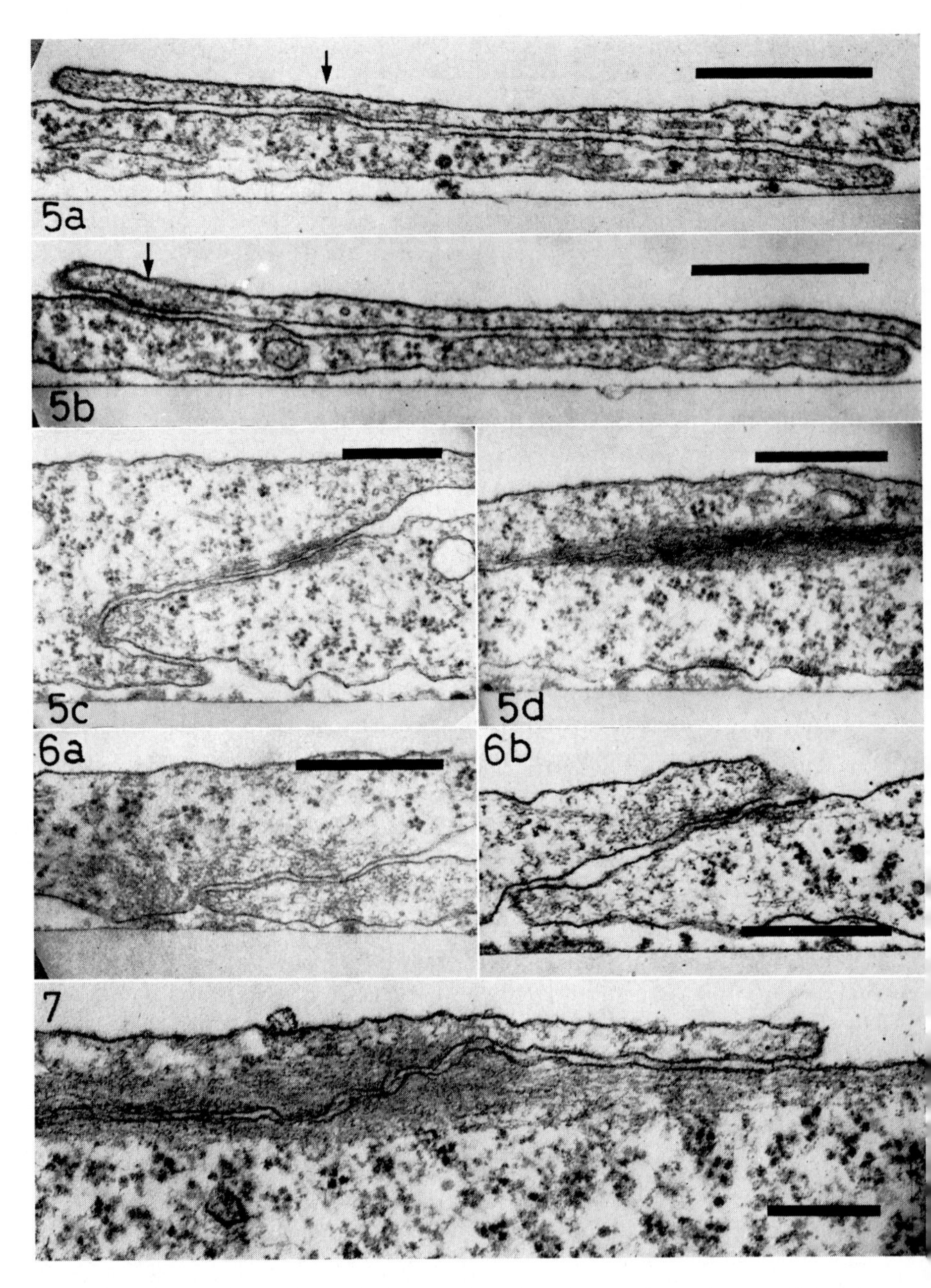

FIG. 5 (Heaysman). Vertical longitudinal sections (a—d) through overlap area 20 s after collision showing differing degrees of specialization (arrows). The bar represents 0.5 μm.

FIG. 6 (Heaysman). (a) Vertical longitudinal section through overlap area 60 s after collision showing similarity between cell–cell specializations and cell–substrate plaque. The bar represents 0.5 μm.

(b) Vertical longitudinal section showing 'lining up' of the microfilaments. The bar represents 0.5 μm.

FIG. 7 (Heaysman). Vertical longitudinal section more than 2 min after collision showing the distortion of the surface shape of the contracting cells and the varying gap between them. The bar represents 0.25 μm.

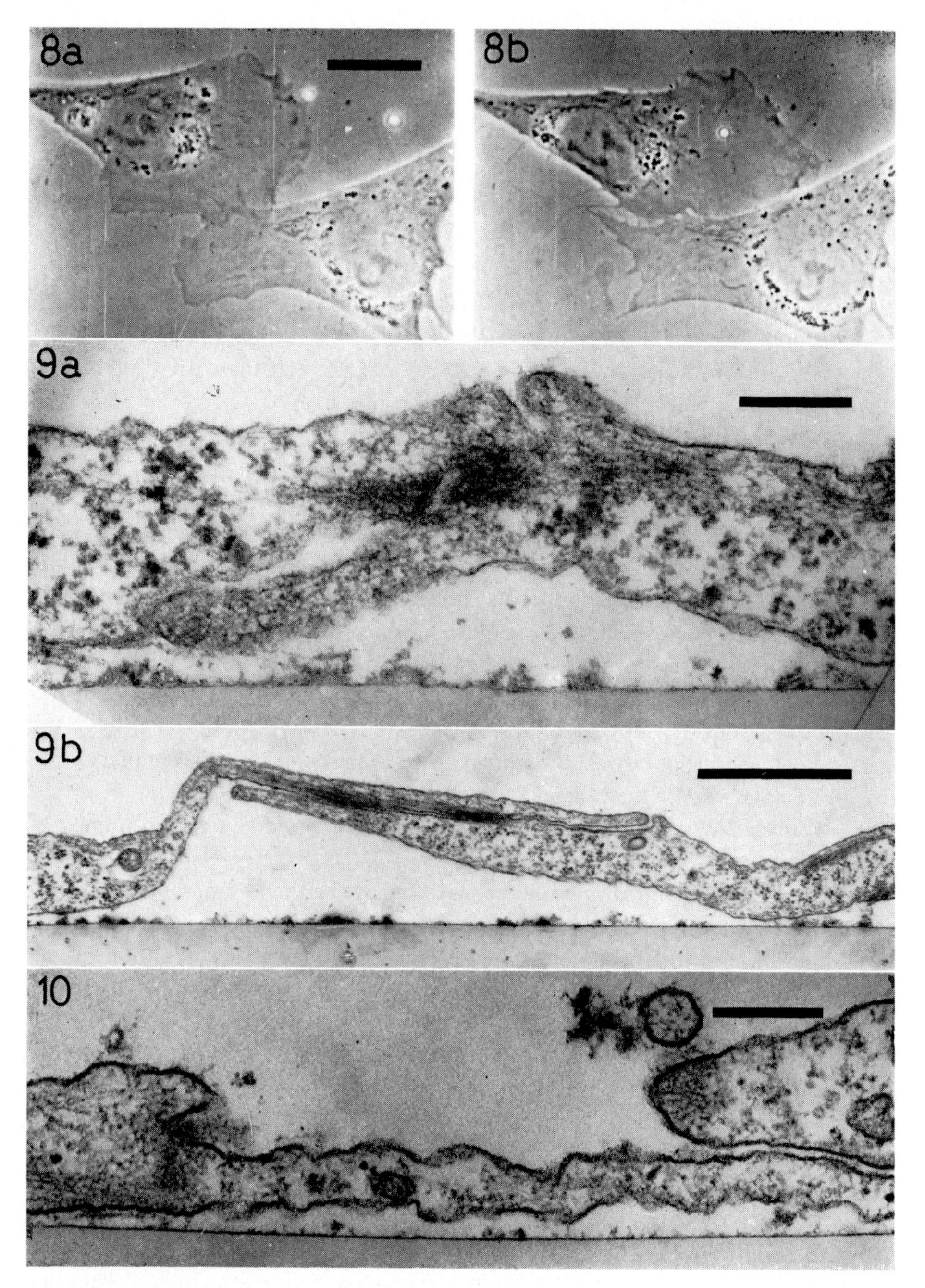

FIG. 8 (Heaysman). Phase contrast light micrographs of two contacting chick heart fibroblasts showing the original underlap and the resultant pulling apart of the cell surfaces. The bar represents 2.0 μm.

FIG. 9 (Heaysman). Vertical longitudinal section of two cells probably about to separate. The orientation of the microfilaments and gross cell distortion suggest tension. The bar represents (a) 0.25 μm and (b) 1.0 μm.

FIG. 10 (Heaysman). Vertical longitudinal sections of two cells just after separation. The bar represents 0.25 μm.

very interesting to know the relation of the apposed plasma membranes in the region of the plaques. Do they form gap junctions or tight junctions?

Heaysman: At high magnification (Fig. 7) we did not find any continuity of the opposing unit membranes; we think there is a continuous gap between the cells, but that this gap varies considerably in width.

When the cells are so close it is very difficult to get a good section of a tight junction. We ought really to do serial sections in both planes and use a tilting stage. But from the evidence to hand, we can say that there are no true tight junctions between these contact inhibiting cells.

Steinberg: Does the retraction you described (Fig. 9) take place once a pseudopod is extended at the other end of the cell which starts to move the other way or immediately after the initial contact?

Heaysman: First a contraction pulls the cells together—we believe this is the result of the tension produced by the filaments—and then the cells separate. Lamellipodia might be produced at other parts of the cell surface and the cells move off, or new lamellipodia might be formed in the same position as before and a new collision occur.

Steinberg: So the retraction does not represent the beginning of movement of the cells, does it?

Heaysman: No, I think it is a breakdown of the contact, rather than a pulling from either end of the cell.

Wolpert: Would you consider this as being part of a general phenomenon? Is there not perhaps a relation between the contacts that you see being made here to the development of microfilaments beneath the membrane at the point of contact in phagocytosis, as Dr Allison described earlier. Doesn't this perhaps confirm the idea that one of the features of cell movement is that contraction is immediately induced at the point of contact with the substratum. In support of this, we have observed a pull when the pseudopods of the sea-urchin mesenchyme contact the substratum (Gustafson & Wolpert 1963, 1967). A good example is that described by Wohlman & Allen (1968) for *Difflugia*, the large amoeba, which pushes out a pseudopod and on contact with the substratum establishes filaments centrally and the pseudopod contracts.

Heaysman: Which way is the contraction?

Wolpert: When the pseudopod touches the substratum it pulls, and the back of the cell follows.

Heaysman: This may be what the transverse fibres are doing. They are often present associated with the plaques. I would like to suggest that the terrific ruffling activity one sometimes sees in cultured cells is to a certain extent a failure of forward movement, the anterior end of the lamella not having made an adhesion to the substrate.

Wolpert: I could not agree more, but I think that one has to sound one note of warning about working on glass. For the cells you have been discussing, the substratum was Araldite, but that is similar to glass. We have looked at fibroblasts of chick limb bud by electron microscopy (Gould *et al.* 1972). This is a three-dimensional situation and here they are not bipolar but multipolar cells. I wonder if your ruffles represent pseudopods which have failed to make contact.

Heaysman: If we grow these cells under methyl cellulose, so that their upward ruffling movement is inhibited, we find a long process going straight along the substrate, where further adhesions must be made.

Gail: Have you found these plaques in cell lines which do not show contact inhibition of locomotion?

Heaysman: We have found them, not in cell lines but in tumour cells, for example MCIM, and in this case they are much the same as in the chick and mouse fibroblasts—if anything I would say we find more of them. Recently we have been looking for cell–cell plaques between colliding S180 (mouse sarcoma) cells and chick heart fibroblasts (Heaysman & Pegrum 1973). These cells do not show contact inhibition (Abercrombie *et al.* 1957). Occasionally the chick cells show signs of the beginning of a response but even ten minutes after collision the S180 cells have no differentiated regions.

Goldman: Martinez-Palomo *et al.* (1969) compared junctional structures in cultures of normal and transformed cells. They found a significant decrease in the number of desmosome-like structures and tight junctions in transformed cells.

Heaysman: In that example, the contact time of the cells was longer than we allow, since we fix the cells at a very early stage. One of the points we wanted to bring out by showing these results now was the fact that these plaques, these electron-dense areas, apparently form and break down very rapidly, possibly from precursors already in the cell.

Porter: Rosenberg (1960) described a cell exudate which tends to form on the substrate in cultures. Do you relate this frosting on the Araldite to what he observes?

Heaysman: I think it is similar.

Porter: If the surface coating is commonly present, would it not affect adhesiveness to the substrate?

Heaysman: I think so—as long as growth is in serum medium. I believe the consequences would be the same on many different substrates.

Porter: It is curious that the cell responds so differently to various substrates if they are all coated with the same cell exudate.

Heaysman: The response is reasonably similar on several different substrates. But we have not studied the electron micrographs of those that differ widely, to know whether because of the substrate or the electrical properties this mat is not laid down.

Allison: Could this be related to the deposition of serum protein on the substratum? This could be checked, for example, by adding labelled serum protein and measuring the concentration by autoradiography. Normally one would expect a layer of serum protein over everything so that the cell makes contact with serum protein rather than with the original substratum.

References

ABERCROMBIE, M., HEAYSMAN, J. E. M. & KARTHAUSER, H. M. (1957) Social behaviour of cells in tissue culture. III. Mutual influence of sarcoma cells and fibroblasts. *Exp. Cell Res.* **13**, 276-291

ABERCROMBIE, M., HEAYSMAN, J. E. M. & PEGRUM, S. M. (1970) The locomotion of fibroblasts in culture. II. Ruffling. *Exp. Cell Res.* **60**, 437-444

ABERCROMBIE, M., HEAYSMAN, J. E. M. & PEGRUM, S. M. (1971) The locomotion of fibroblasts in culture. IV. Electron microscopy of the leading lamella. *Exp. Cell Res.* **67**, 359-367

GOULD, R. P., DAY, A. & WOLPERT, L. (1972) Mesenchymal condensation and cell contact in early morphogenesis of the chick limb. *Exp. Cell Res.* **72**, 325-336

GUSTAFSON, T. & WOLPERT, L. (1963) The cellular basis of morphogenesis and sea urchin development. *Int. Rev. Cytol.* **15**, 139

GUSTAFSON, T. & WOLPERT, L. (1967) Cellular movement and contact in sea urchin morphogenesis. *Biol. Rev. (Camb.)* **42**, 442-498

HEAYSMAN, J. E. M. & PEGRUM, S. M. (1973) Early contacts between fibroblasts—an ultrastructural study. *Exp. Cell Res.* **78**, 71-78

MARTINEZ-PALOMO, A., BRAISLOWSKY, C. & BERNHARD, W. (1969) Ultrastructural modification of the cell surface and intercellular contacts of some transformed cells. *Cancer Res.* **29**, 925-937

ROSENBERG, M. (1960) *Biophys. J.* **1**, 137

WOHLMAN, A. & ALLEN, R. D. (1968) Structural organization associated with pseudopod extension and contraction during cell locomotion in *Difflugia*. *J. Cell Sci.* **1**, 105-114

The growth cone in neurite extension

DENNIS BRAY and MARY BARTLETT BUNGE

MRC, Laboratory of Molecular Biology, Cambridge and The Department of Anatomy, Washington University School of Medicine, St. Louis, Missouri

Abstract The growth cones of elongating nerve fibres* are in many ways similar to the ruffling membranes of migrating fibroblasts. They both move at the same rate and in a characteristic fashion in which surface structures appear to be carried rapidly backwards away from the leading edge. Many of their morphological features seen by light or electron microscopy are the same. It seems likely that these similarities have a common molecular basis and that the two structures have a closely related genetic specification.

Ostensibly their functions are different. Whereas the ruffling membrane appears to be involved in the locomotion of the cell, there is evidence that the growth cone is concerned with neurite elongation.

Observations on isolated neurons in culture show that branch points and fibre surface do not advance with growth, and suggest that these are formed at, or close to, the growth cone.

The apparent paradox is resolved if, in both structures, surface assembly and forward migration are concomitant processes. The model suggested for fibroblast locomotion in which surface membrane is cycled from assembly at the leading edge to uptake at a more proximal region, requires only the modification that rates of assembly exceed rates of uptake to account for nerve growth. Electron microscopy of growth cones has shown structures potentially able to carry out these activities. They contain microfilament networks of the kind often associated with movement, cytoplasmic membrane which shows continuity with the surface membrane, and lysosomal vacuoles which could be involved in the uptake of surface membrane.

INTRODUCTION

It is not necessary for a nerve fibre to have the means of locomotion for it to grow. If, for example, it grew like a hair, by extrusion from the base, then the

* The terms 'nerve fibre' or 'neurite' have been used in this paper to denote all slender processes put out by a neuron, whether axonal or dendritic.

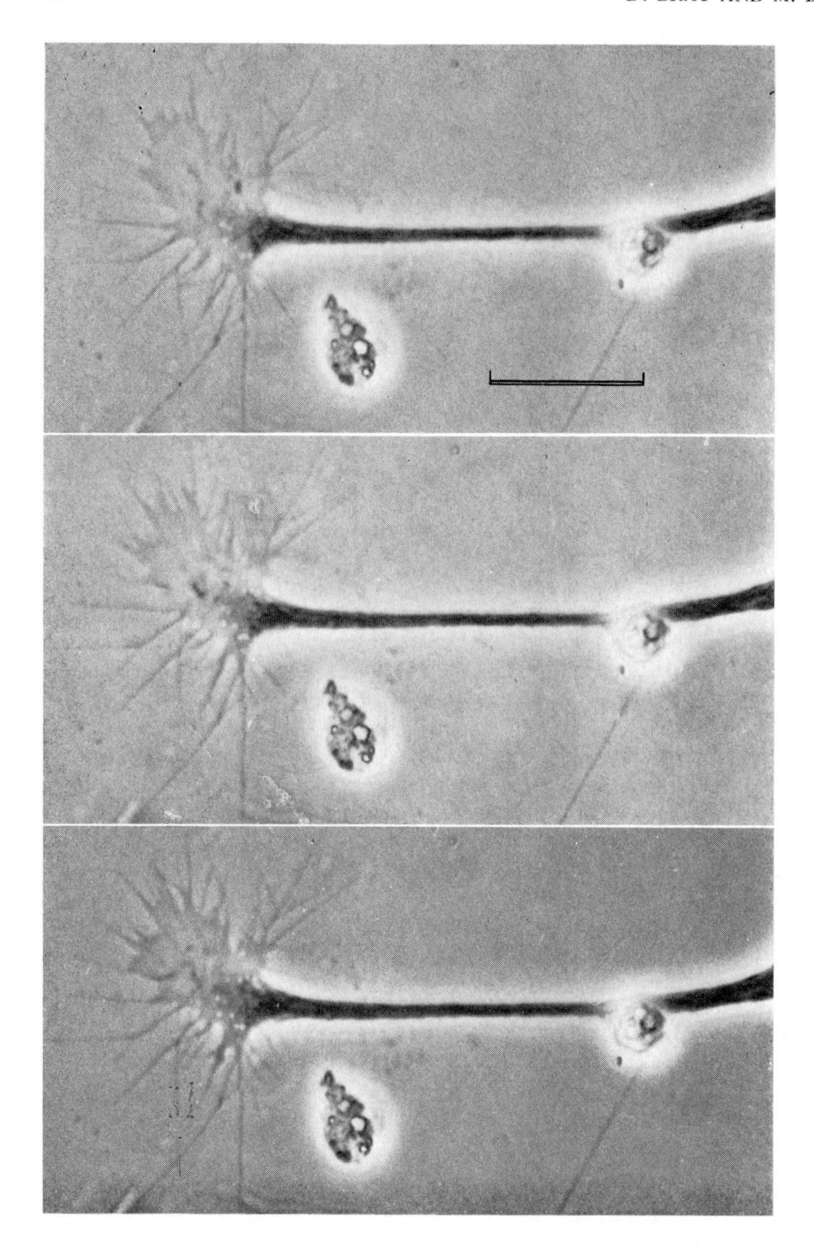

FIG. 1. Growth cone of a chick sensory neuron in culture. Pictures taken at 20 s intervals show its rapid changes in form (calibration bar 20 μm).

tip could be a totally passive object. Every micrometre advanced could be the direct result of the assembly of precisely 1 μm of new fibre at the cell body.

Perhaps the strongest indication that this is not so comes from the nature of the structure found invariably at the tips of growing neurites. Called a growth cone from its appearance in silver-stained preparations, this is seen in living cells as a terminal expansion possessing numerous fine processes (Fig. 1) (Harrison 1910). Its most distinctive feature is the continual motion of filopodia and adjacent regions of membrane which project forwards and then sweep backwards at rates fast enough to be seen by direct observation. In time lapse motion studies this becomes a fluttering movement which has every appearance of an exploratory palpation of the immediate surroundings. Apart from clearly demonstrating that the tip is not a passive structure these movements suggest that part of its action is to propel the fibre across the substrate. This is because a moving growth cone bears an unmistakable resemblance to the ruffling membranes seen at the leading edges of migrating fibroblasts.

LOCOMOTION

Cellular extensions which show ruffling movements are not restricted to nerve cell tips and fibroblasts. Many cells obtained by explantation from a chick embryo show membranous regions at their borders which undergo rapid undulations. Their morphologies vary with the type of cell, being small and localized to the pole of the cell in myoblasts, extensively outspread and with a smooth contour in fibroblasts, and showing a spatulate arrangement in some kinds of epithelial cell. Extensions which are intermediate in form between those of fibroblasts and neurons can be found in Schwann cells derived from sensory ganglia, and often have short processes tipped with small areas of outspread membrane as in a nerve fibre (Fig. 2).

One feature which appears to be unique to neural growth cones is the presence of numerous fine filopodia. These extend radially from the central membranous area, have a diameter of 0.2 μm and a length of 10 μm or more, and are not normally seen in the other cell types in culture. It should be said, however, that the extent of the central membrane is quite variable, and can envelope the filopodia completely to give it a smooth veil-like appearance (Pomerat *et al.* 1967; Nakai 1956). It is also true that fibroblasts can extend microvilli from their surfaces which, although shorter than the filopodia, have a similar diameter and ultrastructural composition (Follett & Goldman 1970).

But it is the similarity of the ruffling movements which provides the strongest evidence of a comparable mechanism in the moving membranes of the fibro-

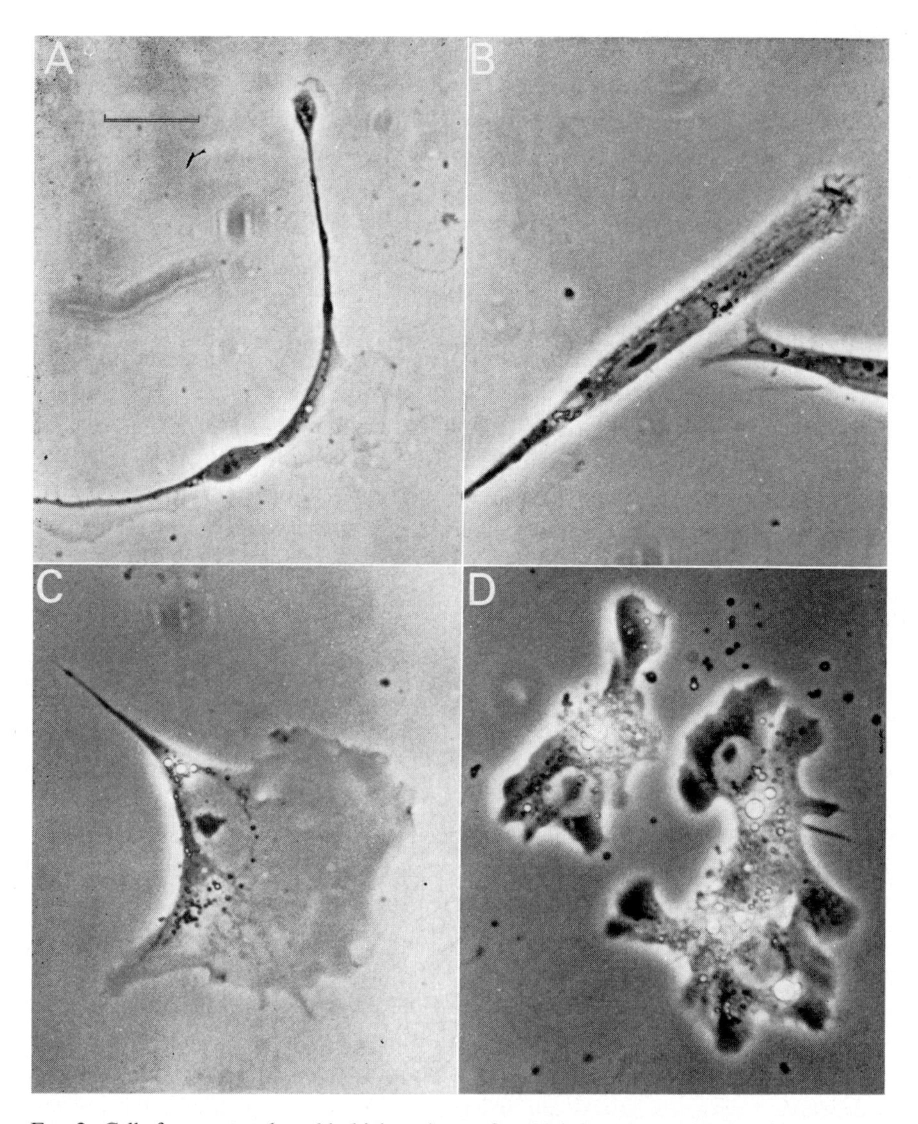

FIG. 2. Cells from a ten-day-old chick embryo after 24 h in culture. Notice the variety of membranous extensions at their periphery: (A) Schwann cell (dorsal root ganglion); (B) myoblast (breast muscle); (C) fibroblast (skin); (D) epithelial cell (liver) (calibration bar 20 μm).

blasts and the growth cone. Whether the smooth membrane or groups of filopodia undulate, the structures all progress slowly forwards to the accompaniment of rapid retrograde waves, and, rather strikingly, the rate of these movements is the same in different cells. Measurement of the growth of chick sensory neurites in culture shows that their tips advance at 26–51 μm/h (Hughes 1953) and exhibit retrograde movements of filopodia or particles on their surface of 70–120 μm/h (Nakai & Kawasaki 1959; Bray, unpublished observations). Chick fibroblasts have been estimated to migrate at an average speed of 36 μm/h and to carry particles backwards on their ruffling membranes at 75 μm/h (Abercrombie *et al.* 1970). Furthermore, the ruffling membrane of fibroblasts and the neural growth cones are similar in their ultrastructure (see later) and in their response to some drugs. In particular, both are immediately paralysed by low concentrations of cytochalasin (Carter 1967; Yamada *et al.* 1970), but are not directly affected by colchicine (Spooner *et al.* 1971; Daniels 1972) or cycloheximide (Spooner *et al.* 1971; Wessells *et al.* 1971).

ASSEMBLY

The unique feature of nerve growth is not the forward migration of the tip, whatever resemblance this may have to fibroblastic movement, but the elongation of the fibre that leaves the rest of the cell behind, and which must involve the continual synthesis and assembly of structural components. It is likely that the centre of most of the macromolecular synthesis of the cell is in the cell body, for this is where the nucleus, the Golgi apparatus and virtually all the ribosomes are located. The products of synthesis might be distributed into new fibres in several ways, of which we can distinguish two extreme cases: either new material is inserted into the fibre base, close to the cell body, or it is transported as some precursor to the tip and incorporated there. These possibilities exist independently of any question of motility, and assembly at either point could be coupled to the locomotion of the tip.

One of the first indications that the cell body is not obligatory for short periods of neurite growth was provided by Hughes (1953). Working on cultures of chick spinal ganglia, he observed that fibres which had been cut close to the explant continued to elongate, often for several hours. This argued against any mechanism in which the fibre is continually pushed outwards from the cell soma, and also indicated that there probably exists a pool of preformed subunits along the fibre; a conclusion since supported by the insensitivity of neurite extension to inhibitors of protein synthesis (Spooner *et al.* 1971; Wessells *et al.* 1971).

More recently, observations of the growth in culture of individually isolated sympathetic neurons have suggested that the site of assembly is close to the growth cone. In the first of these (Bray 1970) the surface of the cells was marked with small particles of glass or carmine, the movements of which were followed as the fibres extended. If surface materials were extruded from the soma in these cells, which electron microscopy has shown to lack Schwann cells and to be bordered by a single plasma membrane (Bunge 1973), then the particles should demonstrate this extrusion. Over long periods, however, whether the particles were close to the soma or several hundred micrometres away, they showed no consistent tendency to move outwards behind the advancing tip.

A second piece of evidence supporting growth at the tip came from an analysis of the often elaborate branching patterns of isolated neurons (Bray 1973). Records of the growth of cells (such as the one shown in outline in Fig. 3) show that almost all the branch points were formed by the bifurcation of growth cones and that, just as with the particles on the surface, those branch points were left behind by the advancing tips. This is a strong indication that at least part of the fibre is deposited in the region of the growth cone although, conceivably, in this case it need not be the surface.

A rather surprising finding from this work was that all the growth cones of a cell grew at about the same rate. Suggested by the constant length of fibre from cell body to growth cones (Fig. 3), this observation was confirmed by direct measurements from time lapse movies, and found to be true at all regions of the cell and at all stages of growth. It was unexpected because the growth cones can vary enormously in their distance from the cell body, in the number of bifurcations they have undergone and in the diameter of their fibre. Since fibroblasts migrate at about the same rate (30–40 μm/h) it is possible that this rate is an intrinsic feature of structures of this sort. However, this leaves unexplained how the cell body can supply precursors over an ever increasing distance, to an ever increasing number of growth cones, and still allow them to advance at a constant speed.

There were two other points of interest. One was that the growth cones in these experiments clearly provided the major attachment of the cell to the culture substrate. If a cell was displaced with a microelectrode, the cell body and the fibres moved as though they were anchored at their tips. Vigorous displacements which broke the fibres often left the growth cones isolated on the substrate. Presumably the greater adhesiveness of the cell at the tips reflects a regional difference in the plasma membrane.

Secondly, the growth cones played a major part in determining the final shape of the cell. They moved in straight lines at a constant rate and branched apparently at random. This behaviour, repeated throughout growth and at an ever increasing number of tips, could have generated all the branching patterns observed. The other parts of the cell played a supportive role, but apart from this, had no obvious influence on the final shape.

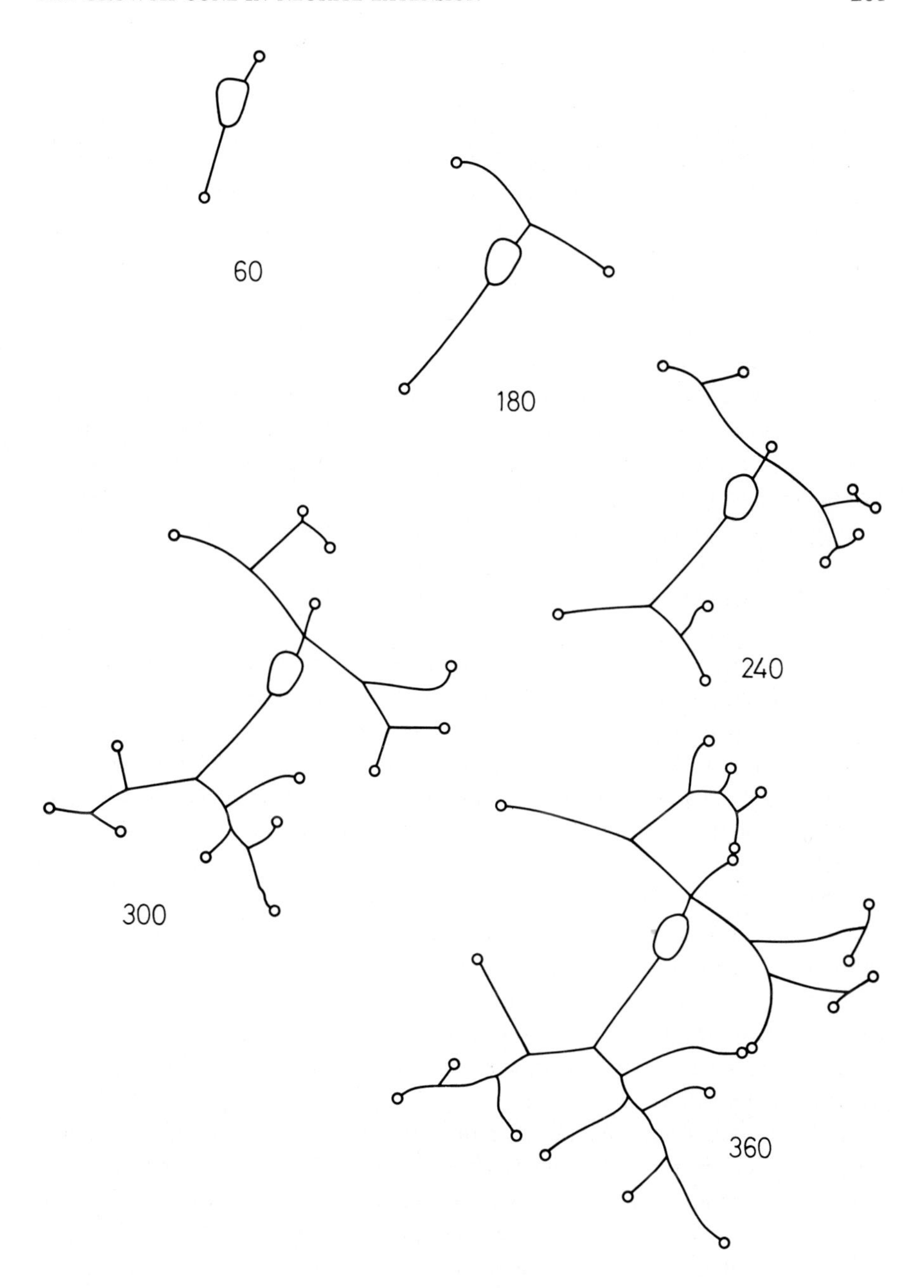

FIG. 3. Outlines of a sympathetic neuron growing in culture at different stages of growth. Most of the branches in the fibre are formed at the growth cones (small circles), but thereafter remain at about the same distance from the cell soma (central oval). Numbers refer to time (in min). For details see Bray (1973).

ULTRASTRUCTURE

Because growth cones of cultured neurons occupy an isolated position, separate both from other cells and from most parts of their parent cell, the correlation between fine structure and activity should be especially clear. If locomotion and cellular assembly do occur in the growth cone, then the structures responsible for these activities should be prominent and perhaps recognizable features of the contents of the growth cone.

The earliest micrographs of structures identified as growth cones were disappointing in this respect for they showed the tips of developing neuroblasts to be little more than bulbous varicosities containing large empty vesicles. More recent work, however, on the tips of sensory neurons in the embryo (Tennyson 1970) and in culture (Yamada *et al.* 1971), as well as our own on cultured sympathetic neurons (Bunge & Bray 1970; Bunge 1973), has shown them to be complicated structures. These reports agree in essence, and show that the growth cones from all three sources have the following characteristics. Within the central regions of the cone most of the organelles of the parent fibre are found, such as microtubules, neurofilaments and mitochondria, but with particularly large amounts of membranous structures such as agranular endoplasmic reticulum, coated and dense-cored vesicles and lysosomal bodies. Towards the periphery of the cone, within and adjacent to the motile filopodia, there are only two kinds of structure. One is a fine and apparently filamentous feltwork, and the other is smooth membrane in tubular or vesicular form (Fig. 4).

The fine network, which at high magnification seems to be made up of elements 5 nm or less in diameter, is often the only structure found in the filopodia (Fig. 5). Filaments of this size have been reported in many vertebrate cells, either in fine networks or in parallel arrays close to the surface, and are currently thought to be involved in cellular movements (Wessells *et al.* 1971). The evidence for this is diffuse but there are, perhaps, two focal points. First, microfilaments are probably made of a protein like muscle actin and function in contractility and secondly they are often in the right place and disposition to be responsible for cell movement.

A protein that is actin-like, or perhaps actin itself, is present in many types of vertebrate cell (Bray 1972; Fine & Bray 1971) and could participate in contraction, particularly since it is often accompanied by a protein with similar properties to muscle myosin (Berl & Puszkin 1970; Adelstein *et al.* 1971). This cytoplasmic actin is probably located in the microfilaments, which are of comparable diameter to F-actin and form the same distinctive arrowhead structures with myosin (Ishikawa *et al.* 1969). These filaments are often present in cells changing shape and occupy positions in which their contraction could

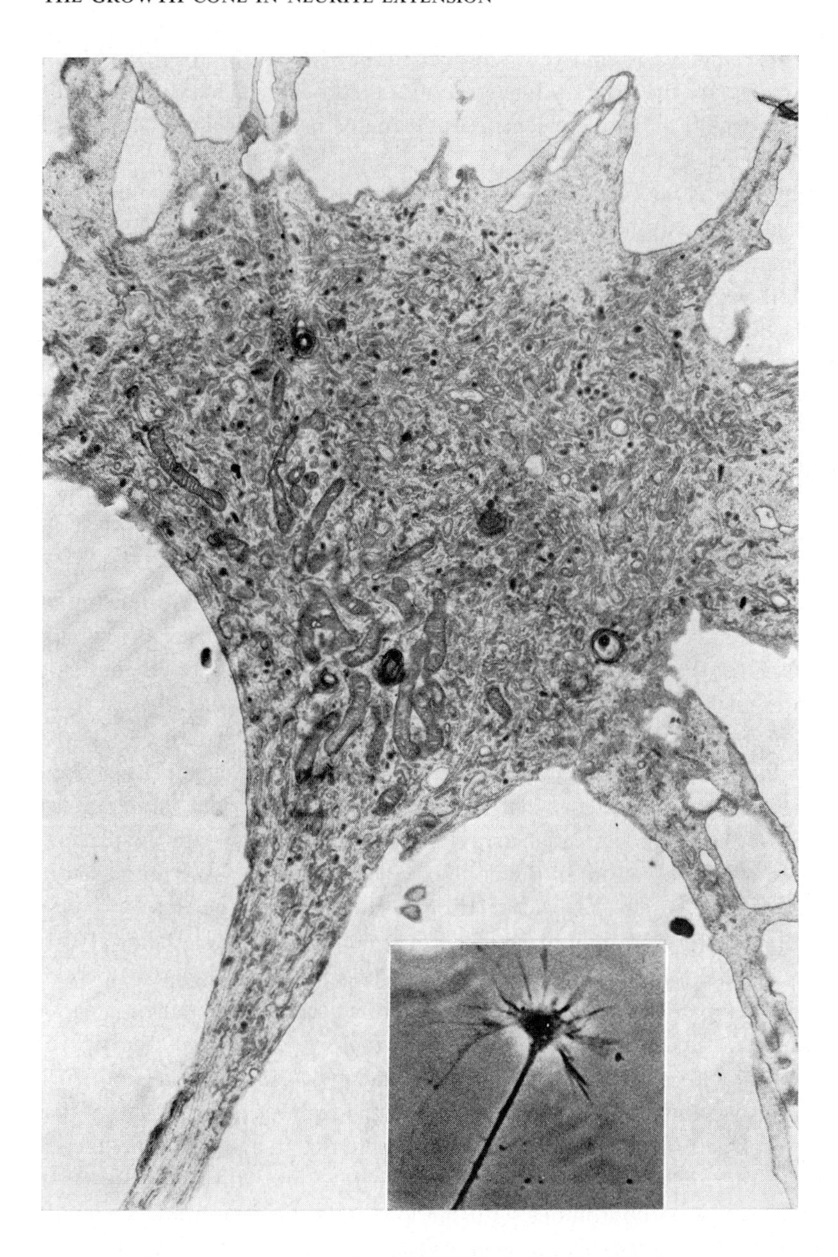

FIG. 4. Low magnification electron micrograph of the growth cone of an isolated rat sympathetic neuron. Mitochondria are clustered in the thickest part of the tip. Membranous organelles occupy the remainder, diminishing near the thin periphery and in the filopodia. × 9 900. *Inset:* light micrograph of the same cone before sectioning. × 756. For details see Bunge (1973).

cause the observed shape changes (Schroeder 1968; Cloney 1966). It has been stated that they are the primary fine structural target of cytochalasin (Wessells *et al.* 1971), which at low concentrations rapidly paralyses many forms of cellular movement (Carter 1967).

So, although it is far from rigorously proven, it is likely that microfilaments in general, and those in the growth cone in particular, are involved in some form of movement. From their location in the filopodia and in the peripheral flanges of the cone, they could produce the fluttering undulations observed in the light microscope. But how an apparently disordered array of filaments produces concerted movements, and how the filaments are related to membrane within and on the surface of the cell remain unanswered questions.

By far the most noteworthy ultrastructural feature of the growth cones we examined was their content of smooth membrane (Fig. 5). This varied in amount, but some cones were almost filled with agranular reticulum. At first sight this appeared to be a jumble of vesicular and tubular forms, but close examination revealed certain recurring distinctive configurations of membrane.

Particularly interesting are the striking arrays of cytoplasmic membrane found close to the lower surface of spread out membranous areas of neurites (Fig. 6). In some sections these anastomotic channels can be seen to open at intervals to the outer plasma membrane, at which points it is possible that surface structures are added. This interpretation is favoured by the appearance of the openings which are clearly different from the coated vesicles normally associated with micropinocytosis in many cell types including nerve cells (reviewed by Holtzman 1969). These distinctive arrays and their continuity with the plasmalemma were not observed in all the cones studied and were sometimes seen in spread out areas in which the exact relation to cones was not clear.

A puzzling feature of endoplasmic reticulum components is their variable occurrence. For although reticulum was abundant in most of the growth cones of cultured sympathetic ganglion cells which we examined, much smaller amounts and no anastomotic membranous arrays were found in the tips of sensory neurons either in culture or *in vivo* (Yamada *et al.* 1971; Tennyson 1970). An even more striking contrast exists between the ruffling membranes of glial cells and fibroblasts. In the former these areas were rich in agranular membrane of precisely the anastomotic kind described before and, furthermore, these tubular elements were continuous with the lower plasma membrane (Yamada *et al.* 1971). In the fibroblasts, however, only occasional vesicles, which had no obvious relation to the outside, were observed (Abercrombie *et al.* 1971).

These fine structural differences are surprisingly large when one considers how similar the two pairs of cell types appear in cultures, and the possibility must be considered that they are artifactual. Films of neurite tips during growth

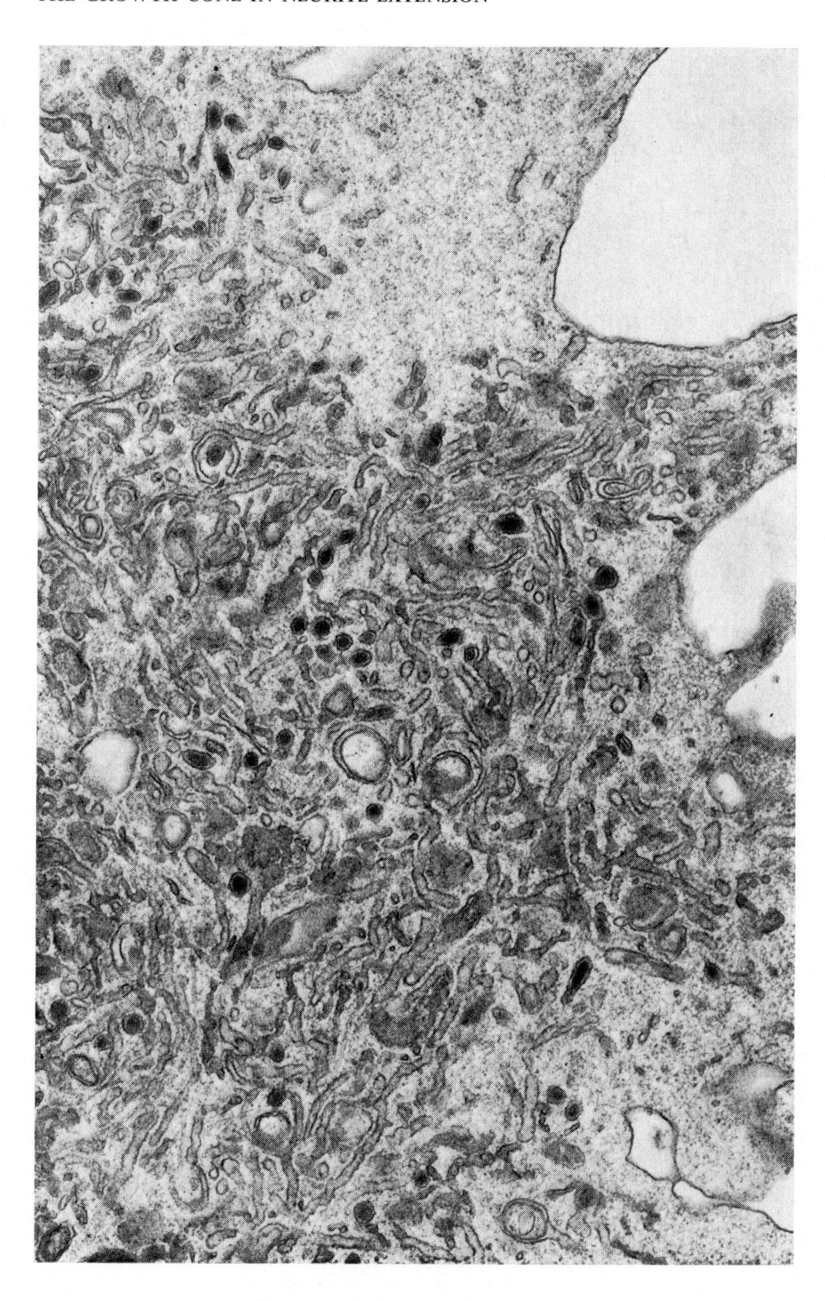

FIG. 5. Higher magnification of part of the growth cone shown in Fig. 4 showing feltwork material at the periphery and a wealth of agranular membranous components more centrally. × 27 000.

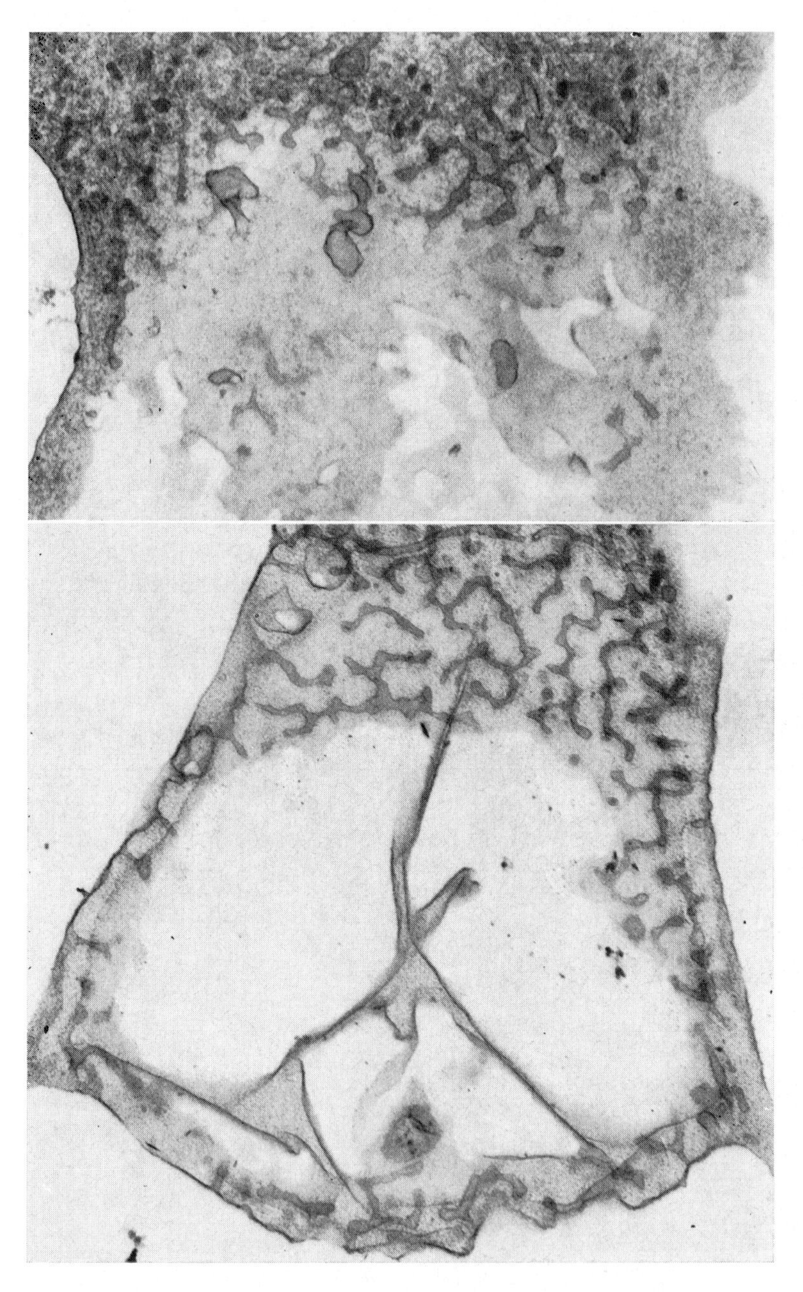

FIG. 6. Spread out areas of nerve fibres sectioned close to the culture surface showing a distinctive arrangement of membranous elements and, in the lower figure, their continuity with the surface membrane. The folds evident in the lower figure suggest that this expansion was ruffling at the time of fixation. × 20 700.

and in the early stages of aldehyde fixation showed blebbing and other signs of morphological change, if the fixative was applied directly to the culture areas. As a consequence we adopted an arrangement in which fixative, at room temperature, was perfused onto the base of the culture dish about 1 cm from the cells under observation, and under these conditions movement was abruptly arrested, usually with complete preservation of form. It seems possible to us that certain fixative conditions (particularly those in which aldehydes are employed initially) permit completion of the addition of membrane to the surface before fixation is complete as has recently been suggested by Heuser & Reese (1972). In this way some changes in the amount and configuration of cytoplasmic membrane could take place during the initial phases of fixation.

DISCUSSION

We have tried to show that neuritic growth cones resemble ruffling membranes of fibroblasts in a number of ways. The outstanding similarities are their motility and the characteristics of the movements which have similar rates and are influenced by the same drugs. Both are membranous extensions of the kind put out by many vertebrate cell types in culture, and at least under some conconditions, have the same appearance under the electron microscope. Even the filopodia, which often mark a growth cone in the light microscope, have the same fine structure as the shorter projections of other cells.

These similarities suggest that the activities of ruffling membranes and growth cones have a common molecular basis in which movement and assembly are closely related. It has been proposed (Abercrombie *et al.* 1970; Harris, pp. 3-20) that the migration of fibroblasts could be produced by the insertion of membranes at the leading edge of the cell and their subsequent resorption close to the nucleus. Such a model could also apply to the assembly of surface material at a growth cone and its rapid backwards movement to the base of the cone. Electron microscopy of growth cones and ruffling membranes has shown structures which could be involved in such a cycle. Both have microfilament networks of the kind often associated with motility, membranous structures which appear in continuity with the outer plasma membrane, and a variety of lysosomal and pinocytic vacuoles which could take up membrane.

Thus, the unique ability of a growth cone to leave behind a nerve fibre will depend on only slight differences of a mechanism common to many cell types. It could be a small step from a component involved in the motility of a migrating neuroblast to one at the tip of a lengthening neuron. If the mechanism discussed here is correct, then one area which is likely to be of importance is the base of the growth cone. For it is here that the resorption of membrane must be balanced against its accumulation into the fibres, and where also the longitudinal elements of the fibre, the neurofilaments and microtubules, diminish in number and lose their parallel orientation. The primary factor which determines growth rather than migration could be a change in membrane properties, an increased rate of microtubule polymerization or the existence of a vectorial transport system. Whatever the first change, control mechanisms could easily exist by which it could trigger the others.

ACKNOWLEDGEMENTS

We wish to thank Dr C. Slater for his advice on the microphotography of cells in culture and Drs S. Brenner and W. M. Cowan for their critical evaluations of this manuscript. This work was supported by Grants NB 03273, NB 02253, NB 04235, and NS 09923 from the N.I.N.D.S., United States Public Health Service.

References

ABERCROMBIE, M., HEAYSMAN, J. E. M. & PEGRUM, S. M. (1970). *Exp. Cell Res.* **62**, 389-398
ABERCROMBIE, M., HEAYSMAN, J. E. M. & PEGRUM, S. M. (1971). *Exp. Cell Res.* **67**, 359-367
ADELSTEIN, R. S., POLLARD, T. D. & KUEHL, W. M. (1971). *Proc. Natl. Acad. Sci. U.S.A.* **68**, 2703-2707
BERL, S. & PUSZKIN, S. (1970). *Biochemistry* **9**, 2058-2067
BRAY, D. (1970). *Proc. Natl. Acad. Sci. U.S.A.* **65**, 905-910
BRAY, D. (1972). *Cold Spring Harbor Symp. Quant. Biol.* **37**, 567-571
BRAY, D. (1973). *J. Cell Biol.* **56**, 702-712
BUNGE, M. B. (1973). *J. Cell Biol.* **56**, 713-735
BUNGE, M. B. & BRAY, D. (1970). *J. Cell Biol.* **47**, 241a
CARTER, S. B. (1967). *Nature (Lond.)* **213**, 261-264
CLONEY, R. A. (1966). *J. Ultrastruct. Res.* **14**, 300-328
DANIELS, M. P. (1972). *J. Cell Biol.* **53**, 164-176
FINE, R. E. & BRAY, D. (1971). *Nat. New Biol.* **234**, 115-118
FOLLETT, E. A. C. & GOLDMAN, R. D. (1970). *Exp. Cell Res.* **59**, 124-136
HARRISON, R. G. (1910). *J. Exp. Zool.* **9**, 787-846
HEUSER, J. & REESE, T. S. (1972). *Anat. Rec.* **172**, 329-330
HOLTZMAN, E. (1969) in *Lysosomes in Biology and Pathology* (Dingle, J. T. & Fell, H. B., eds.), pp. 192-216, vol. 1, Wiley, New York
HUGHES, A. F. (1953). *J. Anat.* **87**, 150-162
ISHIKAWA, H., BISCHOFF, R. & HOLTZER, H. (1969). *J. Cell Biol.* **43**, 312-328
NAKAI, J. (1956). *Am. J. Anat.* **99**, 81-99
NAKAI, J. & KAWASAKI, Y. (1959). *Z. Zellforsch. Mikrosk. Anat.* **51**, 108-122
POMERAT, C. M., HENDELMAN, W. J., RAIBORN, C. W., JR. & MASSEY, J. F. (1967) in *The Neuron* (Hyden, H., ed.), pp. 119-178, Elsevier, New York
SCHROEDER, T. E. (1968). *Exp. Cell Res.* **53**, 272-276
SPOONER, B. S., YAMADA, K. M. & WESSELLS, N. K. (1971). *J. Cell Biol.* **49**, 595-613

TENNYSON, V. M. (1970). *J. Cell Biol.* **44**, 62-79
WESSELLS, N. K., SPOONER, B. S., ASH, J. F., BRADLEY, M. O., LUDUEÑA, M. A., TAYLOR, E. L., WRENN, J. T. & YAMADA, K. M. (1971). *Science (Wash. D.C.)* **171**, 135-143
YAMADA, K. M., SPOONER, B. S. & WESSELLS, N. K. (1970). *Proc. Natl. Acad. Sci. U.S.A.* **66**, 1206-1212
YAMADA, K. M., SPOONER, B. S. & WESSELLS, N. K. (1971). *J. Cell Biol.* **49**, 614-635

For discussion, see pp. 223-232

Extension of nerve fibres, their mutual interaction and direction of growth in tissue culture

GRAHAM A. DUNN

Strangeways Research Laboratory, Cambridge

Abstract The direction of nerve fibre extension in plasma clot cultures has previously been attributed to contact guidance along oriented aggregates of fibrin molecules. Measurement of nerve fibre orientation and density in non-fasciculated outgrowths provided evidence that a contact reaction between nerve fibres is largely responsible for their directional behaviour. Such a contact reaction was observed with time lapse filming techniques and termed 'contact inhibition of extension'. Measurement of the magnitude and direction of stretch which developed within the plasma clots provided further evidence that contact guidance is not important for guiding nerve fibres within plasma clots. Contact guidance does appear to determine the course of fasciculated nerve fibres at clot–liquid interfaces in the cultures although the nature of the guiding structures is not clear.

The radial pattern of nerve fibre outgrowth from an explant in tissue culture was first described by Olivo (1927). He attempted to explain radial outgrowth by a theory of intrinsic guidance whereby each nerve fibre grows continuously in the direction of the 'cellular axis' of the neuron cell body. Weiss (1934) termed the radial outgrowth of nerve fibres from an explant in plasma clot the 'one centre effect' and gave a more satisfactory explanation based on his theory of contact guidance. He showed that nerve fibres in plasma clot cultures tend to follow a course oriented in the direction of stretch of the plasma clot and proposed that the nerve fibres are guided along oriented aggregates of fibrin molecules within the clot. He suggested that a single explant locally dehydrates the plasma clot which consequently shrinks inducing a radial stress pattern in the surrounding clot. This radial tension orients the fibrin micelles of the clot and thus radially orients the direction of extension of the nerve fibres by contact guidance. Furthermore, Weiss showed that when two explants are confronted, nerve fibres grow directly across from one explant to the other, eventually forming a 'bridge' of nerve fibres in the inter-explant zone. He called this pattern the 'two centre effect' and explained it in essentially the same manner as the one centre effect, suggesting that the fibrin micelles would tend to orient strongly between two centres of dehydration. Weiss concluded: 'For the conditions of tissue culture, therefore, the problem of orientation of the nerve fibre reduces itself to the probem of micellar orientation in colloidal substrata which is purely a physicochemical problem'.

I have treated Weiss's theory in some detail because it is generally accepted today to account for the direction of growth of nerve fibres in plasma clot. Since its introduction, little evidence has accumulated against it and in the same paper, Weiss (1934) could find no evidence to support the contemporary theories of 'neurotropism' or chemotropic mechanisms of nerve fibre guidance. However, Levi (1941) observed radial nerve fibre orientation in his cultures while sheath cells, in the immediate vicinity of the nerve fibres, were not oriented. Levi concluded that these plasma clots were not radially oriented since Weiss (1934) has used sheath cell orientation as a method of detecting plasma clot orientation.

In more recent years evidence has been accumulating which strongly suggests that centrifugal migration of fibroblasts from an explant grown on glass is due to contact interactions between cells (Abercrombie & Heaysman 1954; Abercrombie & Gitlin 1965) and it is reasonable to reject Weiss's modified theory of contact guidance by oriented colloidal exudates (Weiss 1945) as an explanation of centrifugal cell locomotion.

The present work was started with the intention of analysing the effects of contact interaction between nerve fibres on their directions of extension in plasma clot cultures.

ONE CENTRE EFFECT

Lumbar dorsal root ganglia were explanted from 8–9-day-old chick embryos onto a coverslip in a medium consisting of two parts of freshly obtained fowl plasma and one part of chick embryo extract containing 3 units/ml of nerve growth factor. The medium was subsequently allowed to clot and the cultures incubated at 37 °C in the hanging drop position for 18–28 h. This procedure produced a typical and repeatable pattern of nerve fibre outgrowth from single explants (Fig. 1). Examination in the light microscope of cultures stained by the Holmes silver impregnation technique revealed that the nerve fibre outgrowth is largely non-fasciculated. The nerve fibres are very strongly oriented in a radial direction and the density of nerve fibres falls off sharply at the border of the outgrowth. These latter two aspects of the pattern seem to depend on the density of the outgrowths since low density outgrowths, obtained with inactive

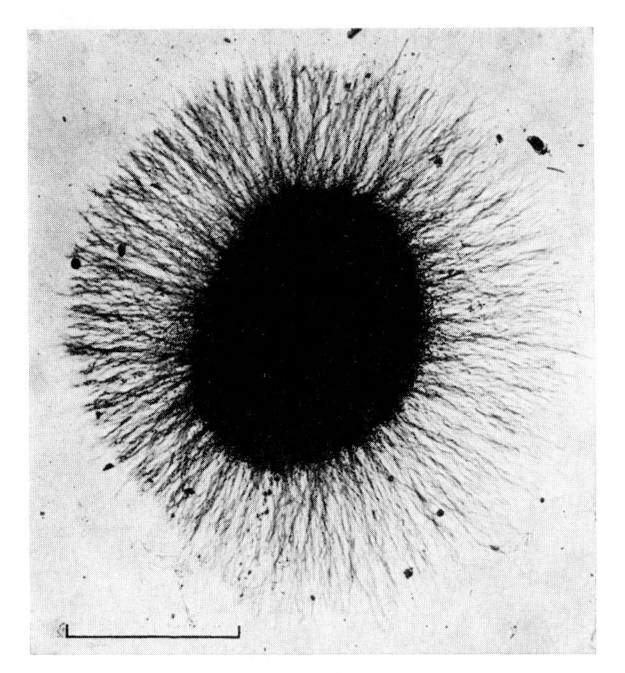

FIG. 1. Eight-day-old lumbar dorsal root ganglion, 27 h *in vitro*, stained by the Holmes silver method. Scale bar 500 μm. (By kind permission of the Editorial Board of the *Journal o, Comparative Neurology*.)

nerve growth factor, show little radial orientation of nerve fibres and no well-defined border to the outgrowth. I found that the magnitude and direction of stresses which developed within the plasma clots could be determined by measuring the optical birefringence of the clot with a polarizing microscope equipped with a λ/20 Brace-Köhler rotating compensator and measuring the thickness of the clot with the calibrated fine-focus control with compensation for the refractive index of the clot. The system was calibrated by measuring the birefringence of a series of clots stretched by a known amount (Fig. 2). The axial ratio of stretch is the ratio of the major to the minor axis of an ellipse formed by stretching the material with a circle drawn on its surface. Birefringence was always positive in the direction of stretch of the plasma clots.

Applying this method to the plasma clot just peripheral to the outgrowth from single living explants, I found that the plasma clot is always slightly stretched in a radial direction with an axial ratio of stretch of about 1.1 (axial ratio is unity when there is no stretch). Harris (pp. 3-20) believes this radial stress to be brought about by mechanical activity of cells within the explant and not by dehydration as Weiss has suggested.

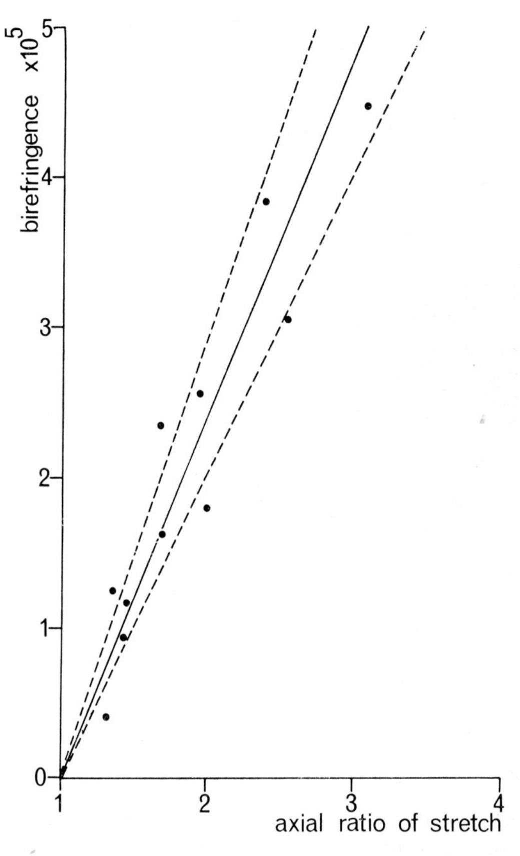

FIG. 2. Optical birefringence plotted against axial ratio of stretch for 11 plasma clots. The solid line is the sample regression passing through the point birefringence = 0, axial ratio = 1. The broken lines indicate the 95% confidence limits for the gradient.

The orientation of nerve fibres was measured as the ratio ($\Sigma\cos\theta/\Sigma\sin\theta$), where θ is the angle between short isometric sections of individual nerve fibres and the direction in which orientation is to be measured, with $0^{\circ} < \theta < 90^{\circ}$. This ratio was selected for three reasons. First, if randomly oriented nerve fibres are embedded in a homogeneous matrix which is subsequently stretched with a given axial ratio of stretch, then the orientation ratio of the nerve fibres in the direction of stretch would be expected to be numerically equal to the axial ratio of stretch. Secondly, the orientation ratio can easily be measured in non-fasciculated nerve fibre outgrowths with a microscope. A rectangular eyepiece grid, with one set of lines oriented in the measuring direction, is superimposed over a 60 μm diameter region of outgrowth. The ratio of the number

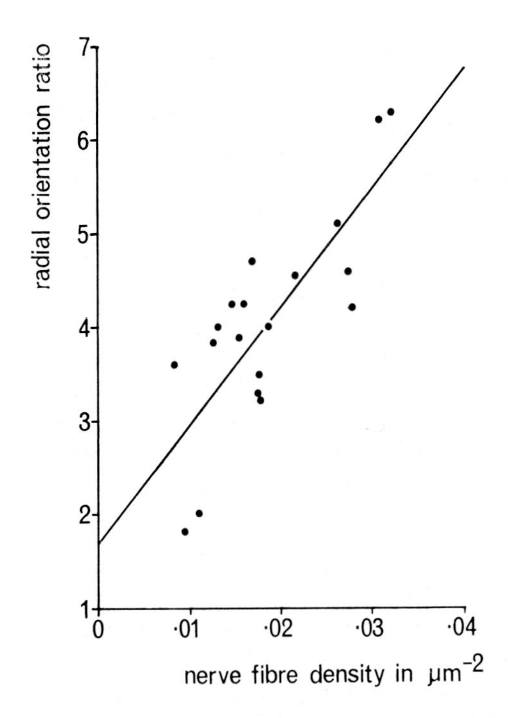

FIG. 3. Radial orientation ratios for 19 nerve fibre fields plotted against their density.

of intersections of nerve fibres and grid lines at right angles to the measuring direction to the number of intersections of nerve fibres and grid lines in the measuring direction is equivalent to the ratio ($\Sigma\cos\theta/\Sigma\sin\theta$). Thirdly, the same orientation ratio can easily be applied to the nuclear orientation of other cell types growing on planar substrata.

Nerve fibre density in the 60 μm diameter regions of the outgrowths was measured by the Curtis modification of the Chalkley random eyepiece array (Curtis 1960). The coincidence of the randomly placed dots with the central axes of the nerve fibres was recorded as a probability (P) that a dot coincides with at least one nerve fibre. Taking account of the fact that at high densities a random dot might coincide with more than one nerve fibre (Dunn 1971) it can be shown that nerve fibre density measured as μm nerve fibre/μm^3 outgrowth region is given by the expression for the optical system used

$$0.0949 \log_e [1/(1-P)].$$

Since the appearance of low density outgrowths suggested that the radial orientation of nerve fibre extension from a single explant is a density-dependent phenomenon, 60 μm diameter fields were blindly selected within a high density outgrowth of a fixed and stained culture. Nerve fibre density and orientation were measured within each field. Fig. 3 shows the results. The solid line is the sample regression of radial orientation ratio on density and has a gradient significantly greatly than zero (t 5.45, d.f.17, $P < 0.001$). The regression line intersects the ordinate at a value of 1.7, but on testing this was not found to be significantly greater than 1.0 (t 1.56, d.f.17, $0.1 < P < 0.2$). There is therefore no evidence to suggest that nerve fibres would be radially oriented at zero density.

For the single explant situation the data indicate that nerve fibre orientation is entirely dependent on local fibre density and we may conclude that it is probable that close-range interactions between nerve fibres are exclusively responsible for their radial orientation. Although the plasma clot is radially stretched to a small extent it is not expected that the magnitude of this stretch (axial ratio 1.1) could be sufficient to account for the strong radial orientation of nerve fibres (mean orientation ratio about 4.0) by a mechanism of contact guidance along oriented fibrin molecule aggregates.

The abrupt fall-off in nerve fibre density at the border of the outgrowth is also best explained in terms of a close-range interaction between nerve fibres since it is not evident in low density outgrowths. The radial orientation of nerve fibres is dependent on local fibre density; therefore, if a growing nerve fibre extends beyond its neighbours, its radial orientation decreases and therefore its radial velocity decreases allowing its neighbours to catch up with it again. A negative feedback system might thus operate to ensure that the distance of a nerve fibre tip from the centre of the explant is similar to that of adjacent nerve fibres.

TWO CENTRE EFFECT

The pattern of nerve fibre outgrowth which results when two dorsal root ganglia are explanted about 0.5 mm apart in the middle of a firm plasma clot is shown in Fig. 4. This is markedly dissimilar to the Weiss two centre effect.

→

FIG. 4. Confronted culture of two lumber dorsal root ganglia, 27 h *in vitro*. Montage of photographs taken with a low-aperture dark-field condenser, Holmes silver method. Scale bar 1 mm. (By kind permission of the Editiorial Board of the *Journal of Comparative Neurology*.)
FIG. 5. Confronted culture of two lumbar dorsal root ganglia just after contact of outgrowths, 23 h *in vitro*, Holmes silver method. Scale bar 250 μm. (By kind permission of the Editorial Board of the *Journal of Comparative Neurology*.)
FIG. 6. Fasciculated outgrowth from confronted ganglia 26 h *in vitro*, showing the 'two centre effect' described by Weiss. Scale bar 500 μm. (By kind permission of the Editorial Board of the *Journal of Comparative Neurology*.)

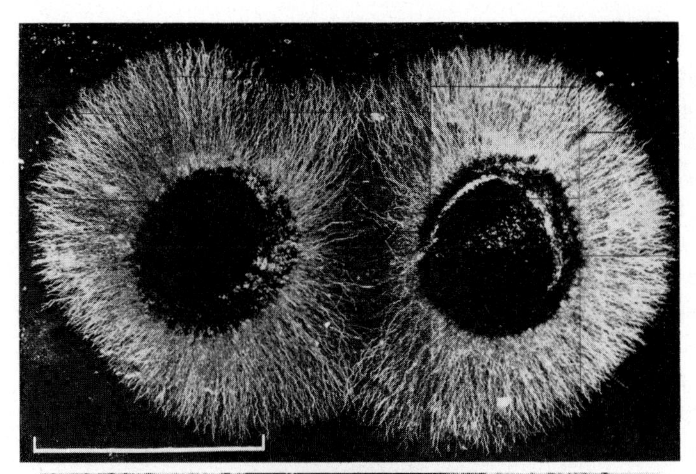

Fig. 4

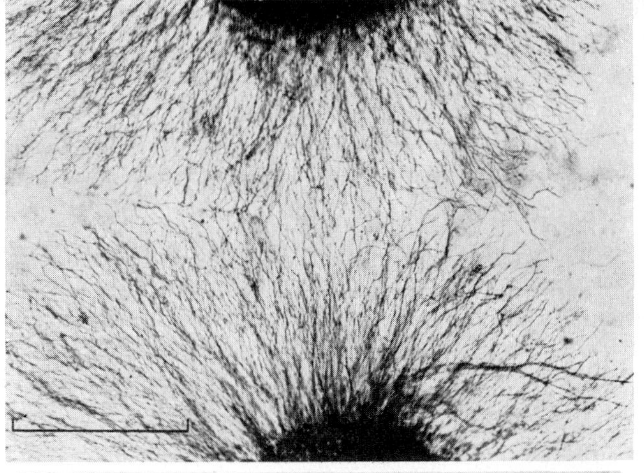

Fig. 5

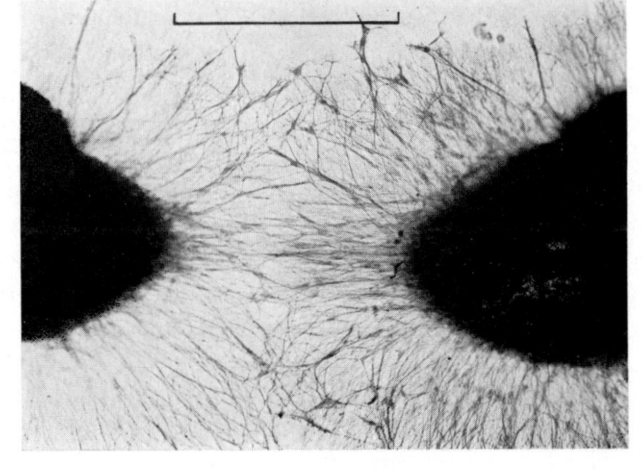

Fig. 6

The nerve fibres between the explants are confused in direction and have a very low density. If the cultures are fixed just before the two outgrowths have made contact, each outgrowth shows a typical radial one centre pattern unaffected by the proximity of the other outgrowth. If the cultures are fixed just after contact of the outgrowths (Fig. 5) the most distal portions of the nerve fibres have deviated sharply from their radial directions where contact has been made. This provides strong evidence that contact between nerve fibres is important in determining these patterns and evidence against a mechanism involving chemotropism away from the explants. The pattern of nerve fibre orientation in these cultures is essentially the same as the pattern of migration of fibroblasts in similar experiments performed by Abercrombie & Heaysman (1954) using chick heart explants grown on glass.

The tensions which developed within the plasma clot were measured and again found to resemble closely the pattern deduced by Weiss. The clots were weakly stretched in a radial direction around the greater part of each explant but in the region between the explants (inter-explant zone) the clot was more strongly stretched in a direction parallel to a line joining the centres of the explants (the low nerve fibre density in the inter-explant zone allowed the measurement of clot birefringence without interference from the intrinsic birefringence of nerve fibres).

Measurements made in six confronted cultures yielded the following results. In the radially oriented parts of the cultures the nerve fibres have a mean radial orientation ratio of 4.25 whereas the clot just peripheral to the outgrowth has an axial ratio of stretch of 1.1. However, in the inter-explant zone the nerve fibres have a much lower orientation ratio of only 1.4 in a direction passing through the centres of the explants, whereas the clot is more strongly stretched with an axial ratio of 1.23. The mean density of nerve fibres is much lower in the inter-explant zone (0.004 μm^{-2}) than it is in the radial part of the outgrowth (0.02 μm^{-2}). These measurements provide further evidence against the possibility of contact guidance being a major contributing factor since the orientation of the nerve fibres is lowest in the most stretched region of the clot.

A typical Weiss two centre effect may be obtained in confronted cultures under certain conditions (Fig. 6), namely wherever the outgrowth is situated at an interface between plasma clot and liquid or when the culture has been incubated for much longer than 24 h and the clot is partially liquefied around the explants. These outgrowths usually lie in a single plane suggesting growth at an interface and they are always at least partially fasciculated which is further evidence that they are not growing within firm plasma clot (Nakai 1960). They will be discussed further under the heading 'contact guidance'.

CONTACT INHIBITION OF EXTENSION

Time lapse cinemicrography was used to film the contact reaction between living nerve fibres extending through a firm plasma clot in a non-fasciculated state. All the contacts observed under these conditions resulted in essentially the same reaction. If the processes (usually filopodial) of the growth cone of one nerve fibre make contact with another nerve fibre anywhere along its length these processes quickly retract, often accompanied by a short retraction of the nerve fibre behind the growth cone (Fig. 7a–c). This retraction usually breaks the filopodial adhesions formed on contact. The total retraction distance measured from the point of contact to the nearest point on the growth cone after retraction was 29.5 $\pm$ 14.6 μm (standard error) for six nerve fibre contacts. This 30 μm long region at the front of an extending nerve fibre is probably the region actively involved in locomotion since Bray (1970) reported that particles are transported backwards on the surface of the growth cone for about 20 μm. After retraction a fan of new lamellar and filopodial processes emerges from the growth cone (Fig. 7d), one of which becomes dominant (Fig. 7e–g) and the nerve fibre continues to extend in the direction of this process. This reaction is very similar to the series of events, termed contact inhibition of locomotion, described when cultured fibroblasts encounter one another on a planar substratum (Abercrombie & Ambrose 1958). I have therefore proposed (Dunn 1971) that this reaction between nerve fibres be called ‘contact inhibition of extension’.

Loeb (1921) proposed that the centrifugal migration of *Limulus* amoebocytes *in vitro* is due to a contact reaction such that contact of one cell with another leads to a resting condition at the point of contact. The cells can subsequently only move in a direction away from each other and cell migration is thus statistically biased in a centrifugal direction. Abercrombie & Gitlin (1965) produced good evidence that contact inhibition of locomotion is responsible for the radial migration of fibroblasts from a chick heart explant on glass. Contact inhibition of extension can similarly account for the radial outgrowth of nerve fibres from a single explant. In a dense outgrowth of nerve fibres the mean free path available to any nerve fibre is greatest in an outward radial direction and each nerve fibre therefore grows in that direction most of the time. Furthermore, if a nerve fibre makes an excursion in any direction, such that the free path available to it is shorter than the retraction distance, then no nerve fibre is established in that direction. This is more likely for non-radial excursions, therefore even less nerve fibre is established in non-radial directions than is accounted for solely on the basis of Loeb’s hypothesis.

Contact inhibition of extension can also account for the pattern of outgrowth in the confronted explant situation. After contact of the outgrowths the mean free path available to nerve fibres in the inter-explant zone is short and approximately the same in all directions. As soon as the mean free path in all directions is shorter than the retraction distance, then further nerve fibre extension is very restricted. The final nerve fibre density in such a randomly oriented portion of the outgrowth is therefore much lower than it is in the radially oriented outgrowth.

CONTACT GUIDANCE

As was mentioned earlier, the Weiss 'two centre effect' sometimes develops in confronted explant cultures. If this pattern is a result of contact guidance, the fact that it only develops at clot–liquid interfaces suggests that contact guidance can only be effective at these interfaces. To test this hypothesis, dorsal root ganglia were grown on the surface of and within plasma clots stretched between two pieces of filter paper on siliconed coverslips. Outgrowths within the clots are largely radially oriented but show a slight tendency to align in the direction of stretch (Fig. 8a). Examination of these cultures in the polarizing microscope shows the clots to be oriented strongly in the direction of stretch with no detectable radial orientation in the vicinity of the explants. Outgrowths on the surface of the clots show little radial orientation but are strongly oriented in the direction of stretch of the clot (Fig. 8b). Control experiments in which the clots were stretched after the cultures had been incubated show that the radial outgrowths are reoriented to the same extent in the direction of stretch both within and on the surface of the clots. Therefore the system of stretching did not stretch the surface more than the interior of the clots.

The strong orientation of nerve fibre extension on the surface of stretched plasma clots is probably due to contact guidance although it is not clear that the nerve fibres are guided by aggregates of fibrin molecules as Weiss has suggested, since these molecules would not be expected to be that strongly oriented. When a clot is stretched it might develop linear surface irregularities of the order of magnitude of the oriented substrata used by Rovensky *et al.* (1971) to demonstrate contact guidance. The situation is further complicated by fasciculation since fasciculating nerve fibres are pulled laterally through the medium (Nakai 1960) and might therefore come to rest in a different direction from that in which they extend. However, I have preliminary results that chick heart fibroblasts also only respond to contact guidance at interfaces of stretched clots with liquid.

→

FIG. 7. Series of photographs taken at three minute intervals showing the contact reaction between two nerve fibres. Phase contrast of living culture 18 h *in vitro*. Scale bar 100 μm. (By kind permission of the Editorial Board of the *Journal of Comparative Neurology*).

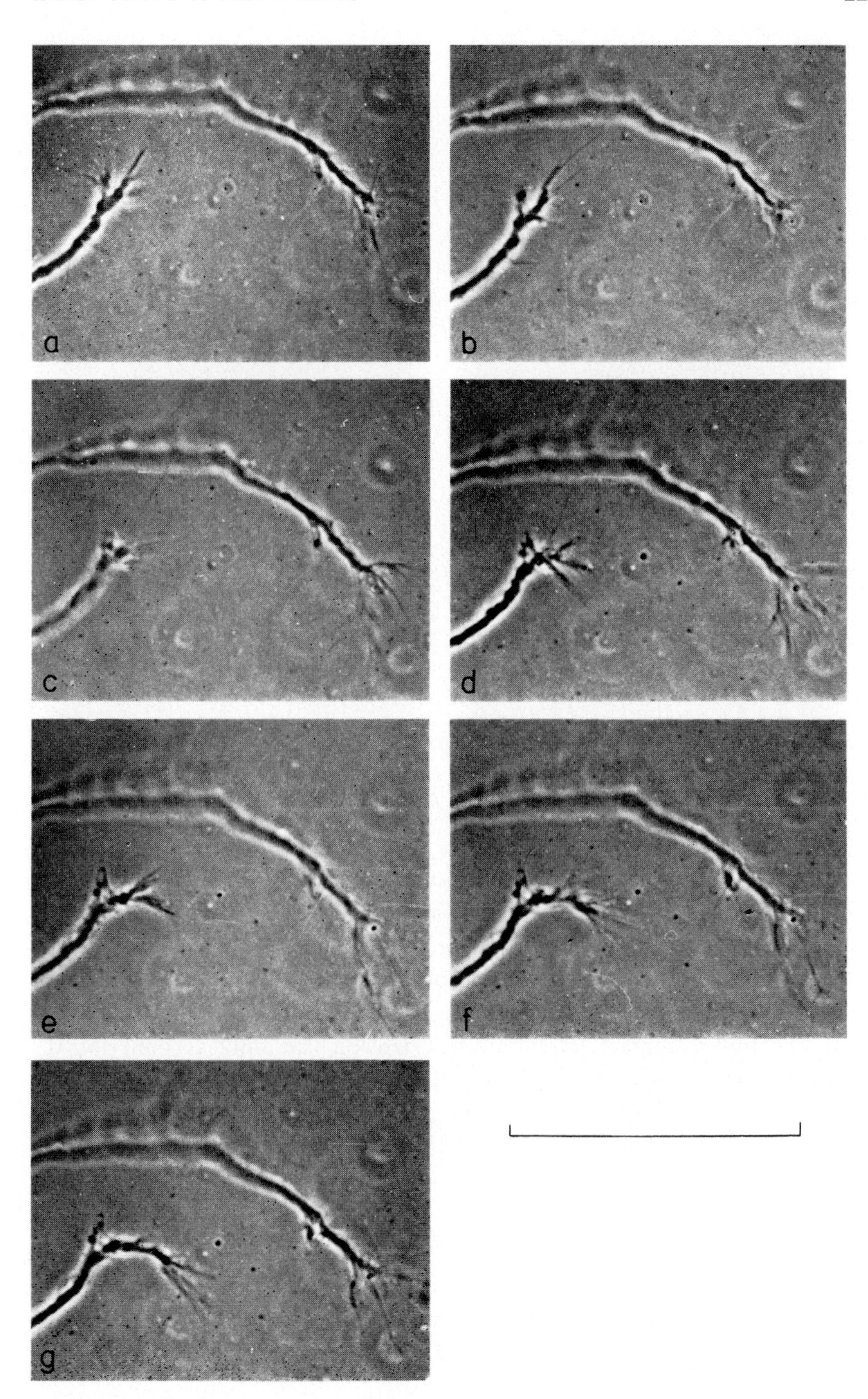
a
b
c
d
e
f
g

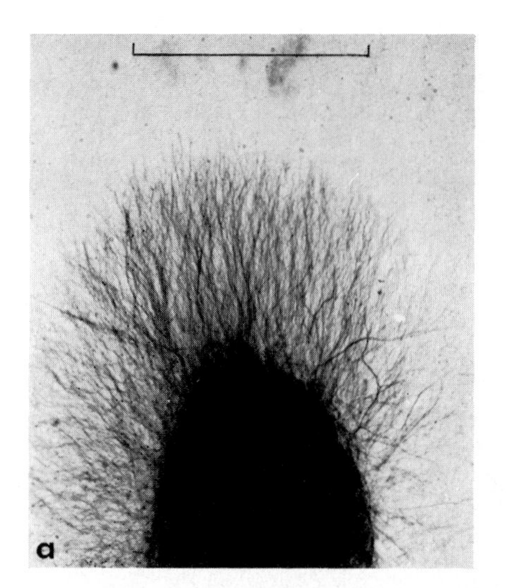

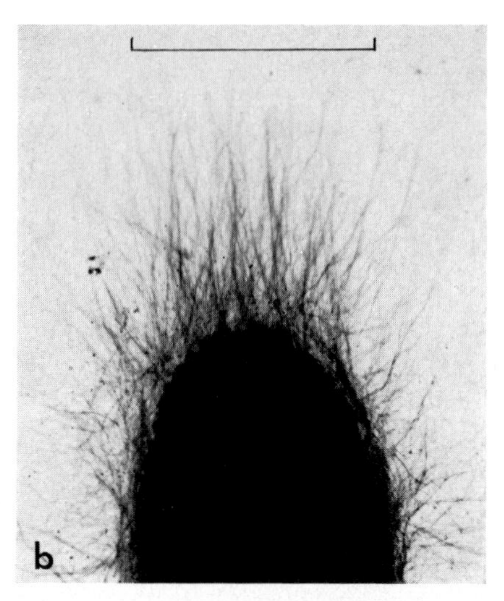

FIG. 8. (a) Non-fasciculated nerve fibre outgrowth from lumbar dorsal root ganglion within plasma clot stretched with an axial ratio of 1.9, 24 h *in vitro;* (b) fasciculated nerve fibre outgrowth from lumbar dorsal root ganglion on surface of plasma clot stretched with an axial ratio of 1.8, 24 h *in vitro*. Both scale bars 500 μm.

SUMMARY

(1) Radial orientation of nerve fibre extension from an explant in non-fasciculating conditions is dependent on local fibre density.

(2) Radial plasma clot orientation is not nearly strong enough to account for radial nerve fibre orientation by contact guidance.

(3) Experiments with confronted explant cultures show that nerve fibre contact is an important mechanism of nerve orientation.

(4) The filmed contact inhibition of extension can fully account for the directional behaviour of nerve fibres under these non-fasciculating conditions.

(5) Nerve fibres may respond to contact guidance at stretched clot–liquid interfaces, possibly along linear surface irregularities at these interfaces.

ACKNOWLEDGEMENTS

I thank Mr Michael Abercrombie of Strangeways Research Laboratory for providing both facilities and advice at all stages of the work. Part of the work was supported by a grant from the Medical Research Council.

References

ABERCROMBIE, M. & AMBROSE, E. J. (1958) Interference microscope studies of cell contacts in tissue culture. *Exp. Cell Res.* **15**, 332-345

ABERCROMBIE, M. & GITLIN, G. (1965) The locomotory behaviour of small groups of fibroblasts. *Proc. R. Soc. Lond. B Biol. Sci.* **162**, 289-302

ABERCROMBIE, M. & HEAYSMAN, J. E. M. (1954) Observations on the social behaviour of cells in tissue culture. II. Monolayering of fibroblasts. *Exp. Cell Res.* **6**, 293-306

BRAY, D. (1970) Surface movements during the growth of single explanted neurons. *Proc. Natl. Acad. Sci. U.S.A.* **65**, 905-910

CURTIS, A. S. G. (1960) Area and volume measurement by random sampling methods. *Med. Biol. Illus.* **10**, 261-266

DUNN, G. A. (1971) Mutual contact inhibition of extension of chick sensory nerve fibres *in vitro*. *J. Comp. Neurol.* **143**, 491-508

LEVI, G. (1941) Nouvelles recherches sur le tissu nerveux cultivé *in vitro*. Morphologie, croissance et relations réciproques des névrones. *Arch. Biol.* **52**, 133-278

LOEB, L. (1921) Amoeboid movement tissue formation and consistency of protoplasm. *Am. J. Physiol.* **56**, 140-167

NAKAI, J. (1960) Studies on the mechanism determining the course of nerve fibres in tissue culture. II. The mechanism of fasciculation. *Z. Zellforsch. Mikrosk. Anat.* **52**, 427-449

OLIVO, O. M. (1927) Migrazione di elementi nervosi coltivati *in vitro*. *Arch. Exp. Zellforsch.* **4**, 43-63

ROVENSKY, Y. A., SLAVNAJA, I. L. & VASILIEV, JU. M. (1971) Behaviour of fibroblast-like cells on grooved surfaces. *Exp. Cell Res.* **65**, 193-201

WEISS, P. (1934) *In vitro* experiments on the factors determining the course of the outgrowing nerve fibre. *J. Exp. Zool.* **68**, 393-448

WEISS, P. (1945) Experiments on cell and axon orientation *in vitro*: the role of colloidal exudates in tissue organisation. *J. Exp. Zool.* **100**, 353-386

Discussion

Albrecht-Bühler: What is the evidence that the material which constitutes the growing fibre is supplied by the cell body?

Bray: The biosynthetic organelles—the Golgi body, the nucleus and by far the majority of the ribosomes—are all in the cell body.

Wolpert: How fast is a filopodium put out?

Bray: Very fast rates have been measured—close to 20 μm/min (Yamada *et al.* 1971).

Porter: How do you interpret the invagination evident at the surface membrane on the growth cone of nerves? Dr Wessells said that Thorotrast (colloidal thorium dioxide) can invade these channels and the morphology suggests that they are indeed invaginations. Do they represent membrane uptake after extensive membrane development?

Bray: They could be sites at which membranes are added to or removed from the surface—both processes probably occur in a growth cone. Their ultrastructural appearance, however, is quite distinct from that of micropinocytic vesicles in other cells, and we believe that they are points at which membrane is put out to the surface. Rounded vesicles, often coated, are seen in other regions of the cone, for example at the base of filopodia, and it is these which we think are forming from the surface.

With regard to Dr Wessell's results with Thorotrast, we have done similar experiments with horseradish peroxidase (Bunge 1973*a*) and find uptake into both rounded and tubular vesicles. I don't think a simple interpretation can be given at this point.

Trinkaus: What is going on at the tip of the so-called growth cone?

Bray: The filopodia appear to be projected forward and then wave back towards the base of the growth cone, like the lamellipodia of the fibroblast. There are thin sheet-like structures between filopodia which are very sensitive to external conditions.

Trinkaus: Do these sheets of membrane spread out on the substratum?

Bray: Yes. Often one sees a nerve fibre that has ceased to grow which has an incredibly spiky appearance. If it is about to start growing again, one almost always sees first the emergence of the more membranous structure between the filopodia.

Wessells: Our time lapse films completely confirm this interpretation (Ludueña & Wessells 1973). There is a continual impression of movement backward along the upper and lateral surfaces of the growth cone.

Bray: I want to ask Drs Wessells and Heaysman about the discrepancy between the ultrastructural appearance of the fibroblasts which Dr Heaysman showed and which had little membranous structure, and that of the glial cells that Dr Wessells and colloborators have studied (Spooner *et al.* 1971); and also between the sensory growth cones which he illustrated in his talk and the sympathetic ganglia which we have shown. It seems to me that these differences are greater than one would expect from the morphological appearances of the two pairs of cells.

Heaysman: The pictures I showed (pp. 187-194) were of the front edge of the lamella and the lamellipodium which do not have these structures. Nearer the nucleus, the structure becomes more like that which you have illustrated.

Wessells: In glial cells and heart fibroblasts we have seen membranous sacs and channels which open to the exterior that are identical to your illustrations, Dr Bray. On treating living cells with Thorotrast just before fixation, we found Thorotrast in every one of the little channels. This is consistent with the idea that they are opening up to the outside as you propose.

I did not know that Abercrombie had measured the rate of movement of cells. Ludueña (unpublished results) has shown that one can change the rate of elongation of axons by changing the substratum on which the nerve cell has been placed. For example, on gelatin the rate of elongation is about 104 μm/h, whereas on plastic it is about 43 μm/h. Together with this increase, the rate of protein synthesis doubles. This latter rate of axon elongation is about the same as that cited by Abercrombie for the locomotion of heart fibroblasts; in contrast, the higher rates observed on gelatin and in conditioned media are closer to rates of regeneration of nerve axons *in vivo*.

Lubińska: What is the time interval between successive photographs?

Wessells: By using sealed flasks on an inverted microscope we are able to photograph the same cell over 24 h, with intervals from 1 s to 5 min.

Gail: What do you know about the adhesiveness of these cells to gelatin substrates?

Wessells: We don't know if there is a difference in adhesiveness between gelatin and plastic. However, one can alter the morphology of nerve axons by varying the concentration of calcium or serum, both procedures apparently operating by increasing the adhesion to the substratum.

Trinkaus: Is the cell morphology different on plastic as opposed to gelatin? Are there any changes in the sites of adhesion?

Wessells: No, axons are straight on both gelatin and plastic, but they are crooked and bent if the concentration of calcium is high or that of serum is low or absent. This effect probably results from increased adhesiveness.

Porter: Dr Dunn, the relation between the number of fibres in the space available in the culture, where the fibre density at the outside is the same as the fibre density close to the explant, would suggest a greater availability of metabolites to the growing fibres. Have you considered doing any perfusion experiments?

Dunn: No, I have not done any, although it would be worth trying. I suspect that the density of these radial outgrowths is limited by mutual contact reactions, as I outlined in my paper. The length of nerve fibre established in any direction is determined by the difference between the free path available in that direction and the retraction distance after contact.

Middleton: How does the fasciculated fibre grow?

Dunn: I refer you to Nakai (1960) who has published a detailed investigation of this subject. With regard to the growth cones, there are more cones along the fascicle behind the leading growth cones.

Bray: We have seen multiple growth cones at the outer limits of the fasciculated fibre in electron micrographs (Bunge 1973*b*).

Wolpert: Is the contact phenomenon with the nerve specific to nerve cells or to any other solid object?

Dunn: Nakai & Kawasaki (1959) found that nerve fibre growth cones seldom retracted on encountering other cells, cell debris or inanimate objects. In my non-fasciculated cultures I have only seen nerve fibres retract spontaneously or on encountering other nerve fibres, never on encountering fibroblasts or other cell types in the culture.

Trinkaus: Some time ago Weiss & Hiscoe (1948) concluded that new materials for the growth of a regenerating axon come from the cell body. How do you reconcile this with your suggestion that synthesis is taking place at the tip?

Bray: In a sense there is no disagreement, since we also would say that precursors of the growing fibre are made in the cell body and must be transported to the tip before being assembled. But I would argue with the view put forward by Weiss & Hiscoe that the entire contents of an axon are continually extruded. They dammed motor nerves by squeezing, and observed the consequences. On the proximal side of the constraint a swelling grew which they interpreted as being due to the interruption of a constant supply of material travelling from the cell body. As Spencer (1972) has suggested, what may be happening is that some of the many fibres of this nerve are damaged, and that the swelling is due to their regeneration.

Trinkaus: So the swelling is due to the presence of more fibres proximally?

Bray: Yes. I think it is due to the regrowth and therefore to the presence of more fibres.

Gail: What about labelling experiments? Are proteins synthesized peripherally?

Bray: I do not know of any such experiments with isolated neurons. Labelling studies on mature neurons show that most protein is synthesized in the cell body and transported down the axon. There is always a small amount of synthesis in the axon itself, but it does not appear to be of a special class of proteins.

Allison: Following up what Dr Porter was saying (p. 225), I would add that the disposition of the cells in the confronted explants looked to me like an example of 'negative chemotaxis'. It is possible that something is produced in the explant which encourages the outgrowth of fibres away from it, so that if there is a relative concentration between two explants fibres would tend to move away from the area of confrontation? A simple explanation would be that something like protease is partially digesting the plasma clot supporting the fibres. Is this possibility adequately excluded?

Dunn: Yes, I think it is. Weiss (1934) could find no evidence for negative chemotropism away from the explant. Furthermore, he tried many possible sources of 'neurotropic' substances and failed to elicit any directional response from the nerve fibres. In my experiments with confronted explants the results argue strongly against negative chemotropism. The outgrowths do not deviate from the radial direction until they have contacted each other and then there is a very sudden and marked deviation. These experiments are very similar to those in which Abercrombie & Heaysman (1954) demonstrated that the directional behaviour of chick heart fibroblasts is due to a mutual contact reaction and not to chemotaxis.

Allison: I am not disputing that you have deviation after contact. But you have an obvious lack of nerve fibres between the two explants. Of course in the early stages of outgrowth the concentration of these substances would be low.

Dunn: When the explants are placed just far enough apart so that the outgrowths fail to make contact during the period of culture, the direction of fibres within each outgrowth is unaffected by the presence of the other explant. When the explants are placed slightly closer together, and cultured for the same time, the direction of fibre extension in the contacted regions of outgrowth is profoundly affected. If these supposed substances have their source in the explants, this small change in distance between explants would not be expected to produce a great change in the concentration of substances in the interexplant zone.

Allison: No; if the explants were placed slightly closer together the concentration of any chemotactic factor would be appreciably greater, since it would fall off with the power of the distance between explants.

Porter: It seems to me that the growth cone is using these filopodia to sense its environment. It extends the filopodium and on contact with another fibre an inhibition of growth in that direction immediately develops. Do you relate this to contact inhibition? What would happen if the filopodium contacted another type of cell?

Dunn: In some respects the contact reaction between non-fasciculating nerve fibres appears to differ from contact inhibition of locomotion in other cell types. However, there are enough similarities between the two reactions that I am convinced that they are mechanistically homologous. I have seen no evidence of this contact reaction when a nerve fibre growth cone encounters a fibroblast.

AXON TRANSPORT

Lubińska: I would like to direct attention to some differences between neurons in culture and neurons *in vivo* which require different methods of analysis of

TABLE 1

Size and shape of axons

	In vivo (adult vertebrates)	*In culture*
Diameter	Up to 20 μm (or more)	A few μm
Length	Up to 1 m (or more)	A few mm
Shape	Regular, almost cylindrical, permanent	Branching, changing continually

intraaxonal movements and axonal elongation under the two experimental conditions. Table 1 sets out the differences in size and shape of axons.

It is not clear why the nerve fibres remain short and thin in tissue culture. In spite of the reduced size they seem to be fully functional electrophysiologically and are able to establish good synaptic connections with other neurons or muscle cells in culture (Crain 1966; Pappas *et al.* 1971). The variability of shape, emission and withdrawal of lateral sprouts, occasional arrests and retractions of the fibre tip do not seem to be due to conditions prevailing in tissue culture. Similar phenomena were observed *in vivo* by Speidel (1933, 1942, 1964) in the transparent fins of tadpoles and seem to characterize the behaviour of growing nerve fibres.

The great length and simplicity of shape of mature (peripheral) axons *in vivo* offer certain advantages for analysis of axoplasmic flow. The surface activity manifested by formation and disappearance of lateral branches in growing fibres is arrested at maturation. The shape of the axons is stabilized: they are generally unbranched except in the preterminal region. The intracellular movements may be studied on stretches of axons at some distance from the main sites of synthetic processes in the perikarya and from the functionally specialized nerve endings. In these stretches the characteristics of migration of axonal components, relatively uncomplicated by other cellular processes, may be determined quantitatively. However, this material requires application of indirect methods of detection whereas in tissue culture the movements of particles may be observed and measured directly in living neurons. The main features of analysis of axoplasmic flow *in vivo* and in culture are listed in Table 2. The general pattern of movements is similar under both conditions. It is bidirectional, various materials moving in the axons both towards and away from the cell bodies. Two ranges of velocity of migration, slow and fast (of the order of 1 and 100 mm/day), have been detected in axons *in vivo*. Various components defined either biochemically or by their known physiological role (such as neurotransmitters, neurosecretions, 'trophic' substances, biochemical markers of small organelles, as well as neurotropic viruses and toxins) move in the

TABLE 2

Analysis of intraaxonal movements

	In vivo (adult vertebrates)	*In culture*
Material	Destroyed cells	Living cells
Observations	At spaced time points	Continuous, good temporal resolution
	Separate portions of the neuron	Whole neuron accessible
	Reconstruction from a mosaic of results obtained on various regions	
	Biochemical, radiometric and morphological analyses of migrating materials	Translocation of particles under light microscope. Poor morphological resolution

axons by fast flow. The range of fast velocities is similar to that of intraaxonal particles measured directly in axons in culture. Much less specific information is available concerning the nature of components migrating slowly and the estimates of slow velocities are less reliable (Lubińska, unpublished results).

Even at a purely descriptive level of the axoplasmic flow many obscure points await clarification. It is not known where the change of direction occurs nor whether the course of flow in the axon is modulated by movements of organelles in the nerve endings during synaptic transmission. The distribution of migrating components at branching points of the axon and the possible changes of flow at the Ranvier nodes have not been investigated. In spite of many attempts (Jankowska *et al.* 1969; Lux *et al.* 1970; Dahlström 1971; and others), no clear-cut influence of nerve stimulation on characteristics of axoplasmic flow has been ascertained, whereas the intracellular movements in other types of cells are strongly affected by stimulation.

Axoplasmic flow is an important factor in the growth of nerve fibres. Since the velocity of fast flow is many times greater than the rate of growth, the elongation of growing fibres cannot be regarded as a simple outpouring of axonal materials. According to Yamada *et al.* (1970) at least two processes are essential for outgrowth of nerve fibres: the axoplasmic flow supplying perikaryal materials to the tip of the fibre and a continual activity of the growth cone where the imported materials, in particular those destined for building of additional stretches of surface membrane, are assembled. When either of these processes is selectively inhibited by specific drugs, the growth of fibres is arrested (Daniels 1968, 1972; Yamada & Wessells 1971; Yamada *et al.* 1971; Chang 1972).

It is interesting to note that axoplasmic flow and growth of regenerating fibres are similarly influenced by temperature. In cold-blooded animals the rates of both processes show a characteristic temperature dependence: they remain

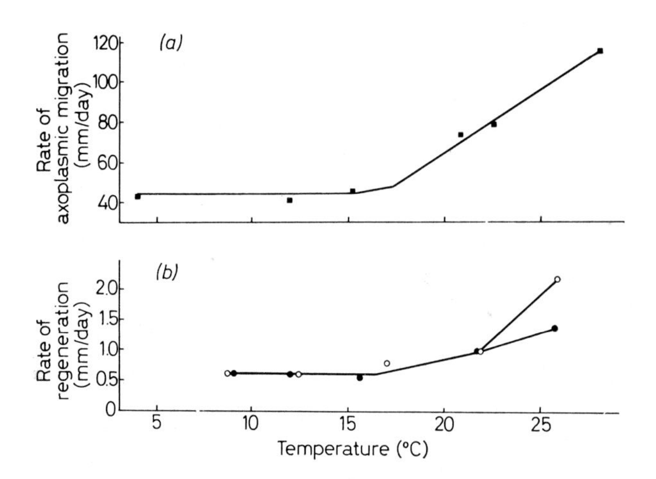

FIG. 1 (Lubińska). Influence of temperature on rates of axoplasmic flow and nerve regeneration. (*a*) Axoplasmic flow of labelled proteins in the connective tissue of *Anodonta*. (*b*) Elongation of regenerating fibres in the sciatic nerves of frog (○) and toad (●).

constant up to about 16 °C and increase at higher temperatures. Fig. 1 shows the influence of temperature on the velocity of axonal migration of labelled proteins in the connective of *Anodonta* (Heslop & Howes 1972) and the rates of elongation of regenerating fibres in the sciatic nerves of frog and toad (Lubińska & Olekiewicz 1950).

Abercrombie: Is there any evidence that membrane components are transported down the axon?

Lubińska: Yes, not only poorly defined membrane elements but apparently some pre-formed organelles also are migrating in the regenerating axons. For example, Matthews & Raisman (1972) found that, in the superior cervical ganglion of rat, ligation of efferent axons causes a significant reduction of the amount of dense-cored vesicles in the perikarya and their accumulation in the axons in front of the ligature. Subsequently these vesicles are found at all levels of the regenerating sprouts including their tips.

Dr Wessells, would a transverse section of a growing axon show any increase in the amount of membranous material compared with a similar cross-section of the non-growing axon? Presumably the material is brought from the perikaryon and assembled at the growth cone. Could you see the material in transit? Is there any information about what happens between the growth cone and the cell body?

Wessells: Many people (for example, Smith *et al.* 1970; Yamada *et al.* 1971) have seen vesicular elements—not long and plate-like, but round—in association with the microtubules and the neurofilament system, but nobody has proved directly that they are moving in either direction. However, in axons we have fixed in osmium tetroxide without initial exposure to glutaraldehyde, the axonal vesicular elements appear as longer sac-like structures, like those described by Dr Bray.

Curtis: Dr Dunn's elegant debunking of the Weiss theory (Weiss 1934) of contact guidance and the neat explanation of why Weiss saw the two-centre effect reminds me that there still remains the curious effect of micro-irregularities of the surface on cell locomotion or its control. Weiss himself using glass fibres, Rosenberg (1962) using behenate (docosanoate) films and other people have shown that there is obviously some guiding effect and change of morphology of the cells in response to shape changes of the substrate.

References

ABERCROMBIE, M. & HEAYSMAN, J. E. M. (1954) Observations on the social behaviour of cells in tissue culture. II. 'Monolayering' of fibroblasts. *Exp. Cell Res.* **6**, 293-306

BUNGE, M. B. (1973*a*) Uptake of peroxidase by growth cones of cultured neurons. *Anat. Rec.* **175**, 280

BUNGE, M. B. (1973*b*) Fine structure of nerve fibers and growth cones of isolated sympathetic neurons in culture. *J. Cell Biol.* **56**, 713-735

CHANG, C. M. (1972) Effect of colchicine and cytochalasin B on axonal particle movement and outgrowth *in vitro*. *J. Cell Biol.* **55**, 37a

CRAIN, S. M. (1966) Development of 'organotypic' bioelectric activities in central nervous tissues during maturation in culture. *Int. Rev. Neurobiol.* **9**, 1-43

DAHLSTRÖM, A. (1971) Axoplasmic transport (with particular respect to adrenergic neurons). *Philos. Trans. R. Soc. Lond. Ser. B Biol. Sci.* **261**, 325-358

DANIELS, M. P. (1968) Colchicine inhibition of nerve process elongation *in vitro*. *J. Cell Biol.* **39**, 31a

DANIELS, M. P. (1972) Colchicine inhibition of nerve fiber formation *in vitro*. *J. Cell Biol.* **53**, 164-176

HESLOP, J. P. & HOWES, E. A. (1972) Temperature and inhibitor effects on fast axonal transport in a molluscan nerve. *J. Neurochem.* **19**, 1709-1716

JANKOWSKA, E., LUBIŃSKA, L. & NIEMIERKO, S. (1969) Translocation of AChE-containing particles in the axoplasm during nerve activity. *Comp. Biochem. Physiol.* **28**, 907-913

LUBIŃSKA, L. & OLEKIEWICZ, M. (1950) The rate of regeneration of amphibian peripheral nerves at different temperatures. *Acta Biol. Exp. (Warsaw)* **15**, 125-145

LUDUEÑA, M. A. & WESSELLS, N. K. (1973) Cell locomotion, nerve elongation, and microfilaments. *Dev. Biol.*, in press

LUX, H. D., SCHUBERT, G., KREUTZBERG, G. W. & GLOBUS, A. (1970) Excitation and axonal flow: autoradiographic study on motoneurons intracellularly injected with a ^{3}H-amino acid. *Exp. Brain Res.* **10**, 197-204

MATTHEWS, M. R. & RAISMAN, G. (1972) A light and electron microscopic study of the cellular response to axonal injury in the superior cervical ganglion of the rat. *Proc. R. Soc. Lond. B Biol. Sci.* **181**, 43-79

NAKAI, J. (1960) Studies on the mechanism determining the course of nerve fibres in tissue culture. II. The mechanism of fasciculation. *Z. Zellforsch. Mikrosk. Anat.* **52**, 427-449

NAKAI, J. & KAWASAKI, Y. (1959) Studies on the mechanism determining the course of nerve fibres in culture. I. The reaction of the growth cone to various obstructions. *Z. Zellforsch. Mikrosk. Anat.* **51**, 108-122

PAPPAS, G. D., PETERSON, E. R., MASUROVSKY, E. B. & CRAIN, S. M. (1971) Electron microscopy of the *in vitro* development of mammalian motor end plates. *Ann. N.Y. Acad. Sci.* **183**, 33-45

POMERAT, C. M., HENDELMAN, W. J., RAIBORN, C. W., JR. & MASSEY, J. F. (1967) Dynamic activities of nervous tissue *in vitro* in *The Neuron* (Hydén, H., ed.), pp. 119-178, Elsevier, Amsterdam

ROSENBERG, M. D. (1962) Long-range interactions between cell and substratum. *Proc. Natl. Acad. Sci. U.S.A.* **48**, 1342-1349

SMITH, D. S., JARLFORS, U. & BERANEK, R. (1970) The organization of synaptic axoplasm in the lamprey. *J. Cell Biol.* **46**, 199-219

SPEIDEL, C. C. (1933) Studies on living nerves. II. Activities of amoeboid growth cones, sheath cells, and myelin segments, as revealed by prolonged observation of individual nerve fibers in frog tadpoles. *Am. J. Anat.* **52**, 1-80

SPEIDEL, C. C. (1942) Studies of living nerves. VII. Growth adjustment of cutaneous terminal arborizations. *J. Comp. Neurol.* **76**, 57-74

SPEIDEL, C. C. (1964) *In vivo* studies of myelinated nerve fibres. *Int. Rev. Cytol.* **16**, 173-231

SPENCER, P. S. (1972) Reappraisal of the model for 'bulk axoplasmic flow'. *Nat. New Biol.* **240**, 283-285

SPOONER, B. S., YAMADA, K. M. & WESSELLS, N. K. (1971) Microfilaments and cell locomotion. *J. Cell Biol.* **49**, 595-613

WEISS, P. (1934) *In vitro* experiments on the factors determining the course of the outgrowing nerve fibre. *J. Exp. Zool.* **68**, 393-448

WEISS, P. & HISCOE, H. B. (1948) Experiments on the mechanism of nerve growth. *J. Exp. Zool.* **107**, 315-316

YAMADA, K. M. & WESSELLS, N. K. (1971) Axon elongation: effect of nerve growth factor on microtubule protein. *Exp. Cell Res.* **66**, 346-352

YAMADA, K. M., SPOONER, B. S. & WESSELLS, N. K. (1970) Axon growth: roles of microfilaments and microtubules. *Proc. Natl. Acad. Sci. U.S.A.* **66**, 1206-1212

YAMADA, K. M., SPOONER, B. S. & WESSELLS, N. K. (1971) Ultrastructure and function of growth cones and axons of cultured nerve cells. *J. Cell Biol.* **49**, 614-635

Modes of cell locomotion *in vivo*

J. P. TRINKAUS

Yale University, New Haven, USA

Abstract The surface activity and locomotion of deep cells of the *Fundulus* blastoderm were studied *in vivo* with time lapse cinemicrography. During late cleavage, the surfaces of the blastomeres begin to undulate gently. By early blastula, these undulations gradually increase in amplitude and hemispherical surface protrusions called *blebs* appear. These blebs form and retract rapidly and at mid blastula some may be seen adhering to the surfaces of other cells. At the same time, they often expand into elongate *lobopodia.* Cell locomotion is first evident in mid blastula and continues throughout gastrulation. During locomotion, the leading edge of a deep cell behaves in various ways. When blebs and lobopodia adhere to a substratum (other deep cells, the undersurface of the enveloping layer or the periblast) and retract, the cell may move in the direction of the shortening cell process. Alternatively, blebs and lobopodia may adhere, but not shorten. Locomotion is accomplished rather by protoplasmic flow into the protrusion. Blebs and lobopodia may also flatten and spread on the substratum as *lamellipodia.*

During the past few years, concerted study of tissue cell locomotion under the ideal optical conditions of cell culture has brought us perceptibly closer to an understanding of how certain tissue cells move *in vitro.* In contrast, our knowledge of cell locomotion within the organism remains in a primitive state. Indeed, much that is already known for cells *in vitro* is not only unknown for cells *in vivo*, but has not even been studied. This discrepancy is no doubt due in large part to the technical impediments that most *in vivo* situations present. However, whatever the reasons, it is most regrettable. An understanding of how cells move within organisms is obviously essential if we are finally to confront two of the major problems of cell biology: how cells move during normal morphogenesis and during neoplastic invasion. *In vitro* studies are of course helpful, in that they reveal how cells *might* move *in vivo.* However, they do not necessarily reveal how cells actually *do* move *in vivo*, where the environment is largely three dimensional and where the substrata available to moving cells are mainly other cells or their products, and not a plane surface composed of glass or plastic (see Elsdale & Bard 1972). To learn how cells move *in vivo* they must be studied *in vivo*.

The only extensive studies of cell locomotion within an organism are the pioneering investigations of Speidel (1933, 1935) on neuron outgrowth in amphibian larvae and of Gustafson & Wolpert (1967) on mesenchyme cell movement during sea urchin gastrulation. The present investigation represents the initial steps of a similar study of cell locomotion in the embryo of the teleost *Fundulus heteroclitus*.

During blastula and gastrula stages of *Fundulus* there are two general categories of cells (Trinkaus & Lentz 1967; Bennett & Trinkaus 1970). In the first, the cells lie at the surface of the blastoderm, where they are tightly joined to each other to form a cohesive monolayer termed the *enveloping layer*. Those of the second category are found in the segmentation cavity, enclosed above by the enveloping layer and below by the syncytial periblast. These are the so-called *deep cells*. During gastrulation they engage in extensive morphogenetic movements to form the germ ring, the embryonic shield, and, eventually, the embryo itself. They also contribute to the yolk sac, along with the enveloping layer (see Trinkaus 1969). The normal surface activity and locomotion of *Fundulus* deep cells were studied *in vivo* in the intact blastoderm by time lapse cinemicrography. The lucidity of the *Fundulus* egg makes this possible. For a full outline of the method and detailed illustrations of deep cell behaviour, see Trinkaus (1973).

RESULTS

The beginning of surface activity: blebbing

Deep blastomeres in the late cleavage and early blastula blastoderm show distinct surface activity, well before the onset of cell locomotion. This begins around the 128-cell stage with barely perceptible undulations of the cell surface. As cleavage continues, these undulations increase in amplitude, rate of formation and frequency, and soon become converted into hemispherical blebs. These blebs, which have a diameter between a quarter and half of a cell diameter, form relatively slowly at first, in about 60 s, on only a small percentage of the cells. As development continues, however, they form more rapidly, usually in about 10–20 s, and on more and more cells. In general, protrusion of a bleb is more rapid than its retraction. In some instances, blebs do not retract directly, but bulge around the cell surface in a propagated wave. Although this 'circus' or 'limnicola' movement has often been seen in culture (e.g., Holtfreter 1944; Johnson 1970; Harris 1973*a*), this appears to be the first time it has been described *in vivo*. Blebs are characteristically hyaline in appearance, apparently because they lack cytoplasmic organelles (Trinkaus & Lentz 1967). In spite of this rapid blebbing, all deep cells of early blastulae are stationary, that is, they do not engage in translocation. This seems to be due in part to the fact that at this stage blebs do not adhere to the surface of other cells. They retract freely and do not lose their rounded contour upon retraction.

In view of the fact that blebbing has often been observed on cultured cells (e.g., Price 1967) (particularly after some time in culture), many have considered it to be an abnormal *in vitro* phenomenon. It therefore must be emphasized that blebbing of *Fundulus* deep cells is a normal activity *in vivo* that is constantly in evidence throughout blastula and gastrula stages, both before and concurrently with cell locomotion, and indeed often as a part of the process of locomotion. Blebbing ('pulsatory activity') is also a normal activity of primary mesenchyme cells in sea urchin embryos (Gustafson & Wolpert 1961). The presence of blebbing in embryos of two such widely separated forms suggests that it is a common feature of normal cell surface activity during blastula and gastrula stages of embryos generally. In view of the importance of blebbing in deep cell locomotion in *Fundulus* (see later), the possibility that it occurs in other forms is worth careful attention.

The presence of surface undulations in late cleavage blastomeres and of blebbing in early blastulae of *Fundulus* is clear evidence that, contrary to some popular notions, cells begin to differentiate at an early stage—not only before the completion of gastrulation but, indeed, before it begins. Although the changes at the cell surface and within the cell that lie at the basis of this differentiation are not known, three possibilities are increased hydrostatic pressure within the cell, increased capacity to form new membrane rapidly (Harris 1973*a*), and increased surface deformability (Weiss 1965). It is perhaps significant that these surface changes happen at about the time that primary nucleoli (Karasaki 1965) first become evident (Lentz & Trinkaus 1967).

The advent of locomotion

Deep cells begin locomotion in mid blastula, before the onset of gastrulation. Cell displacement begins gradually, involving at first only an occasional cell, until by late blastula and early gastrula it is commonplace. Still, at any one time, the majority of cells do not seem to move, although almost all show much surface activity: undulations, blebbing, and limnicola activity. The initial displacement observed varies considerably from about 60 μm (two or three times the diameter of a cell) to about 150 μm. During gastrulation, however, the upper limit of the displacement is much greater. Some cells in the germ ring move as much as a quarter to a third of the circumference of the egg when they converge toward the embryonic shield, as shown by marking with vital dyes (Oppenheimer 1936) and carbon particles (Trinkaus, unpublished results); in other words, some cells must translocate about 800 μm. Mesenchyme cells in the yolk sac of late gastrulae have also been observed to move large distances. Apart from the cells in the germ ring and the yolk sac near the embryonic shield that move directionally toward the shield, deep cell movements, both in blastulae and gastrulae, appear to be random, in so far as direction is concerned. It should be emphasized, however, that no quantitative analysis has been made to see if this is really so (see Lucy & Curtis 1959; Elton & Tickle 1971; Gail & Boone 1970).

The organs of locomotion

Coincident with the appearance of cell translocation is the elongation of certain blebs into finger-like *lobopodia* and the flattening of the apical portion of some blebs and lobopodia to form broad *lamellipodia*.

Lobopodia begin as blebs, but instead of retracting immediately they persist and elongate. This elongation is relatively slow, compared to initial bleb formation, and may take up to 2–3 min. Like bleb formation, however, lobopodial formation is apparently not true growth, where the mass or volume of the cell increases. When a large lobopodium appears, the rest of the cell visibly diminishes in size. It seems that the lobopodium forms, in part, at least, as a result of flow of cytoplasm from the cell body into the bleb. Although lobopodia may act as organs of locomotion, they do not always do so. Lobopodia sometimes elongate to form long, straight filopodia.

Lamellipodia first appear as a flattening of the leading edge of a bleb or lobopodium. Typically, a bleb or lobopodium is thrust out, its leading edge momentarily flattens against the substratum and spreads forward at the same time, its edge showing delicate scalloping. Then it rounds up again. Often it flattens for only a few seconds. The scalloping suggests adhesion at several points, with concave non-adhering regions in between, as in fibroblasts and epithelial cells in culture (Harris 1973*b*; DiPasquale 1973). The substratum observed in time lapse cinemicrography is usually the undersurface of the enveloping layer, but it may also be the upper surface of the periblast or the surface of another deep cell. Lamellipodia appear more frequently during gastrulation than in the blastula stage, involve a greater area of the cell surface and persist longer. These broad lamellipodia may form either by the flattening of blebs or lobopodia or *de novo*, directly from the cell surface.

Regardless of the organ of locomotion dominant at the time, the cell body of deep cells observed in locomotion usually retains its almost spherical form during the stages studied. Its changes in shape are relatively small and, in any case, it has not been observed to flatten or elongate greatly.

The adhesiveness of deep cells

It seems likely that the appearance of locomotory activity in mid blastula stems in part from an increase in surface adhesiveness of the non-adhesive or lowly adhesive deep cells of early blastulae, so that they are able to gain the traction necessary for translocation. Several observations support this hypothesis. (1) When blebs and lobopodia contact the surface of another cell, they often spread on it. (2) The trailing edge of a moving cell frequently drags behind the front end, becoming stretched, with elongated, straight contours. (3) When the stretched rear end of such a cell is about to break its contact, it is often pulled to a point, which rounds off only when the contact is visibly broken. (4) The trailing edge then retracts quite rapidly. (5) If a retracting cell preserves its contact with another cell, the latter will often be dragged along in the retraction. (6) Cells in contact tend to conform to each other's surfaces, not remaining semi-spherical, as in early blastulae. (7) A deep cell with three or more lamellipodia is immobilized and assumes a fibroblast-like form with straight or concave contours between the processes. In fibroblasts, lamellipodia have been demonstrated by microdissection to be loci of attachment to the substratum (Goodrich 1924; Harris 1973*b*). And, as with fibroblasts, when deep cell lamellipodia detach and round off, the whole cell rapidly retracts and rounds up.

Evidence which corroborates the increase in surface adhesiveness during the blastula stage is the behaviour of deep cells in culture (Trinkaus 1963) and their junctional contacts, seen by electron microscopy (Trinkaus & Lentz 1967). In culture, dissociated cells of early blastulae remain rounded, form blebs and aggregate very little with each other. Late blastula and early gastrula cells, in contrast, tend to spread on the glass substratum and aggregate more readily. Observations of the fine structure of blebs of early blastulae confirm the lack of contacts with the surfaces of other cells. In late blastula and early gastrula, however, surface protrusions conform closely to the surfaces of other cells, forming junctions 15–20 nm wide, focal tight junctions, and apparently some gap junctions as well (T. Betchaku & Trinkaus, unpublished results).

Thus it would appear that the differentiation of locomotory activity among *Fundulus* deep cells depends in part on two distinct events: (1) the formation of blebs and lobopodia, the primary organs of locomotion, and (2) a change in surface properties which causes blebs and lobopodia to begin to adhere to other cells. This provides the traction necessary for locomotion.

Although the evidence that there is indeed an increase in surface adhesiveness is impressive, the results could also be explained in part by an increase in deformability (Weiss 1965). It is conceivable, for example, that blebs do not form lobopodia or lamellipodia and are retracted in a rounded state in early blastulae partly because of high rigidity of the cell surface and cytoplasm and that in mid and late blastulae blebs do form lobopodia and begin to spread over the surfaces of other cells to form lamellipodia partly because of a decrease in this rigidity. We have recently tested this possibility by applying negative pressure to a portion of the cell surface and have found that deep cells of early gastrulae are indeed more readily deformed (more extensible) than are deep cells of early blastulae (Tickle & Trinkaus 1973).

Modes of deep cell locomotion

All these surface protuberances—blebs, lobopodia and lamellipodia—can act as organs of locomotion. It would be artificial, however, to designate a mode of locomotion by the type of cell extension at any particular time, since that extension might transform into another in the next instant. Blebs can extend to form lobopodia and spread to form lamellipodia; lamellipodia can round out to form blebs; and lobopodia can spread to form lamellipodia and retract to form blebs. In view of this, I propose simply to describe the various ways deep cells have been observed to move, leaving until later the problem of interpretative classification.

A common mode of locomotion in late blastula (and in gastrula, though seemingly less frequently) depends on the alternation of blebs and lamellipodia. A bleb forms and flattens on the substratum into a lamellipodium, which invariably spreads forward as it forms. The lamellipodium then fills with cytoplasm, presumably flowing from the cell body, to form a new bleb, which thrusts further forward. This new bleb in turn flattens to form a new leading lamellipodium, which spreads still further forward, and so the cycle continues. Cytoplasmic flow and adhesive spreading on the substratum alternate to effect translocation.

Two variants of this mode of translocation have been observed: in one, blebs predominate, in the other, lamellipodia. In late blastula and throughout gastrulation one often sees deep cells engaging in a sporadic, random, bumbling form of locomotion that appears to depend exclusively on blebs or short lobopodia. Cells seem to displace other cells by thrusting out blebs and flowing into them. Locomotion occurs *as the bleb is forming*, in the direction of the long axis of the cell, with the bleb leading the way. It is as if the whole cell flows forward. The reason for the absence of flattening or other obvious evidence of adhesion is not known. It seems possible, however, that the leading edge of the blebs adheres just enough to prevent their retraction, but not enough to cause flattening. Consequently the formation of a bleb, transient adhesion of its tip, flow of cytoplasm into it, and de-adhesion of the trailing edge of the cell suffice as a mechanism of locomotion. There are two direct pieces of evidence that the leading blebs are adhesive: (1) the delicate scalloping occasionally observed at the leading edge of a leading bleb and (2) the lack of retraction of a leading bleb. There is also evidence that the trailing edge of such a cell is adhesive. Frequently, when the rear end retracts and catches up with the advancing leading edge, it does so suddenly, as if it had just broken an adhesion. Although this mode of locomotion superficially resembles classical amoeboid movement, because of the protrusion of a 'pseudopodium' (a bleb) and the flow of cytoplasm into it, I prefer not to term it 'amoeboid'. Such blebs have been observed to flatten momentarily, showing scalloping. And even though there is undoubtedly protoplasmic flow, the resolution of our films is not sufficient to reveal details of its nature. It is not known, in particular, whether a fountain zone (Allen 1961) or a cylinder of gel containing sol (Mast 1926) exists in deep cells. It is semantically imprecise to designate a mode of cell movement as 'amoeboid' simply because protoplasmic flow is involved. Mammalian polymorphonuclear leucocytes, for example, show obvious protoplasmic flow but differ strikingly from amoebae in that there is no cylinder of gel (Ramsey 1972). Similar study of the nature of protoplasmic flow of deep cells is needed, but will probably have to be done in culture.

The other variant, where the lamellipodium predominates as the organ of locomotion, has been observed only during gastrulation. The lamellipodium may form either from the flattening of a bleb or a lobopodium, or *de novo*. In either case, such lamellipodia are broader, distinctly more prominent, and longer lasting than those formed in blastulae. Typically, a fan-like lamellipodium forms, spreading forward at the same time. It then may extend further, stretching the rounded cell body, or retract somewhat only to spread forward again. Eventually, the cell body is dislodged, rounds up more, and catches up with the forward spreading lamellipodium. When a lobopodium forms a lamellipodium, the following sequence is characteristic. The tip of the lobopodium spreads out on the substratum to form a broad lamellipodium. With this, the leading edge advances and the whole cell extension lengthens. This places the cell body under increased tension and it may also advance at this time. Typically, such a lamellipodium continues to spread laterally and forward, with the result that most or all the lobopodium becomes flattened. This causes a considerable shortening of the cell extension and invariably results in much forward displacement of the cell body. From time to time, these lamellipodia apparently break their adhesions, for they round up and become bleb-like.

Another mode of locomotion distinctive of the gastrula stage is by the use of long lobopodia. Typically, a cell extends a lobopodium which is 30–100% longer than the diameter of the cell body. The tip adheres and the lobopodium immediately begins to shorten. As this occurs, the tip remains in place and the cell body moves forward a distance corresponding almost exactly to the amount of shortening, responding immediately and totally to the pull which is exerted. This suggests that the cell body is either non-adherent or only lightly adherent to its substratum. As such a long lobopodium shortens there might be some cytoplasmic flow into it, for sometimes the cell body appears to decrease in size. Another long lobopodium may then be thrust out and the process repeated. This is the so-called 'inchworm' method of cell translocation. All modes of deep cell locomotion are spasmodic, in that the rate of cell movement varies greatly, and sporadic, in that a cell may move for a while and then cease movement completely, only to begin again. But the inchworm method is especially so. It tends to be an all or nothing phenomenon. When the cell body is moving forward in response to the pull of the shortening lobopodium its rate of movement is very high. But then it frequently does not move at all for a time.

There are a number of variations on this mode of locomotion. A lobopodium may continue to advance without shortening, the cell body moving with it, or a lobopodium may continue to advance *as it is shortening* and moving the cell body. Sometimes, in this latter movement, the cell body advances a distance which corresponds closely to the sum of the lobopodial shortening and the forward movement of the tip, which again suggests that in such a case the cell body is not adherent or only lightly adherent to the substratum. On other occasions, the cell body does not respond immediately to the shortening of the lobopodium; instead the cell elongates somewhat before advancing, thus implying that its adhesion to its substratum is strong enough to resist momentarily the pull of the shortening lobopodium. In other cases, the tip of the lobopodium appears to be less adherent than the cell body; for as the lobopodium retracts it is quickly resorbed into the cell body. In such instances, there is of course no cell displacement. A final variation on this mode of movement is rather spectacular and appears only in mid and late gastrulae. The long lobopodium adheres not only at its tip but extensively along its length as well. This results in obvious branching and bending of the lobopodium and varying movement of the cell body, related to both the shortening of the extension and to its several changing branched loci of adhesion. Analysis of the locomotion of such cells is rather complicated; nevertheless the amount of translocation is still roughly predictable, if the shortening and changing adhesions are studied by frame-by-frame film analysis.

In spite of these variations in precise mode of locomotion of deep cells, there is a basic similarity. In each case, (*a*) a cell protrudes a process; (*b*) the process adheres to the substratum and tension develops within the process. This tension is usually gross and therefore obvious. When it is not evident, perhaps it is too small to be readily detected (it could, for example, be localized at or near the tip); (*c*) the cell body is pulled forward, or simply flows into the process; (*d*) the process pushes out further, or another process is extended; *et cetera.* Because of these similarities and since individual deep cells may change from one mode of locomotion to another, it seems possible, or even probable, that the same basic mechanism is at work in all. If this were so, the manner of locomotion at any given moment could depend on variation in one essential aspect of the locomotory process, such as, for example, adhesiveness of the cell body. If it adheres weakly, the slight pull of an adhering bleb could be sufficient to displace it and the bleb would become an organ of locomotion. If the cell body adheres more firmly, it would not be displaced immediately and as a consequence the bleb might continue spreading against the substratum, forming a lamellipodium. If, finally, the cell body adheres even more strongly, it would not be dislodged by either adhering blebs or spreading lamellipodia and the bleb or lamellipodium might continue to move further ahead, pulling out a long lobopodium. Of course, it is also possible that the differences among the three modes of locomotion lie as well, or even primarily, in the behaviour of the leading edge.

DISCUSSION

A primary result of this study is the discovery that cell locomotion in the *Fundulus* blastoderm begins during the blastula stage, well in advance of gastrulation. There are at least two distinct stages in the differentiation of this locomotory activity: formation of the primary organs of locomotion and a change in surface properties, so that cells adhere to various substrata. It would be of interest to know if locomotion begins in a similar way in other organisms. In this regard, the observations of Gustafson & Wolpert (1961, 1967) on sea urchin embryos are apropos. The first detectable activity of the primary mesenchyme cells is their blebbing. The blebs are replaced by long filopodia, which serve as organs of locomotion. It is not clear whether these filopodia actually form from blebs, but, in view of the *Fundulus* results, such a sequence is most plausible. Significantly, the rate of extension of sea urchin filopodia, about 10 μm/min (Gustafson 1964), is about the same as for *Fundulus* lobopodia (Trinkaus 1973).

A dominant feature of the locomotion of *Fundulus* deep cells is movement by the shortening of long cell protrusions. This mode of movement is of special interest because, in its essential features, it has been found in a number of other forms: sea urchin gastrulae (Gustafson 1964; Gustafson & Wolpert 1967); regenerating ascidians (Izzard 1971); mixed aggregates of chick retinal pigment cells with heart cells (Trinkaus & Lentz 1964); and probably in the chick blastoderm as well (Trelstad *et al.* 1967). Its occurrence in such diverse systems suggests that this mode of movement is widespread during development.

Incidentally, it is not known whether the shortening of extended cell processes during locomotion (in *Fundulus* or in any of these systems) is due to active contraction (as is usually thought) or to the elastic recoil of a visco-elastic element returning from a stretched state to a more stable configuration (a rounded cell) (Francis & Allen 1971).

Although translocation by long lobopodia or filopodia seems quite different from classical fibroblast movement, deep cell movement by means of broad lamellipodia is superficially much like that of fibroblast-like cells. There are at least two important differences, however. (1) In contrast to fibroblasts, the bulk of such moving deep cells remains rounded. The reason for this is not known, but it seems likely that it is due to weak adhesion of the cell body to the substratum (Trinkaus 1973). (2) Unlike fibroblasts and indeed unlike cells of the *Fundulus* enveloping layer (Lentz & Trinkaus 1971), deep cells appear not to be subject to contact inhibition of movement. They are typically invasive and move readily over each other, over the periblast and over the undersurface of the enveloping layer. In this regard, we have recently detected ruffling activity in lamellipodia of deep cells cultured *in vitro* on a glass substratum and *in vivo* moving on the periblast with Nomarski interference optics (Trinkaus, W. S. Ramsey & T. Betchaku, unpublished data). Although details of this ruffling remain to be analysed, it appears to be much like that of fibroblast cells, in that many ruffles form at the leading edge of the lamellipodium and propagate backward, away from the edge. In view of the invasive, non-contact inhibiting characteristics of *Fundulus* deep cells, it will be of much interest to observe the effect of cell contacts on this ruffling. Incidentally, I believe this to be the first time that individual cells moving *in vivo* have been found to show ruffling. This is an encouraging observation, for it suggests that cell locomotion *in vivo* indeed shares many features with cell movement *in vitro*.

ACKNOWLEDGEMENTS

I am indebted to Richard Ebstein for his invaluable assistance during an early phase of this investigation. This research was supported by Grants GB-6880 and GB-7828 from the National Science Foundation.

References

ALLEN, R. D. (1961) A new theory of amoeboid movement and protoplasmic streaming. *Exp. Cell Res.* (Suppl.) **8**, 17-31

BENNETT, M. V. L. & TRINKAUS, J. P. (1970) Electrical coupling between embryonic cells by way of extracellular space and specialized junctions. *J. Cell Biol.* **44**, 592-610

DIPASQUALE, A. (1973) *Contact Relations and Locomotion of Epithelial Cells in Culture*, Ph.D. Dissertation, Yale University, New Haven, USA

ELSDALE, T. R. & BARD, J. (1972) Cellular interactions in mass culture of human diploid fibroblasts. *Nature (Lond.)* **236**, 152-155

ELTON, R. A. & TICKLE, C. A. (1971) The analysis of spatial distributions in mixed cell populations: a statistical method for detecting sorting out. *J. Embryol. Exp. Morphol.* **26**, 135-156

FRANCIS, D. W. & ALLEN, R. D. (1971) Induced birefringence as evidence of endoplasmic viscoelasticity in *Chaos carolinensis*. *J. Mechanochem. Cell Motility* **1**, 1-6

GAIL, M. H. & BOONE, C. W. (1970) The locomotion of mouse fibroblasts in tissue culture. *Biophys. J.* **10**, 980-993

GOODRICH, H. B. (1924) Cell behavior in tissue cultures. *Biol. Bull. (Woods Hole)* **46**, 252-262

GUSTAFSON, T. (1964) The role and activities of pseudopodia during morphogenesis of the sea urchin larva in *Primitive Motile Systems in Cell Biology* (Allen, R. D. & Kamiya, N., eds.), pp. 333-349, Academic Press, New York

GUSTAFSON, T. & WOLPERT, L. (1961) Studies on the cellular basis of morphogenesis in the sea urchin embryo. Directed movements of primary mesenchyme cells in normal and vegetalized larvae. *Exp. Cell Res.* **24**, 64-79

GUSTAFSON, T. & WOLPERT, L. (1967) Cellular movement and contact in sea urchin morphogenesis. *Biol. Rev. (Camb.)* **42**, 442-498

HARRIS, A. K. (1973*a*) Cell surface movements related to locomotion. *This volume*, p. 3-20

HARRIS, A. K. (1973*b*) Location of cellular adhesions to solid substrata. *Dev. Biol.* in press

HOLTFRETER, J. (1944) A study of the mechanism of gastrulation. Part II. *J. Exp. Zool.* **95**, 171-212

IZZARD, C. S. (1971) Cell movement by filopod contraction in the tunic of *Botryllus Schlosser*. XI Meeting Am. Soc. Cell Biol. abstract, p. 157

JOHNSON, K. E. (1970) Role of changes in cell contact behavior in amphibian gastrulation. *J. Exp. Zool.* **175**, 391-427

KARASAKI, S. (1965) Electron microscopic examination of the sites of nuclear RNA synthesis during amphibian embryogenesis. *J. Cell Biol.* **26**, 937-958

LENTZ, T. L. & TRINKAUS, J. P. (1967) A fine structural study of cytodifferentiation during cleavage, blastula, and gastrula stages of *Fundulus heteroclitus*. *J. Cell Biol.* **32**, 121-138

LENTZ, T. L. & TRINKAUS, J. P. (1971) Differentiation of the junctional complex of surface cells in the developing *Fundulus* blastoderm. *J. Cell Biol.* **48**, 455-472
LUCY, E. C. A. & CURTIS, A. S. G. (1959) Time lapse film study of cell aggregation. *Med. Biol. Illus.* **9**, 86-93
MAST, S. O. (1926) Structure, movement, locomotion, and stimulation in Amoeba. *J. Morphol.* **41**, 347-425
OPPENHEIMER, J. M. (1936) Processes of localization in developing *Fundulus*. *J. Exp. Zool.* **73**, 405-444
PRICE, Z. H. (1967) The micromorphology of meiotic blebs in cultured human epithelial (HEp) cells. *Exp. Cell Res.* **48**, 82-92
RAMSEY, W. S. (1972) Locomotion of human polymorphonuclear leucocytes. *Exp. Cell Res.* **72**, 489-501
SPEIDEL, C. C. (1933) Studies of living nerves. II. Activities of amoeboid growth cones, sheath cells, and myelin segments, as revealed by prolonged observation of individual nerve fibers in frog tadpoles. *Am. J. Anat.* **52**, 1-80
SPEIDEL, C. C. (1935) Studies of living nerves. III. Phenomena of nerve irritation and recovery, degeneration and repair. *J. Comp. Neurol.* **61**, 1-80
TICKLE, C. A. & TRINKAUS, J. P. (1973) Change in surface extensibility of *Fundulus* deep cells during early development, in press
TRELSTAD, R. L., HAY, E. D. & REVEL, J. P. (1967) Cell contact during early morphogenesis in the chick embryo. *Dev. Biol.* **16**, 78-106
TRINKAUS, J. P. (1963) The cellular basis of *Fundulus* epiboly. Adhesivity of blastula and gastrula cells in culture. *Dev. Biol.* **7**, 513-532
TRINKAUS, J. P. (1969) *Cells into Organs: the forces that shape the embryo*, Prentice-Hall, Englewood Cliffs, New Jersey
TRINKAUS, J. P. (1973) Surface activity and locomotion of *Fundulus* deep cells during blastula and gastrula stages. *Dev. Biol.* **30**, 68-103
TRINKAUS, J. P. & LENTZ, J. P. (1964) Direct observation of type-specific segregation in mixed cell aggregates. *Dev. Biol.* **9**, 115-136
TRINKAUS, J. P. & LENTZ, T. L. (1967) Surface specializations of *Fundulus* cells and their relation to cell movements during gastrulation. *J. Cell Biol.* **32**, 139-153
WEISS, L. (1965) Studies in cell deformability. I. Effect of surface charge. *J. Cell Biol.* **26**, 735-739

Discussion

Wessells: What is the current view on Kessel's observation (1960) that colchicine has no effect on epiboly?

Trinkaus: We have just reinvestigated that. If an egg is placed in colchicine just as the epiboly is about to begin, epiboly does continue, almost always to completion. Thus, a large part of this enormous spreading is unaffected by colchicine. Moreover, as Kessel (1960) observed, the deep cells apparently do not divide; they remain quite large, and as a consequence one gets a strange diffuse germ ring and embryonic shield. I am sure that organogenesis does not proceed normally but we cannot say more yet. However, the formation of an embryonic shield indicates that convergence of deep cells continues. Thus cell locomotion continues in the presence of a concentration of colchicine that blocks cell division, presumably by disrupting microtubules.

de Petris: Is the speed of spreading the same in the presence of colchicine as in normal conditions?

Trinkaus: The rate of epiboly is the same for the first few hours. Then it slows down a little in the presence of colchicine.

Curtis: Does colchicine get inside these cells?

Trinkaus: It must, it blocks cell division.

Middleton: As I understand it, at the start of epiboly the enveloping layer is a single layer of epithelial cells which in the course of epiboly spreads to cover the whole egg. Are more cells introduced into this layer as it spreads?

Trinkaus: As far as we know none is introduced. We have never seen cells move into the layer in our time lapse films. The layer becomes thinner, covers a wider area and is put under enormous tension as epiboly approaches completion. When you sever the marginal connections of the blastoderm with the periblast, the blastoderm immediately retracts for a considerable distance. Even with a local severance there is a local retraction in a matter of seconds.

Goldman: I understand there is a conspicuous lack of fibres in the cytoplasmic processes in your electron micrographs. Could this be due to poor fixation?

Trinkaus: We don't know. This puzzles us as much as it does you, and I certainly am not prepared to say that there are no microtubules or microfilaments there.

Porter: What is the nature of the covering?

Trinkaus: This extraneous membrane, the chorion, is removed mechanically before exposure to colchicine; but, in any case, it is permeable to small molecules.

Goldman: I don't think that the arrest of mitosis is a good criterion for establishing whether colchicine also breaks down other cytoplasmic microtubules. We have found that we can prevent mitotic spindle formation in BHK-21 cells treated with 0.05 μg/ml colchicine, but that the microtubules found in the fibroblastic processes of the cells remain (Dickerman & Goldman, unpublished observations). It is essential to use electron microscopy to verify that microtubules are destroyed.

Trinkaus: We have looked at untreated deep cells with the electron microscope (Lentz & Trinkaus 1967; Betchaku & Trinkaus, unpublished results) and do not find microtubules in non-mitotic cells. Since we find microtubules in the spindle of mitotic cells, we take this as evidence that the fixation was adequate for microtubules. Our fixation was also adequate for microfilaments of the 5–6 nm variety. Cells of the enveloping layer, especially the marginal ones, have abundant microfilaments (as does the periblast). Significantly, these cells too appear to lack microtubules.

Porter: In the isolation of cells from amphibian embryos, the flask cells which border on a blastopore in epiboly are loaded with microtubules. Thus the development of the asymmetry in these flask cells is associated with tubule assembly (Perry & Waddington 1966).

Trinkaus: Possibly the process of elongation depends on an increase in the adhesiveness of the surface. The adhesive spreading of a cell along the surfaces of other cells into the space between them could conceivably result in much elongation.

Wohlfarth-Bottermann: I can exemplify how difficult it can be to reveal contractile filaments in thin sections. Fig. 1*a* shows a portion of the groundplasm of *Physarum*, where numerous ribosomes and a network of filaments are visible. We believe that this network is the contractile substance. Fig. 1*b* is a section of the fine structure of a 2% actomyosin-gel thread model which contracts to half its initial volume on addition of ATP. The picture represents the fine structure before addition of ATP. In Fig. 1*c*, Fig. 1*a* and *b* are superimposed. Note how difficult it is now to identify the truly contractile structures in the groundplasm (Komnick *et al.* 1972).

Trinkaus: As previously mentioned, we see an abundance of filaments in *Fundulus* eggs, especially in the marginal cells of the enveloping layer and in the periblast. I described our experiments on their possible contractile properties before (p. 147).

Porter: Mr Abercrombie, you have described the progress of fibroblasts over a surface as a flooding out and then a slight contraction. Do you relate any of that behaviour to any of the spasmodic behaviour seen in the cells Professor Trinkaus has described?

Abercrombie: Yes, indeed I do. I think all these kinds of movement grade into each other. They are all basically dependent on some sort of projection mechanism, which might be due to a squeezing or to a conduction by microtubules, and this is combined with a retraction mechanism probably based on an actomyosin system.

Trinkaus: I should emphasize that the three modes of movement are all interchangeable, that is, an individual cell may move alternatively by means of one or another. A cell may bleb. The bleb may flatten to form a lamellipodium and then bulge out again to form another bleb, and then maybe send out a lobopodium from another part of the cell, or that same bleb might elongate into a lobopodium, attach at its end, shorten and thus by this means move the cell forward. Equally, the tip of a lobopodium might flatten to form a lamellipodium and fill again with cytoplasm to form a bleb or lobopodium. The fact that the same cell may shift from one mode to another does indeed suggest that the fundamental features are the same, with varying emphasis on dif-

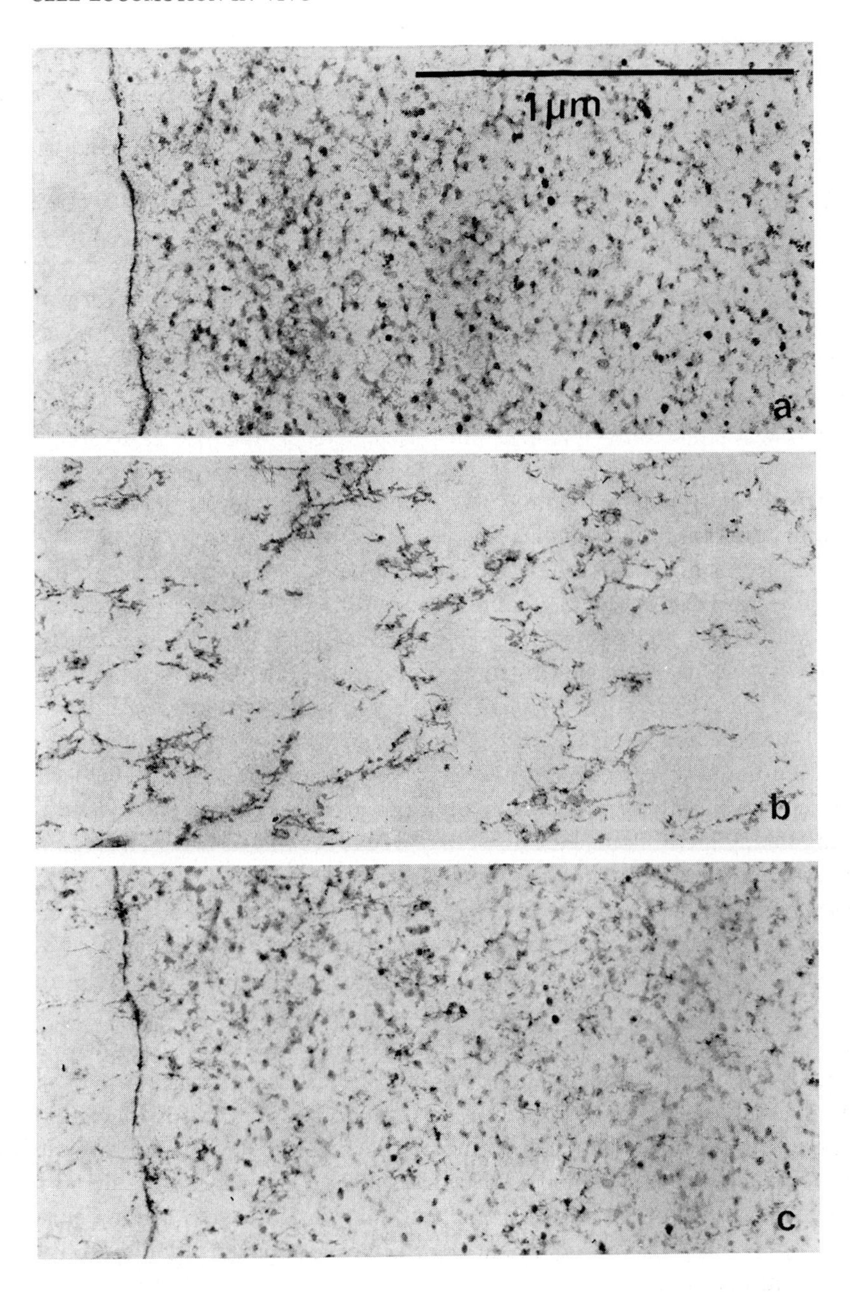

FIG. 1 (Wohlfarth-Bottermann). (*a*) Groundplasm of *Physarum*, (*b*) 2% actomyosin gel, (*c*) superimposition of pictures (*a*) and (*b*). (from Komnick *et al.* 1972.)

ferent aspects. Of course, we do not understand at all what causes a cell to shift from one of these activities to another; nor do we understand why a fibroblast not in contact with other cells will suddenly form a new ruffling lamellipodium and move out in a new direction.

Heaysman: You said that your cells only show ruffling if their upper surface was free of other cells.

Trinkaus: It was only under this condition that we saw it happen.

Heaysman: This agrees with our observation that when cells are grown underneath methyl cellulose, where their upper surface is not free to move, they do not ruffle.

Trinkaus: Often when a cell underlaps another cell in culture, it is ruffling as it approaches, but stops ruffling as it moves under the other cell. This seems to be the same phenomenon.

Heaysman: For many years, we have referred to the ruffling membrane as the locomotory organ as though ruffling was necessary for locomotion. We now believe that the ruffling is a corollary of the situation in which we are growing the cells, and is probably due to the failure of adhesion of the moving point of the lamella to the substrate. In his terminology, Professor Trinkaus is using lamellipodium where we use lamella, and ruffle where we would use lamellipodium.

Trinkaus: No. Our use of the term ruffle is the same as yours—an upward protrusion of the upper surface of the flattened margin of the cell which arises usually, but not always, by uplift of the margin (Ingram 1969; Harris 1969). It is true, however, that we use lamellipodium in a broader sense than you do, referring to any broad flattened extension of a cell (Trinkaus 1973). You would restrict the term to apply only to the advancing leading edge of such a large flattened region (Abercrombie *et al.* 1970). In my opinion, this is too restrictive. It would, for example, exclude use of this term for many cells moving *in vivo*, possessing definite flattened marginal regions that are clearly involved in locomotion, because the poor resolution of the *in vivo* situation prevents detailed observation of the leading edge. We need a general term with wide application for broad flattened cell protrusions, just as we do for hemispherical bulges (bleb), elongate finger-like processes (lobopodium) and for long, thin extensions (filopodium). Lamellipodium would seem to be the appropriate word. (The term lamella would also be satisfactory, but less so.) I humbly suggest that where you have used lamellipodium you substitute leading edge or leading lamella. Of course, where there is any doubt one should always explain what one means.

Harris: An alternative term is pseudopod, but this suggests that tissue cells move in the same way as the large, free-living amoebae, whereas their mode of locomotion is completely different.

Trinkaus: I consider pseudopodium to be particularly unfortunate when applied to tissue cells. This term has long been applied to the locomotory extensions of free-living amoebae and implies protoplasmic flow, with gelated ectoplasm and solated endoplasm, and, of course, a fountain zone at the front end. In most tissue cells we simply do not know whether such a situation exists or not—in many cases it definitely does not. Because of this, the use of pseudopodium for tissue cells, such as fibroblasts, mesenchyme cells and epithelial cells, is misleading and has caused much confusion. Simple morphological terms such as lamellipodium, filopodium etc., are clearly preferable for the present.

Abercrombie: Professor Trinkaus and I concluded that the most satisfactory solution is always to explain what we meant.

References

ABERCROMBIE, M., HEAYSMAN, J. E. M. & PEGRAM, S. M. (1970) The locomotion of fibroblasts in culture. II. 'Ruffling'. *Exp. Cell Res.* **60**, 437-444

HARRIS, A. K. (1969) Initiation and propagation of the ruffle in fibroblast locomotion. *J. Cell Biol.* **43**, 165A

INGRAM, V. M. (1969) A side view of moving fibroblasts. *Nature (Lond.)* **222**, 641-644

KESSEL, R. G. (1960) The role of cell division in gastrulation of *Fundulus heteroclitus. Exp. Cell Res.* **20**, 277-282

KOMNICK, H., STOCKEM, W. & WOHLFARTH-BOTTERMANN, K. E. (1972) Ursachen, Begleitphänomene und Steuerung zellulärer Bewegungserscheinungen. *Fortschr. Zool.* **21**, 1-74

LENTZ, T. L. & TRINKAUS, J. P. (1967) A fine-structural study of cytodifferentiation during cleavage, blastula, and gastrula stages of *Fundulus heteroclitus. J. Cell Biol.* **32**, 121-138

PERRY, M. M. & WADDINGTON, C. H. (1966) Ultrastructure of the blastopore cells in the newt. *J. Embryol. Exp. Morphol.* **15**, 135-154

TRINKAUS, J. P. (1973) Surface activity and locomotion of *Fundulus* deep cells during blastula and gastrula stages. *Dev. Biol.* **30**, 68-103

THE LOCOMOTION OF PRE CELLS

Time lapse film observations show that early in the life of a given culture the majority of cells are isolated, poorly spread upon the substrate and display vigorous blebbing activity. These isolated cells move only relatively slowly and owing to frequent changes in direction tend to oscillate over relatively small areas of substrate. Collisions between adjacent cells apparently lead to the development of stable adhesions since, with time, there is a gradual decrease in the number of isolated cells and a concomitant increase in the number of cells incorporated into islands of cells. The result of this process is that after about 12 h the culture consists either of an unbroken confluent sheet of cells or of a network of interlinked islands, depending on the initial cell inoculum.

In comparison with the rounded blebbing isolated cells, the cells that are incorporated into islands are extensively spread upon the substrate and those cells at the periphery of an island display typical leading lamellae. Similar behavioural differences between isolated cells and those in contact with other cells have been noted in cultured *Fundulus* gastrula cells (Trinkaus 1963) and in epithelioid Limpet haematocytes (Partridge, unpublished findings). The reasons for these differences are obscure but my observations of cultured PRE cells suggest the following explanation.

The mean speed of locomotion of isolated PRE cells is low [about 7 μm/h (Middleton, unpublished data)] when compared with similar data for other cell types: Abercrombie & Heaysman (1953) observed a mean speed of 82 μm/h for isolated chick heart fibroblasts and Bereiter-Hahn (1967) obtained values of up to 1200 μm/h for isolated amphibian ectodermal cells in culture. The low speed of PRE cells is apparently due to the cells being essentially unpolarized. The majority of the cells show no distinct leading lamella and such locomotion as they undergo appears to result from undirected blebbing. In contrast both chick heart fibroblasts and amphibian ectodermal cells show distinct leading lamellae which polarize their direction of locomotion. PRE cells can, however, become polarized. Shortly (within about 30 min) after colliding with another cell or island of cells the previously isolated blebbing cell spreads extensively on the substrate, produces a leading lamella from that part of its margin not in contact with other cells and attempts to move off in a direction other than that which led to the collision (Fig. 1). It seems likely, therefore, that the differences observed between isolated cells and those in contact with other cells can be explained in terms of the degree to which the cells are polarized. It appears that the establishment of a cell–cell contact polarizes the locomotion of a cell away from the direction of the contact and leads to the observed behavioural changes. However, the situation must be more complex than this since in

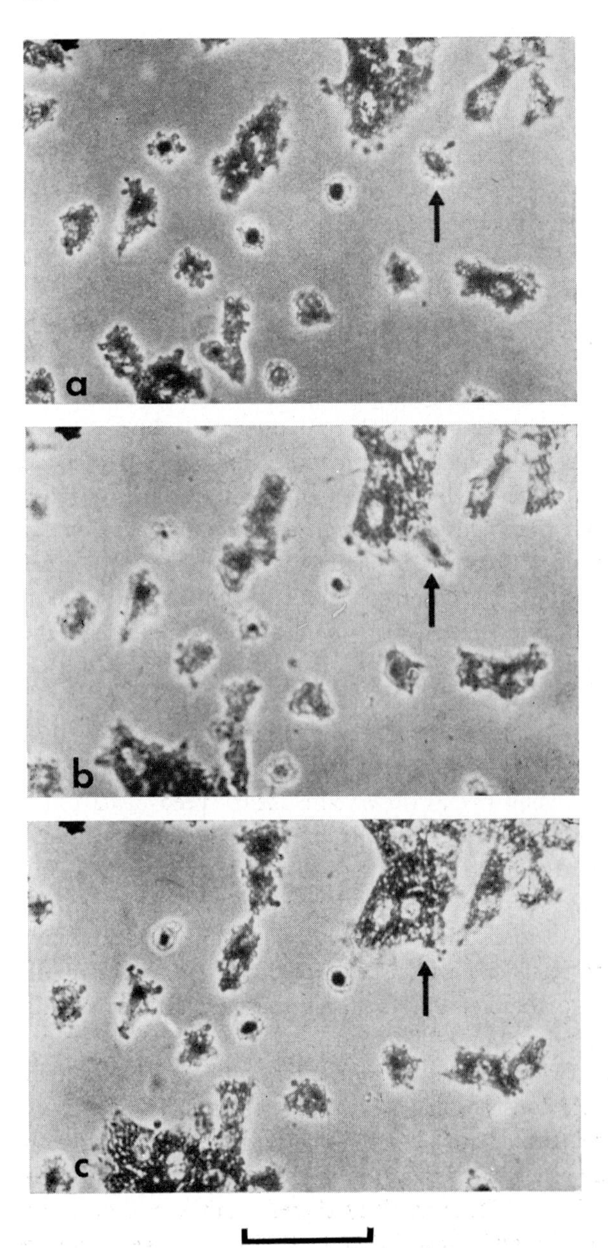

FIG. 1. Stills taken from a time lapse film to illustrate the spreading of an isolated pigmented retina epithelial cell (arrow) after contact with an island of cells: (*a*) after 4 h in culture; (*b*) after 5 h in culture; and (*c*) after 6 h in culture (bar equals 20 μm).

islands of PRE cells composed of less than about 30 cells, mean cell spread area (i.e. the average area of substrate occupied by a single cell) increases linearly with an increase in the number of cells constituting the island (Middleton, unpublished results).

MONOLAYERING AND CONTACT INHIBITION OF PRE CELLS

Quantitative analysis of confluent PRE cultures fixed after 24 h growth shows the cells to be strongly monolayered (Middleton 1972). This, together with the knowledge that the cells are capable of locomotion and that such locomotion brings about collisions between them, indicates that the cells are restricted from moving over the surfaces of others. Thus, PRE cells in common with fibroblasts exhibit contact inhibition of locomotion. The mean overlap index [observed number of nuclear overlaps expressed as a percentage of the number expected if the cells were randomly arranged (Abercrombie *et al.* 1968)] of the analysed PRE cultures was only about 2.0% (Middleton 1972), which is only about a quarter that of embryonic chick heart fibroblasts [8.4% (Abercrombie *et al.* 1968)] and a ninth that of neonatal mouse muscle fibroblasts [18.3% (Abercrombie *et al.* 1968)]. Therefore these epithelial cells show a stronger tendency to monolayer and thus more stringent contact inhibition than either of the fibroblast types.

The mechanism of contact inhibition is unknown (see Abercrombie 1970 for discussion) but two of the possible hypotheses are (*a*) the inability of the dorsal surface (the surface of the cell furthest from the substrate) of one cell to provide a suitable substrate for the locomotion of another cell (Carter 1967), and (*b*) the development of strong lateral adhesions between colliding cells such that further movement in the direction of the adhesion is impossible (Curtis 1967). Both these possibilities have been investigated in cultures of PRE cells.

The spreading of PRE cells on a monolayer of the same cell type

A substrate to which cells can adhere has long been known to be a prerequisite for cell spreading and locomotion in culture. It has therefore been suggested that monolayering could result from the dorsal surfaces of the cells being non-adhesive, or only weakly adhesive to other cells of the same type, since if this were the case one cell would be incapable of moving over another (Carter 1967). This theory can be tested by determining whether or not freshly added cells are capable of spreading and moving on a pre-existing monolayer of cells. Such an experiment is difficult to perform with fibroblasts since even in confluent cultures small areas of clear substrate remain, onto which freshly added cells may settle. However, confluent cultures of PRE cells form a gapless, strongly monolayered sheet and thus present a continuous substrate of their dorsal surfaces to added cells. When freshly dissociated PRE cells, at a concentration sufficient to form another confluent monolayer, are added to such a culture the majority of the added cells fail to spread. Cultures of this type fixed after 24 h show the added cells to have formed aggregates lying on top of the original cell sheet (Middleton, unpublished results). Thus the added cells have demonstrated 'associative behaviour' (Abercrombie 1967) in that their ability to adhere to each other is now greater than their ability to adhere to the substrate provided. Quantitative observations confirm that the added cells have failed to spread since the observed number of nuclear overlaps is two orders of magnitude less than the number expected from two superimposed layers of PRE cells (Middleton, unpublished results).

These observations strongly suggest that the dorsal surface of one PRE cell does not provide a suitable substrate for the locomotion of another cell of the same type. However, it is not clear whether or not this phenomenon underlies monolayering. Carter (1967) has proposed a theory of contact inhibition based on such a suggestion: the contact of the leading lamella of one cell with the non-adhesive, or relatively less adhesive, dorsal surface of another cell will prevent the forward extension of that leading lamella and thus bring about a cessation of movement in that direction and contact inhibition. However, Abercrombie (1970) considers that this theory is incompatible with the phenomenon of contact inhibition as it has been observed in fibroblasts. He notes that when fibroblasts collide the leading lamella of one cell frequently passes some distance ventrally (i.e. between the cell and its substrate) to the other cell before it displays the contraction and cessation of ruffling activity characteristic of contact inhibition. Since in this case the leading lamella is passing beneath the other cell no change of substrate is involved and thus the adhesive properties of the dorsal surface of the cell can be playing no part in contact inhibition.

Although contact inhibition between PRE cells has not yet been observed in sufficient detail to determine whether the processes occurring during the reaction are identical to those of fibroblasts, it seems likely that the phenomenon is fundamentally similar in both cell types. If this is so the adhesive properties of the dorsal surfaces of PRE cells might not be important so far as contact inhibition is concerned.

The stability and ultrastructure of contacts between PRE cells

It has been suggested that contact inhibition and monolayering are the consequences of the formation of lateral adhesions between colliding cells since further movement in the direction of the adhesion might be prevented (Curtis 1967). This possibility has been investigated in cultures of PRE cells.

Time lapse film observations suggested that collisions between isolated PRE cells or between islands of PRE cells led to the development of stable contacts between the cells or islands (see before). This was confirmed by detailed film analysis of 25 collisions between PRE cells. Isolated PRE cells were followed for periods of up to 16 h after a collision with another cell or island of cells, but in only one case was a cell seen to move away again after a collision. The majority of cells, once having established contact with another cell or island of cells, did not subsequently become isolated again upon the substrate (Middleton, unpublished results). It should be stressed that sub-confluent cultures were used for this analysis and thus there was always a large area of the substrate unoccupied by cells. Fibroblasts colliding under similar conditions normally exhibit contact inhibition and, after a delay of 10–20 min, move away on an unoccupied area of substrate (Abercrombie 1970). In this respect, therefore, there is a marked contrast between the behaviour of PRE cells and fibroblasts.

Whilst the formation of an adhesion appears to be an integral part of contact inhibition between fibroblasts (Abercrombie 1970) it is clearly not of sufficient strength to prevent the fibroblasts involved from moving apart again. In contrast the adhesions developed between colliding PRE cells are normally retained once formed. This suggests either that the adhesions formed between these cells are stronger than those formed between fibroblasts or that the cells can only exert a relatively weak locomotory force against the developed adhesion.

In the light of this information it was of interest to look at the ultrastructure of the contacts developed between cultured PRE cells. In collaboration with Miss S. Pegrum, I cultured PRE cells for 24 h on an Araldite substrate and prepared sections vertical to the substrate for electron microscopy by a method similar to that of Abercrombie *et al.* (1971). Our observations revealed the presence of an apically situated (i.e. close to the dorsal surface of the cells) contact specialization between every PRE cell in every section examined. These specializations consist of localized regions in which the membranes of adjacent cells are closely apposed and they are always associated with a localized accumulation of cytoplasmic fibrils (Fig. 2). Similar, but more highly differentiated contact specializations consisting of gap junctions and *zonula adhaerentes* have recently been described between the cells of three-week-old

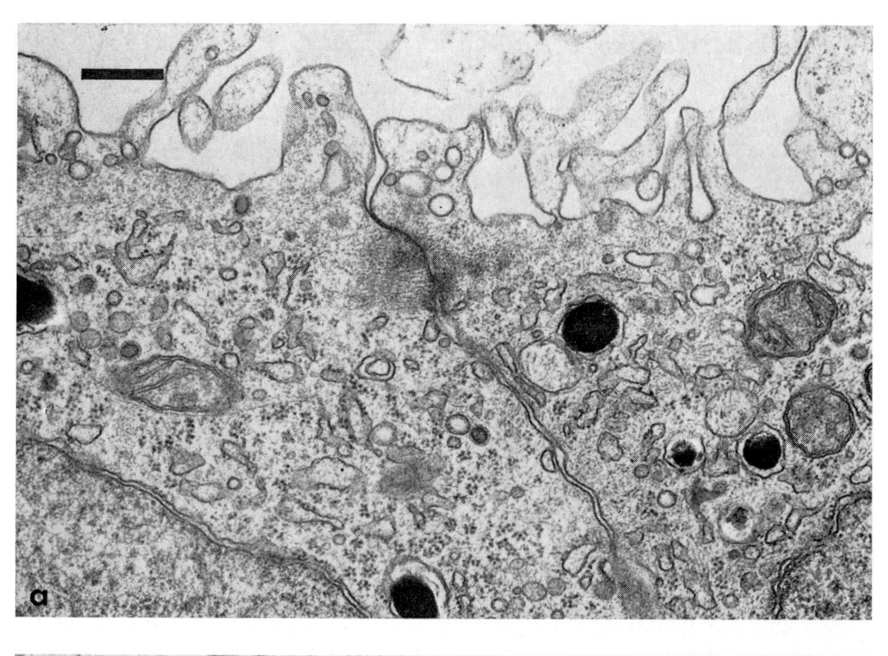

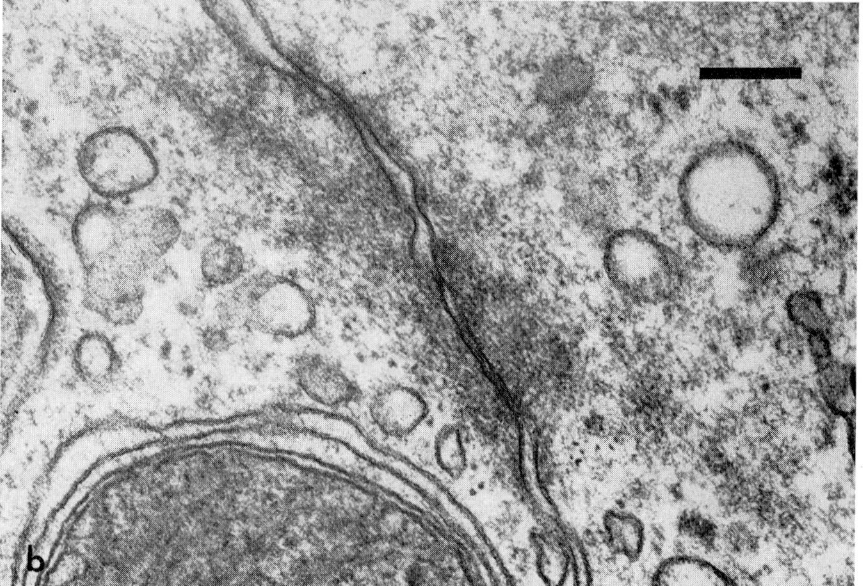

FIG. 2. (*a*) Electron micrograph showing a specialized contact between cultured pigmented retina epithelial cells. Scale bar represents 0.5 μm.
(*b*) Higher power electron micrograph of a similar contact specialization to that shown in (*a*). Scale bar represents 0.2 μm.

clones of chick embryo PRE cells (Crawford *et al.* 1972). The contacts that we have observed, however, differ from most of the described contact specializations between cells in that their intermembrane gap of 6.0–8.0 nm is larger than that found in gap junctions [about 2.0 nm (Revel & Karnovsky 1967; Brightman & Reese 1969)] and smaller than that found in desmosomes [about 24.0 nm (Farquhar & Palade 1963)]. However, contact specializations with a similar intermembrane gap (6.0–7.5 nm) have been described between the cells of the cohesive epithelium that constitutes the enveloping layer of *Fundulus* blastulae and gastrulae (Trinkaus & Lentz 1967; Lentz & Trinkaus 1971). For the sake of convenience the nomenclature of Trinkaus & Lentz (1967) will be followed and the contact specializations found between PRE cells will be referred to as apical junctions.

The fact that these apical junctions are found in every section of PRE cells suggests that they form a continuous band running completely around the cells. It seems possible, therefore, that this band could provide a localized region of strong adhesion between the cells and thus be responsible for the stable contacts formed between the cells in culture. Some support for this proposal comes from the suggestion that the apical junctions in *Fundulus* are at least partially responsible for maintaining the cell contacts in the enveloping layer despite the considerable tension placed upon them during epiboly (Lentz & Trinkaus 1971).

If apical junctions provide the basis for the stable contacts seen between PRE cells then fibroblasts which do not normally form stable contacts would not be expected to show such contact specializations. The evidence on this point has until recently been equivocal. Devis & James (1964) described the appearance of occasional tight junctions between cultured guinea-pig fibroblasts and Flaxman *et al.* (1969) obtained essentially similar results with L-strain fibroblasts. In contrast, Harris (1970) found no sign of specialized cell contacts in cultures of embryonic chick heart fibroblasts and 3T3 cells. Recently, however, Heaysman & Pegrum (1973) have followed the development of specialized contacts between the leading lamellae of colliding chick heart fibroblasts. These contacts start to develop within 20 s of the collision, become more differentiated with time and are retained until the cells separate. The contact specializations they describe appear to be similar to those found between PRE cells, in that they have an intermembrane gap of 5.0–10.0 nm and are always associated with cytoplasmic microfilaments.

Since both PRE cells and fibroblasts form specialized adhesions after contact, the presence of such specializations cannot alone explain the stability of PRE cell contacts. There must be, as yet, undetected differences in the adhesive or locomotory properties of the two cell types.

The similarity of the contacts formed between PRE cells and fibroblasts suggests that the development of a specialized contact or adhesion is an integral part of contact inhibition in both cell types. Certainly, with fibroblasts there appears to be an appropriate temporal relationship between the development of the adhesion and the behavioural changes seen in contact inhibition (Heaysman & Pegrum 1973). However, the time at which the specializations develop between PRE cells has not yet been determined.

The precise role of such a contact specialization remains obscure. There is no doubt that the development of a sufficiently strong lateral adhesion could, by effectively preventing further locomotion, lead to monolayering. Indeed Curtis (1967) developed his theory of contact inhibition around this idea. Whilst in the light of present knowledge such a theory could account for monolayering in PRE cells, Abercrombie (1970) argues that it cannot account for the contraction that is seen when fibroblasts display contact inhibition.

Alternatively, it is possible that the development of a specialized region of adhesion on the leading lamella of a cell leads to contact inhibition by interference with the locomotory mechanism of the cell. Recent evidence has suggested that the synthesis or assembly of membrane components in the leading lamella is important in the movement of fibroblasts (Abercrombie *et al.* 1970). It is thus not unlikely that the development of a specialized adhesion in this region could lead to the paralysis of the leading lamella and cessation of movement seen in contact inhibition. However, this suggestion again fails to account for the contraction.

Another possibility is that these contact specializations represent sites of intercellular communication and thus provide a pathway for the exchange of a 'signal' which triggers contact inhibition. Indeed it is known that both fibroblasts (Furshpan & Potter 1968; Borek *et al.* 1969) and epithelial cells (Borek *et al.* 1969) can show electrical coupling in culture. The contact specializations seen between PRE cells and between chick heart fibroblasts are, however, unlike the gap junctions to which electrical coupling is normally ascribed [see Bennett & Trinkaus (1970) for discussion].

Whilst the idea that the exchange of a signal through specialized contacts causes contact inhibition is attractive, it is difficult to see how such a mechanism could operate. Trinkaus *et al.* (1971) have criticized the hypothesis on the grounds that it cannot explain the local inhibition of ruffling of the leading lamella seen during fibroblast contact inhibition. Also, Abercrombie (1970) has pointed out that it is difficult to see how the 'signal' could avoid inhibiting the movement of any cell producing it, although the theoretical approach of Loewenstein (1968) could overcome this objection.

If the formation of contact specializations is fundamental to the process of contact inhibition, cells showing a lack of contact inhibition would not be expected to form such specializations. The evidence on this point is inconclusive. In general cells lacking contact inhibition seem to form specialized contacts with each other less frequently than cells that do show contact inhibition (Trinkaus & Lentz 1967; Martinez-Palomo *et al.* 1969). However, it is clear that more detailed behavioural and ultrastructural studies will be required to elucidate the nature of any correlation between the lack of specialized contacts and the lack of contact inhibition.

In conclusion, the common occurrence of specialized contacts between both fibroblasts and PRE cells colliding in culture suggests that such specializations are fundamental to the phenomenon of contact inhibition. The precise nature and role of these specializations, however, remains obscure, but it seems possible that the behavioural differences observed between fibroblasts and PRE cells are explicable in terms of quantitative or qualitative differences in the adhesions that the different cell types form.

ACKNOWLEDGEMENTS

I am most grateful to Michael Abercrombie and Joan Heaysman for their advice and encouragement. I am also indebted to Miss Melanie Holmes for her expert technical assistance and Miss Eva Crawley for her photographic expertise.

References

ABERCROMBIE, M. (1967) *Natl. Cancer Inst. Monogr.* **26**, 249-264
ABERCROMBIE, M. (1970) *In Vitro* **6**, 128-142
ABERCROMBIE, M. & HEAYSMAN, J. E. M. (1953) *Exp. Cell Res.* **5**, 111-131
ABERCROMBIE, M. & HEAYSMAN, J. E. M. (1954) *Exp. Cell Res.* **6**, 293-306
ABERCROMBIE, M. & MIDDLETON, C. A. (1968) in *Epithelial Mesenchymal Interactions* (Fleischmajer, R. & Billingham, R. E., eds.), pp. 56-63, Williams and Wilkins, Baltimore
ABERCROMBIE, M., LAMONT, D. M. & STEPHENSON, E. M. (1968) *Proc. R. Soc. Lond. B Biol. Sci.* **170**, 349-360
ABERCROMBIE, M., HEAYSMAN, J. E. M. & PEGRUM, S. M. (1970) *Exp. Cell Res.* **62**, 389-398
ABERCROMBIE, M., HEAYSMAN, J. E. M. & PEGRUM, S. M. (1971) *Exp. Cell Res.* **67**, 359-367
BENNETT, M. V. L. & TRINKAUS, J. P. (1970) *J. Cell Biol.* **44**, 592-610
BEREITER-HAHN, J. (1967) *Z. Zellforsch. Mikrosk. Anat.* **79**, 118-156
BOREK, C., HIGASHINO, S. & LOEWENSTEIN, W. R. (1969) *J. Membrane Biol.* **1**, 274-293
BRIGHTMAN, M. W. & REESE, T. S. (1969) *J. Cell Biol.* **40**, 648-677
CARTER, S. B. (1967) *Nature (Lond.)* **213**, 256-260
CRAWFORD, B., CLONEY, R. A. & CAHN, R. D. (1972) *Z. Zellforsch. Mikrosk. Anat.* **130**, 135-151
CURTIS, A. S. G. (1967) *The Cell Surface: its Molecular Role in Morphogenesis*, Academic Press, London

DEVIS, R. & JAMES, D. W. (1964) *J. Anat.* **98**, 63-68
FARQUHAR, M. G. & PALADE, G. E. (1963) *J. Cell Biol.* **17**, 375-412
FLAXMAN, B. A., REVEL, J. P. & HAY, E. D. (1969) *Exp. Cell Res.* **58**, 438-443
HEAYSMAN, J. E. M. & PEGRUM, S. (1973) *Exp. Cell Res.* **78**, 71-78
FURSHPAN, E. J. & POTTER, E. D. (1968) *Curr. Top. Dev. Biol.* **3**, 95-127
HARRIS, H. (1970) *Am. Zool.* **10**, 324 (abstr.)
HARRISON, R. G. (1910) *J. Exp. Zool.* **9**, 787-846
HOLMES, S. J. (1913) *Univ. Calif. Publ. Zool.* **11**, 155-172
LENTZ, T. L. & TRINKAUS, J. P. (1971) *J. Cell Biol.* **48**, 455-472
LOEWENSTEIN, W. R. (1968) *Dev. Biol.* **2** (Suppl.), 151-183
MARTINEZ-PALOMO, A., BRAISLOVSKY, C. & BERNHARD, W. (1969) *Cancer Res.* **29**, 925-937
MIDDLETON, C. A. (1972) *Exp. Cell Res.* **70**, 91-96
REVEL, J. P. & KARNOVSKY, M. J. (1967) *J. Cell Biol.* **33**, C7-C12
TRINKAUS, J. P. (1963) *Dev. Biol.* **7**, 513-532
TRINKAUS, J. P. & LENTZ, T. L. (1967) *J. Cell Biol.* **32**, 139-153
TRINKAUS, J. P., BETCHAKU, T. & KRULIKOWSKI, L. S. (1971) *Exp. Cell Res.* **64**, 291-300
VAUGHAN, R. B. & TRINKAUS, J. P. (1966) *J. Cell Sci.* **1**, 407-413
WEISS, P. (1961) *Exp. Cell Res.* **8** (Suppl.), 260-281

Discussion

Trinkaus: DiPasquale (1972) has observed, as you have, this same rapid blebbing all over the cell on the part of isolated corneal epithelial and epidermal cells from the chick embryo. So, it looks as if generalized blebbing of isolated cells is especially characteristic of epithelial cells. He has also observed that these cells possess few if any microtubules. Perhaps isolated cells lacking microtubules bleb more readily.

Porter: We have been studying the surfaces of CHO cells (Porter *et al.* 1973), which we have synchronized. The blebs are retained through the G_1 phase but not into the S phase. We thought that this was a surface feature we could rely on to indicate the phases. We discovered, however, that if the cells were plated out very sparsely at the outset of culturing, they retained their blebs all the way through the cycle. They do not flatten out till the next generation when the population density forces them into contact.

Middleton: A low percentage of isolated pigmented retina epithelial (PRE) cells does occasionally spread fully upon the substrate (Middleton, unpublished results). It occurs to me that in these cases the cells might have become polarized by minor irregularities in the substrate rather in the same way that fibroblasts show contact guidance (Weiss 1961).

Trinkaus: DiPasquale & Bell (1972) have discovered that the upper surface of an epithelial sheet does not support spreading or active migration. They filmed individual fibroblastic cells which they had seeded onto the surface of a sheet and onto the plastic substratum beside the sheet. Within one hour, the cells on plastic had flattened and begun active movement, whereas those on the sheet remained round and did not spread, without exception. If the dish was moved, rounded cells on the upper surface of the epithelial sheets moved freely with the flow of medium, indicating an apparent lack of adhesion.

Middleton: I found that the aggregates of added cells adhered firmly enough to the underlying epithelial sheet to withstand a fairly vigorous washing and a melanin bleaching technique.

Trinkaus: P. B. Bell (unpublished results) seeded retinal pigment cells onto a sheet of heart cells in order to find out how the former moved in a sorting out mechanism. The results were disappointing, because the retinal pigment cells behaved in exactly the same way on the fibroblast sheet as the retinal pigment cells did on a retinal pigment cell sheet. This suggests that the adhesiveness of the upper surface is too low to allow spreading.

Middleton: I have done the reverse experiment, seeding fibroblasts prepared from embryonic chick heart cultures onto a confluent sheet of PRE cells. The result was somewhat unexpected: after 24 h the fibroblasts had in some way penetrated the epithelial sheet and come to lie, apparently normally spread, on the substrate beneath the sheet (Middleton, unpublished results). Presumably the added fibroblasts must have adhered initially to the upper surfaces of the PRE cells, for it is difficult to see how they could penetrate the sheet without first attaching to it.

Curtis: With a layer of epithelial cells on an epithelial sheet, how do you know that some of the epithelial cells from the top layer do not insert themselves in the epithelial sheet underneath?

Middleton: I think it is very likely that some of the added cells do force their way into the existing sheet.

Wessells: If one dissociates sensory ganglia and plates out the cells, after 24 h many of the single nerve cell bodies are on top of glial cells, and a much lower percentage is directly on the substratum. Ludueña (unpublished results) mixed a suspension of heart fibroblasts prelabelled with tritiated thymidine with a suspension of dissociated unlabelled ganglion cells. After 25 h, just as many nerves with axons are on top of heart cells as on glial cells. Thus, any 'trophic' effect of foreign cells on nerve cells is apparently non-specific.

Trinkaus: Do the nerve cells spread out?

Wessells: No, the nerve cell body never spreads in response to different sorts of living cells as substratum. But the nerve cell body will sometimes spread if serum concentrations are dropped to 0.5% or less.

Porter: What about the behaviour of epithelial cells when they divide? They seem to leave the sheet to go through mitosis and then re-insert themselves into the sheet.

Middleton: I'm sure that you are right, but in these cases the cells are re-spreading on their original collagen or glass substrate. I was trying to determine whether the cells could spread upon a substrate of other cells.

Porter: Have you mixed pigmented with non-pigmented cells?

Middleton: Not with non-pigmented epithelial cells, but I have mixed PRE cells with non-pigmented fibroblasts. If a mixed suspension of chick PRE cells and chick heart fibroblasts is seeded onto a collagen substrate the cells initially settle down with a fairly random distribution. However, observations after 24 h show that a two-dimensional 'sorting out' process has taken place. The PRE cells form typical epithelial islands surrounded by a network of fibroblasts. Fibroblasts are never incorporated into the epithelial islands and very few isolated PRE cells are found within the fibroblast network (Middleton, unpublished results).

Trinkaus: We also observed this in mixed aggregates of retinal pigment cells and heart cells (Trinkaus & Lentz 1964).

Steinberg: Our observations (D. Garrod & Steinberg, unpublished findings, 1970) on the two-dimensional aggregation and sorting out of liver cell populations, both alone and in combination with limb bud cells, are very similar to yours. The liver parenchymal cells sort out from the liver fibroblastic cells and from the limb bud cells in two dimensions on the substratum. As you described, isolated cells do not move but stay rounded up and bleb. Since they only jig about on one spot, we cannot measure a speed of translocation. But when they encounter another group of cells through these oscillations, they quickly adhere, flatten and begin to ruffle. Curiously, islands of liver cells can translocate, whereas the individual liver cells do not.

Abercrombie: When one seeds cells on top of another layer of cells, should one not distinguish between simple failure to spread and failure to spread plus 'clumping'? Contact inhibition could stop the spreading of each cell. The cell is after all in contact with another cell all the way round its periphery, and this is supposed to stop its movement. It seems to me, however, that Professor Curtis' evidence is more suggestive of a non-adhesive surface, because the cells not only do not spread but pull together into clumps.

Trinkaus: The retinal pigment cells in Bell's experiment cohere tightly when they come into contact on the fibroblast substratum.

Steinberg: With Dr L. Wiseman, we have studied (unpublished results, 1971) the movements of chick embryonic cells labelled with tritiated thymidine after attachment to the surfaces of already formed spherical aggregates, to see whether these cells would 'diffuse' or be immobilized on the surfaces.

Trinkaus: Is this similar to the Roth–Weston experiment (Roth & Weston 1967; Roth 1968)?

Steinberg: It is similar only in the sense that individual cells are being put onto aggregates. We put on either cells of the same kind, or cells that, if present in large numbers, would sort out either externally or internally to the cells of the aggregate. We have used both reaggregates and intact fragments which have never been dissociated but have been allowed to round up and become spherical. In all cases, the results were identical. Regardless of cell type and state of dissociation, the cells that started on the outside 'diffuse' and move about, some of them leaving the surface and moving inside. Within 24 h some of them can be found in the centre of the aggregates. Electron micrographs of comparable aggregates, which have had no cells placed on them, show no broad gaps between the cells; there is no gaping apart of the membranes. This is not readily explained in terms of channels for the cells to creep into. It seems, rather, that the labelled cells must be squeezing into areas where all the cells were originally in adhesion. So, the conclusion that cells surrounded in three dimensions by other cells are incapable of moving seems to me to be thrown into serious question.

Harris: Roth & Weston (1967) found that more cells of a given tissue type adhered to aggregates of that type than to aggregates of another tissue type. Did you find such differences?

Steinberg: Unexpectedly, we did not.

Wolpert: Are you saying that a cell will pass through another cell mass in which the cell gaps are about 20 nm?

Steinberg: It certainly seems that way, although we were not able to follow the path of any given cell. Sections through these liver aggregates do not reveal any cavities. So I see no other explanation, except that the cells burrow their way in.

Middleton: A similar process must be taking place in the experiments I described in which fibroblasts penetrate a confluent sheet of PRE cells to reach the underlying substrate (see p. 263): they must be able to break down the apical junctions between the epithelial cells in order to burrow through the sheet. The only alternative explanations are either that the fibroblasts create gaps in the sheet by lysing individual epithelial cells or that the fibroblasts are able actually to pass through intact epithelial cells, rather as lymphocytes move out of blood vessels.

Trinkaus: Or else this is the first proof that a cell can shrink to 20 nm in width.

Wolpert: This is important, because I have always assumed that the principal component of the motile force of cells is pulling (Gustafson & Wolpert 1967). This phenomenon is based on pushing.

Abercrombie: Earlier evidence did not show this. When Weston and I (1967) put together two lots of disaggregated cells, we got no intermixing.

Steinberg: The only explanation I can see at present is that a large group of cells behaves differently from a single cell of the same type, judging from their locomotory behaviour on the plastic (cf. preceding paper and discussion). If we fuse two aggregates of the same kind, one of which is labelled, they will round up to form the hemispheres of a spheroid, as described by Weston & Abercrombie (1967). If we section such a spheroid, we find only a little intermixing at the boundary (Wiseman 1970).

Curtis: I suspect it is more complicated than that. Some time ago I separated amphibian cells and deliberately constructed, by layers, an aggregate which was already sorted out artificially (unpublished results). Thereupon the cells intermixed randomly and then re-sorted. Dr Middleton, are the cells you add, which presumably have been freshly separated, carrying over properties which make them behave abnormally?

Middleton: The added PRE cells have been freshly dissociated with papain (Steinberg 1962; Middleton 1972) before being seeded onto the existing sheet of PRE cells. To this extent they are certainly abnormal in comparison with their *in vivo* state. However, the fact remains that the added cells are unable to spread upon the upper surface of a cultured PRE sheet but can, in control cultures, spread perfectly normally on a collagen, glass or plastic substrate.

Curtis: Might this not explain the surprising intermingling, even of unlike types, which Professor Steinberg describes?

Steinberg: It might. We have found that recently trypsinized cells have quite different assembly properties from cells which have recovered from the effects of trypsinization (Wiseman 1970; Wiseman *et al.* 1972). Aggregates of more recently trypsinized chick embryonic heart or liver cells enveloped other aggregates that were identical except for having been trypsinized earlier, just as though they were composed of cells from two different tissues. Recovery seems to take as long as two or three days. So freshly dissociated cells can show quite altered behaviour.

de Petris: A physiologically important case where cells can penetrate inside other cells is that in which lymphocytes cross the endothelial capillary membrane. Electron microscopy shows a gap of 20 nm surrounding the invading lymphocytes. Also, in macrophages, the lymphocyte can move inside the cell; no gap, apart from the usual one of about 20 nm, is seen.

Steinberg: Do the lymphocytes actually penetrate the cytoplasm?

Allison: It seems improbable that they actually penetrate the cytoplasm, but they could be within membrane-bounded cisternae.

de Petris: The cytoplasm of the two cells remains of course always separated by the two plasma membranes.

Curtis: The theory that they do go through another cell has been challenged recently by Schoefl (1972).

Abercrombie: Even so, I have seen films of lymphocytes unmistakably inside another cell, and moving actively. Their movement is hard to understand.

Trinkaus: When the retinal pigment cells divide in a sheet and round up, do they pull entirely away from their neighbours?

Middleton: Time lapse film observations suggest that a PRE cell in mitosis pulls away from and breaks its contacts with neighbouring cells. However, as it rounds up it appears to leave retraction fibres that anchor it to the substrate (Middleton, unpublished results).

Trinkaus: I am extremely puzzled by this. Occasionally in time lapse films of the enveloping layer of *Fundulus* we see dividing cells round up, pull away from their neighbours and leave retraction fibres, yet we know from electro-physiological studies that this is a highly impermeable layer (Bennett & Trinkaus 1970). How can physiological continuity persist in the face of these apparent anatomical gaps?

Middleton: Until we look at this with the electron microscope I don't think we can be certain whether a mitosing cell breaks all its contacts with its neighbours. Possibly the retraction fibres that we see terminate in tiny point contacts on other cells.

Huxley: Is the total area of material involved in the specialized contact points of any one cell in a sheet of cells constant? Suppose a certain limited amount of material was available for making contact; then, if the cell used all this in establishing contacts with the substrate and neighbouring cells, there would be none left for sustaining additional adhesions necessary for the continuation of movement. If it could be verified, this might explain the inhibition of movement by contact with surrounding cells, without any elaborate mechanisms for switching off the formation of contacts itself.

Abercrombie: That is an attractive theory. One does not know which way up these cells have landed in culture; presumably they have landed on the substrate in all orientations.

Heaysman: For contact adhesions between cells and their underlying substrate we need only concern ourselves with the cells that are settling down onto the layer below. They should not have used up all their adhesive material and should be able to attach to the underlying sheet of cells as they would to any normal substrate.

Huxley: Maybe the cell normally needs similar attachment sites on the other cells in order to attach and grow, that is, those sites which have been used up in cell–cell and cell–substrate contacts.

Heaysman: The substrate does not need to form a reciprocal attachment.

Curtis: On the question of using up adhesions, I would like to observe that in some work of Dr J. G. Edwards (unpublished) on heavily trypsinized BHK cells, the size of the aggregates formed was extremely small. They limit themselves, even though adhesiveness seems to be great: this implies that they have few adhesive sites per cell. Two possible interpretations are: either only one cell needs an adhesive site to form an adhesion, or both cells must possess one adhesive site each to form a contact. The evidence suggests that only one adhesive site on one cell is necessary for the formation of one adhesion in this situation. The less heavily trypsinized cells do not behave in this way, and appear to have many sites.

Abercrombie: Dr Middleton, did you do any ultrastructural studies on the difference between the blebbing cells and the cells which have become polarized by attachment to an island?

Middleton: We have not looked at this in any detail. Well-spread PRE cells at the edge of an island show cell–substrate plaques often associated with oblique, dorsally directed fibre tracts similar to those you described in fibroblasts (Abercrombie *et al.* 1971). We have not as yet seen any such structures in isolated blebbing PRE cells (Middleton & S. Pegrum, unpublished results).

Harris: If you explanted the neural retina so that the apical surface (the side with the specialized contacts) was downward, would these contacts move to the upper side, or would new ones form there, or what?

Middleton: I can't answer that directly. However, the explants of PRE tissue that I use are composed of much folded entangled sheets of the tissue much of which must have lost its normal orientation. Despite this, ultrastructural observations of the epithelial sheets that migrate out from such an explant show that the cells are correctly orientated, having apically situated contact specializations (Middleton & S. Pegrum, unpublished results). It seems likely therefore that as the initially disoriented cells start to migrate they either reorientate themselves or their contacts appropriately with respect to the substrate.

Allison: Surely it is not just a matter of the amount of contact which is made, but that these contacts are discriminating. The epithelial cells, for example, that approach a fibroblast obviously do not make contact with it. They prefer company of their own kind, as do nerve cells. As Dr Wessells pointed out (pp. 72-76), nerve cells elongate perfectly well on other cell types, so that they are not contact inhibited by other cell types although they are contact inhibited by other nerve cells.

Trinkaus: DiPasquale (1972) has also studied this and has found that fibroblasts do make adhesive contacts with the edge of an epithelial sheet, but these adhesions are less durable than epithelial–epithelial and fibroblast–fibroblast adhesions.

Wessells: Sengel (1958) took primitive chick ectoderm when it was two cell layers thick, inverted it and noticed a complete switch in morphology of the two cell types. This is consistent with the idea of rearrangement in an epithelial population.

Steinberg: The foregoing comments stress how the term 'contact inhibition' has been used by many people, in a variety of ways that by no means imply the action of a common mechanism. I do not mean just 'contact inhibition of growth' as compared with 'contact inhibition of movement', but contact inhibition in relation to various aspects of cell movement itself. The term has been used to refer to the behaviour of small regions of cell membrane when one cell touches another; to the behaviour of a whole cell when it collides with another cell, in terms of the positional changes of that whole cell; and even to behaviour at the level of whole populations of cells (e.g. monolayering, parallel alignment). The underlying causes of the behaviour in each of these cases are different, so that the term is changing meaning, sometimes subtly, when used in these different contexts.

Middleton: Certainly I would agree that monolayering and contact inhibition are not necessarily synonymous. The former may be induced by processes other than contact inhibition, although it is commonly used as a measure of contact inhibition.

References

ABERCROMBIE, M., HEAYSMAN, J. E. M. & PEGRUM, S. M. (1971) The locomotion of fibroblasts in culture. IV. Electron microscopy of the leading lamella. *Exp. Cell Res.* **67**, 359-367

BENNETT, M. V. L. & TRINKAUS, J. P. (1970) Electrical coupling between embryonic cells by way of extracellular space and specialized junctions. *J. Cell Biol.* **44**, 592-610

DIPASQUALE, A. (1972) *An Analysis of the Contact Relations and Locomotion of Epithelial Cells*, Ph. D. Dissertation, Yale University, New Haven, Connecticut, USA

DIPASQUALE, A. & BELL, P. B., JR. (1972) The cell surface and contact inhibition of movement. *J. Cell Biol.* **55**, 60A

GUSTAFSON, T. & WOLPERT, L. (1967) Cellular movement and contact in sea urchin morphogenesis. *Biol. Rev. (Camb.)* **42**, 442-498

MIDDLETON, C. A. (1972) Contact inhibition of locomotion in cultures of pigmented retina epithelium. *Exp. Cell Res.* **70**, 91-96

PORTER, K. R., PRESCOTT, D. & FRYE, J. (1973) Changes in surface morphology of CHO cells during the cell cycle. *J. Cell Biol.*, in press

ROTH, S. A. (1968) Studies on intercellular adhesive selectivity. *Dev. Biol.* **18**, 602-631

ROTH, S. A. & WESTON, J. A. (1967) The measurement of intercellular adhesion. *Proc. Natl. Acad. Sci. U.S.A.* **58**, 974-980

SCHOEFL, G. I. (1972) The migration of lymphocytes across the vascular endothelium in lymphoid tissue. *J. Exp. Med.* **136**, 568-588

SENGEL, P. (1958) La différenciation de la peau et des germes plumaires de l'embryon de poulet en culture *in vitro*. *Année Biol.* **34**, 29-52

STEINBERG, M. S. (1962) The role of temperature in the control of aggregation of dissociated embryonic cells. *Exp. Cell Res.* **28**, 1-10

TRINKAUS, J. P. & LENTZ, J. P. (1964) Direct observation of type-specific segregation in mixed cell aggregates. *Dev. Biol.* **9**, 115-136

WEISS, P. (1961) Guiding principles in cell locomotion and cell aggregation. *Exp. Cell Res.* **8**, (Suppl) 260-281

WESTON, J. A. & ABERCROMBIE, M. (1967) Cell motility in fused homo- and heteronomic tissue fragments. *J. Exp. Zool.* **164**, 317-324

WISEMAN, L. L. (1970) *Experimental Modulation of the Assembly Properties of Embryonic Cell Populations*, Ph. D. thesis, Princeton University

WISEMAN, L. L., STEINBERG, M. S. & PHILLIPS, H. M. (1972) Experimental modulation of intercellular cohesiveness: reversal of tissue assembly patterns. *Dev. Biol.* **28**, 498-517

Effects of drugs on morphogenetic movements in the sea urchin

T. GUSTAFSON

Wenner-Grens Institute, University of Stockholm

Abstract During a study of the morphogenesis of the muscular system of sea urchin larvae it was observed that the contractile strands around the upper part of the intestine and also projecting down from the coelom are formed by coelomic pseudopods, that is by elements similar to the contractile elements responsible for much of the morphogenetic activity during earlier stages of embryonic development such as gastrulation. This observation suggested to us that investigations of the control mechanisms of muscular activity would clarify their possible role in the initiation of morphogenetic movements in this system. Extensive investigations of the larval muscular physiology indicated that monoamines, in particular serotonin and acetylcholine, participate in the control of swallowing, in the pumping movements of the coelom and in movements of the lower part of the intestine. Accordingly, many agents known for their affinities to various receptors, either as agonists or antagonists, or for their ability to interfere with the synthesis, storage or breakdown of biogenic amines were studied with particular emphasis on the effects on gastrulation. Antiserotonins and lipid-soluble anticholinergic agents had pronounced inhibitory effects, and some phase specificity could be noticed in the sense that antiserotonins were particularly active against early gastrulation whereas the anticholinergic agents were more effective against the second phase of gastrulation in which pseudopods are dominant. In several cases, the effects of antiserotonins could be counteracted by an excess of exogenous serotonin. Agents supposed to interfere with the biosynthesis of serotonin are strongly inhibitory, but the effect of a presumed inhibitor of tryptophan hydroxylase could be counteracted by an exogenous supply of 5-hydroxytryptophan. It is suggested that serotonin and acetylcholine initiate different types of morphogenetic cell movements in the sea urchin embryo.

My collaborators and I have spent about ten years studying the morphogenesis of the sea urchin larva by time lapse cinematography (see Gustafson & Wolpert 1967). We can summarize the main events in the development of the sea urchin larva from the late blastula to an early pluteus stage in Fig. 1. Much of the shaping of the embryo is due to morphogenetic movements, in some cases

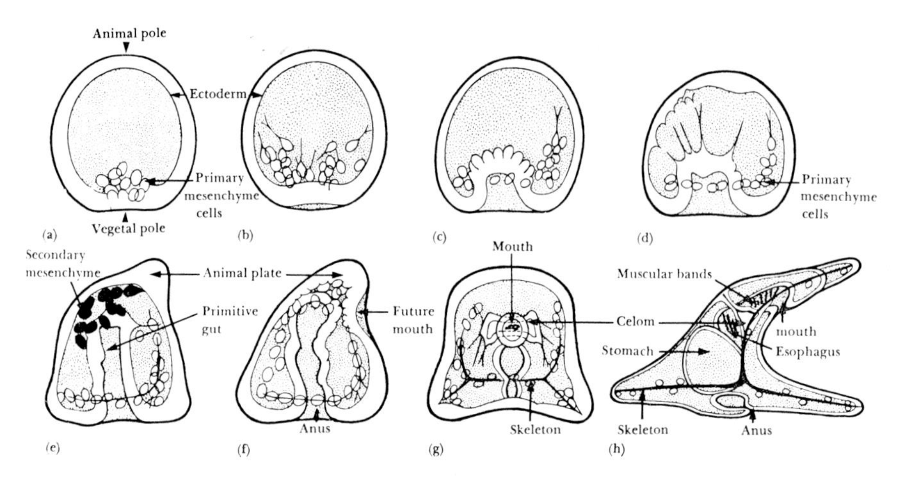

FIG. 1. Some stages in early development of the sea urchin larva *(Psammechinus miliaris)*; (a), (b): the primary mesenchyme cells are released from the blastula wall at the lower (vegetal) pole and enter the blastula cavity. The infolding of the lower part of the wall (the formation of the rudiment of the primitive gut) has two steps; the end of the first step is illustrated in (c) and the beginning of the second step in (d). An early pluteus stage is shown from the mouth side (g) and in lateral view (h).

movements of individual cells but more often movements of cells still attached to each other and still forming a sheet only one cell layer thick. Before discussing these movements let us a consider the blastula wall of the sea urchin. The development of the different zones of the blastula wall is depicted diagrammatically in Fig. 2. In the wall, the cells are mutually attached to each other at two levels and in addition attached to the surrounding hyaline membrane by microvilli (Fig. 3a). If the cells try to round up, thereby breaking some of their mutual contacts, the cell sheet tends to curve inwards (Fig. 3b). An invagination would occur unless the cells also break their contacts with the hyaline membrane, in which case the cells would be released into the blastula cavity. This model appears to have a practical relevance several times during the morphogenesis of the lower zones of the blastula (Fig. 1) although the exact details of the contractile events leading to the so-called 'rounding up' remain to be elucidated. The detachment of the cells in zone 1, the future primary mesenchyme (cf. Fig. 2), is one example, the invagination of the succeeding zones (zones 2–6) forming the archenteron rudiment another and the delineation of the coelom rudiment (zone 3) from the rest of the archenteron is an third example. After the early events in the lower part of the larva, a new type of cellular motility appears in each of the zones 1–3 manifested by a more or less pronounced pulsatory activity of the cells, which might be one aspect of their attempts to round up or

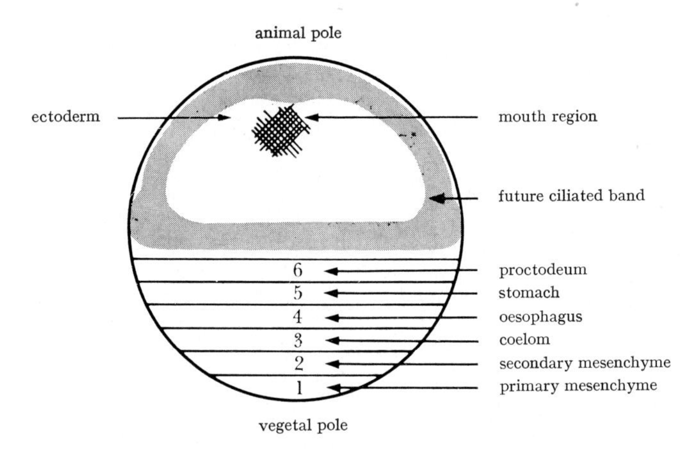

FIG. 2. The diagram shows the development of the different parts of the wall of the sea urchin blastula. The relative proportions of the zones 1–6 are enlarged for convenience. The shaded zone indicates the approximate site of the future ciliated band, including the animal plate.

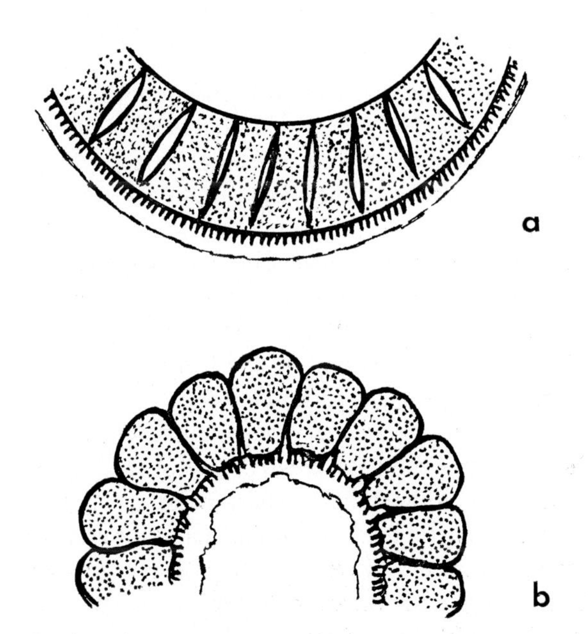

FIG. 3. Diagrammatic representation of the mechanism of early infolding of the rudiment of the primitive gut: (a) the cells are all attached to a supporting membrane (the hyaline layer or membrane); (b) if the cells tend to break their contacts and round up, the cell sheet, fixed at its ends, curves inwards.

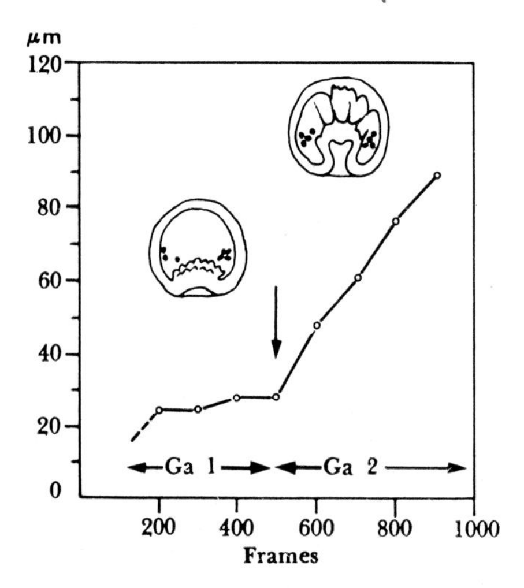

FIG. 4. The progress of formation of the primitive gut. The ordinate shows the extent of the infolding and elongation of the rudiment and the abscissa is the number of frames passed (time lapse film, frame intervals 20 s). The arrow indicates the time of the first appearance of pseudopods at the tip of the rudiment.

even an important step in this process. The cells form thin pseudopods which explore the blastula cavity and the inner wall of the larva, attach to the wall and shorten. After the subsequent migration of the primary mesenchyme, either the archenteron rudiment extends into a tube or the coelom rudiment extends into two coelomic sacs (cf. Fig. 1).

Morphogenesis of the zones 1–3 is thus a two-step process (see Fig. 4). The first step depends on some process associated with pulsatory movements of the cells and a tendency to round up, whereas contraction of attached pseudopods is involved in the second step. A further generalization is that the movements concerned conform to a simple time–space pattern (cf. Fig. 2). Motility is first noticed in zone 1, becomes more apparent in zone 2 until it becomes intense in zone 3, as if some factor were released or formed in each of these three zones. We have focused our attention on the nature of this 'factor'.

From the beginning of our studies we were struck by the fact that the muscular strands associated with the oesophagus are derivatives of the former pseudopods of the coelom (Fig. 5). We therefore assumed that we could begin to understand the initiation of the morphogenetic motility by studying the control of muscular contractility. Eight years ago we therefore started a study of the pharmacology of the muscular activity [see Gustafson *et al.* (1972*a*) for details]. In brief, we trapped a pluteus larva in a mesh of a nylon net inserted between the slide and

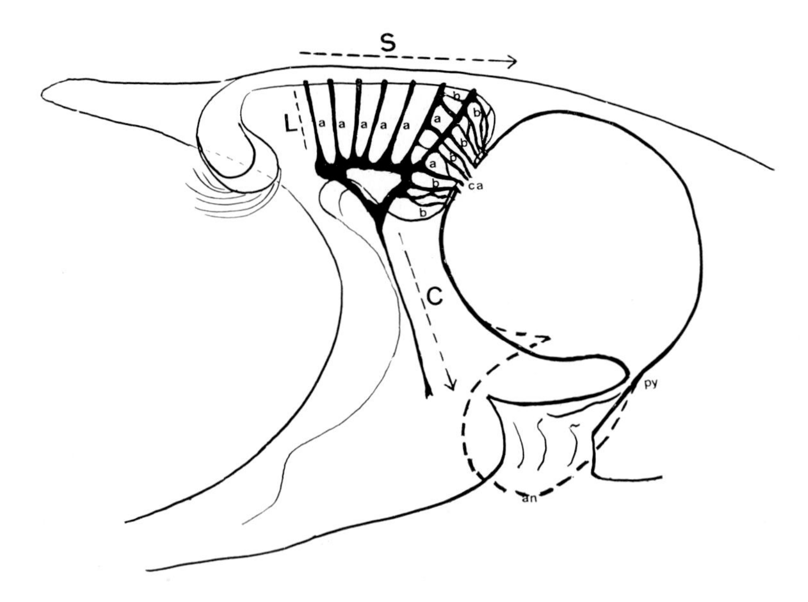

FIG. 5. The muscular system of the coelom–oesophagus complex, formed by pseudopodal elements of the coelom. Of the contractile activities swallowing (S) is due to concerted, consecutive contractions of the elements marked a and b, the C-movements (C) correspond to contractions of the muscular elements directed downwards, and the L-movements correspond to contractions of one or a few a-elements: contour of the proctodeum during filling is shown by a dashed line; ca, cardium; py, pylorus; an, anus. [From Gustafson *et al.* (1972*a*).]

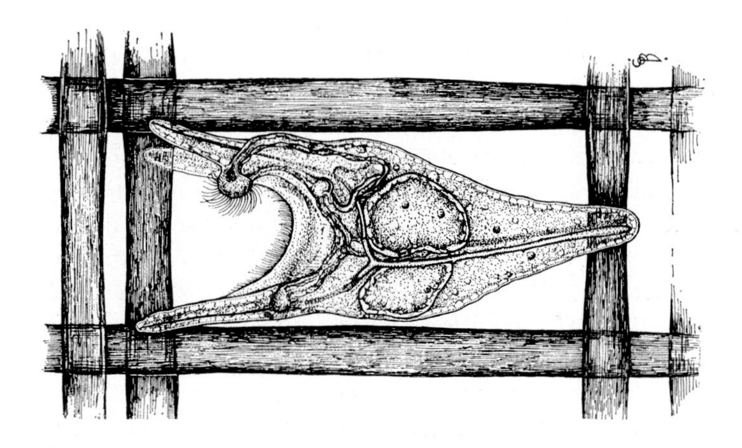

FIG. 6. An echinopluteus trapped within a mesh of a nylon net inserted between the slide and the coverslip (Gustafson & Kinnander 1956). [From Gustafson *et al.* (1972*a*).]

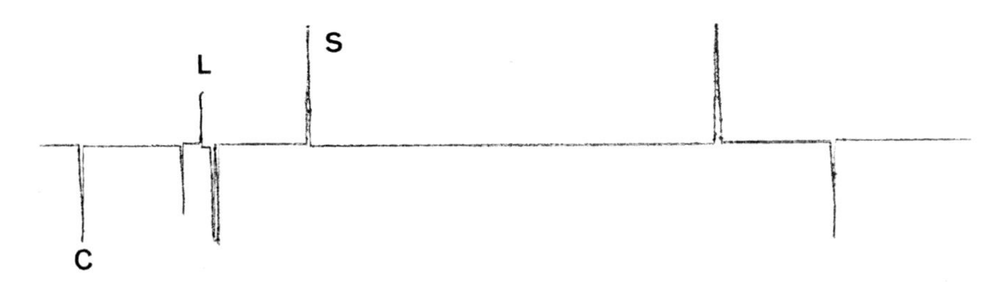

FIG. 7. Muscular (C, L and S) movements indicated on the moving paper strip (see Gustafson *et al.* 1972*a*). This example represents the actvity of an undisturbed larva in pure sea water and the time of observation is 7 min. [From Gustafson *et al.* (1972*a*).]

the coverslip (Fig. 6). The muscular contractions in response to solution of various chemical agents were recorded manually on a moving paper strip (Gustafson *et al.* 1972*a*). The activity of an undisturbed larva is illustrated in Fig. 7. We found that treatment with monoamine oxidase inhibitors eventually resulted in an intense activity. Some catecholamines or indoleamines might therefore be involved. After many experiments we found that 5-hydroxytryptamine (serotonin) has a strong stimulatory effect on muscular activities (Fig. 8) which is unsurpassed by any other indoleamines or catecholamines studied so far in this system. The specificity of the response to serotonin is corroborated by experiments with serotonin antagonists, while further support for the presumed role of serotonin is provided by the dramatic increase in the activity of the muscular system elicited by precursors of serotonin (i.e. tryptophan and 5-hydroxytryptophan). Accordingly, we may conclude with some confidence that serotonin to some extent governs muscular contractility in the pluteus larva. We cannot rule out the control by the catecholamines, however. In fact, tyrosine (the precursor of catecholamines) has a tryptophan-like effect although this only slowly develops and is comparatively moderate.

After this excursion into the pharmacology of the pluteus larva we found it worthwhile to investigate the action of serotonin in the morphogenetic activity, and treated larvae with various agents. We always started treatment in the late blastula stage and always before hatching [since hatching can be assumed to be correlated with decreased permeation of a series of agents (Gustafson & Toneby 1970)]. More than 300 agents were tested. It was evident from our results that many agents known as serotonin antagonists retarded gastrulation, and also that the effects of these agents were focused mainly on early gastrulation. The specificity of the antiserotonin effects was supported in a number of cases (cf. Fig. 9) where the effect of the antiserotonins could be considerably attenuated by the presence of serotonin in the medium. The function of

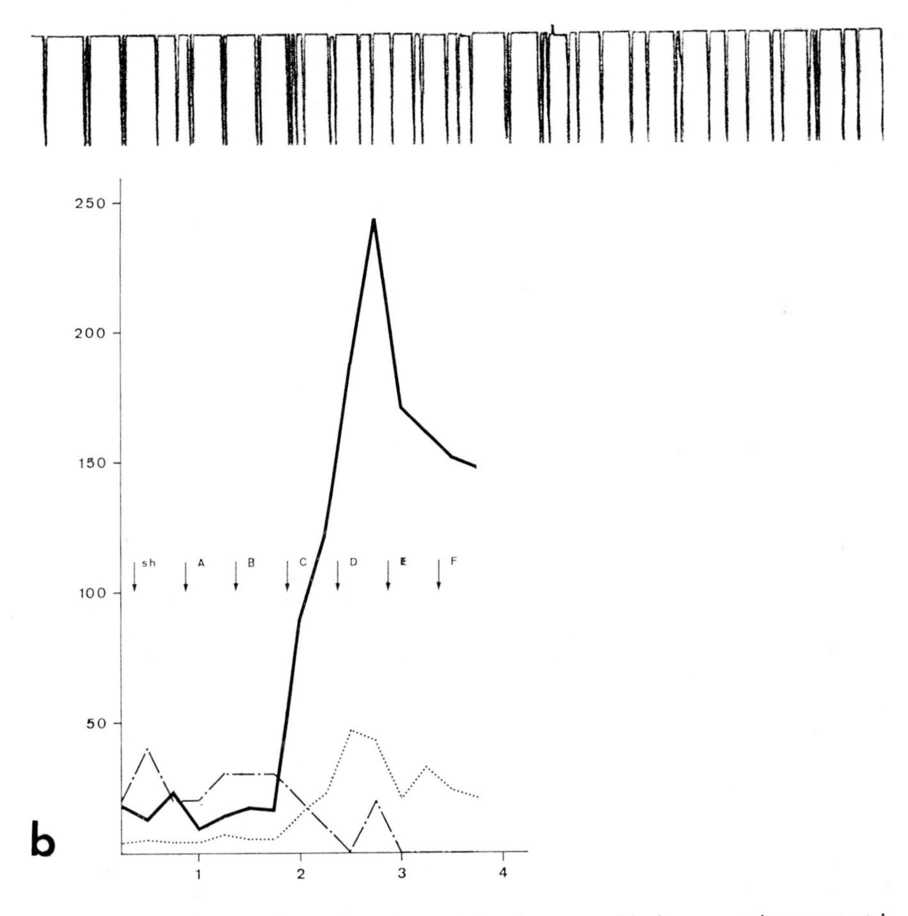

FIG. 8. Effects of serotonin on muscular activity of a sea urchin larva: moving paper strip records of the effect of 50μM-serotonin after 3 h (owing to the enormous activity the speed of the moving paper strip had to be increased threefold); (b) effect of stepwise ($\times$ 10) increase in concentration from 8.5nM (A) to 0.85mM (F); sh denotes changes from sea water to sea water, i.e. a 'sham operation'; abscissa shows time in hours and ordinate shows frequency of movements (i.e. number of movements/15 min). Solid line represents C-movements (cf. Figs. 5 and 7); — · — · — · — S-movements ($\times$10); . . . . L-movements. [From Gustafson *et al.* (1972*a*).]

serotonin is backed up by the observation that agents interfering with the formation of serotonin have a marked effect on gastrulation. For instance, diethyl dithiocarbamate, an agent known to block certain hydroxylations of aromatic compounds, strongly retarded gastrulation. This retardation could be

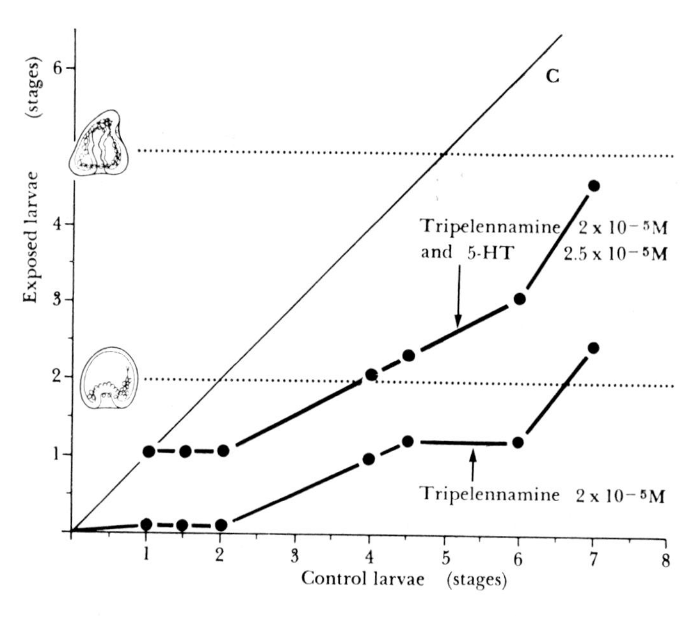

FIG. 9. Inhibitory effects of an antiserotonin (antihistamine), and its alleviation by serotonin (5-hydroxytryptamine, 5-HT). Exposed larvae and control larvae (in pure sea water) are compared, with respect to the progress of morphogenesis from the late blastula without mesenchyme (at zero) to an early pluteus stage (at 6). If the exposed larvae were unaffected the points of comparison with the control would all fall on the line C. Induced retardation would depress the line below C. The ends of primary and secondary invagination are marked by lower and higher dotted line, respectively. Note retardation of formation of the primary mesenchyme and early invagination of the archenteron rudiment by tripelennamine and alleviation of this inhibition by serotonin (5-HT) [see Gustafson & Toneby (1970), from which this is reproduced].

counteracted if 5-hydroxytryptophan was added to the medium together with the inhibitor (Fig. 10). Furthermore, inhibitors of the next step in the metabolism of tryptophan to serotonin also had an inhibitory effect, particularly directed against the first phase of gastrulation.

But all these results would be rather uninteresting if serotonin or some related substance was not present in the embryo at the right time. The results of our Russian colleagues Buznikov *et al.* (1964) show that serotonin is present in the larva notably during early gastrulation, the phase most sensitive to serotonin antagonists (Fig. 11). Fluorescence histochemical studies by Buznikov and by ourselves also show the occurrence of some monoamines in the archenteron. Our own analyses (M. Toneby, unpublished results) by, for example, thin-layer chromatography reveal serotonin in the pluteus larva where, according to Buznikov, it reappears after being more or less absent during secondary invagination.

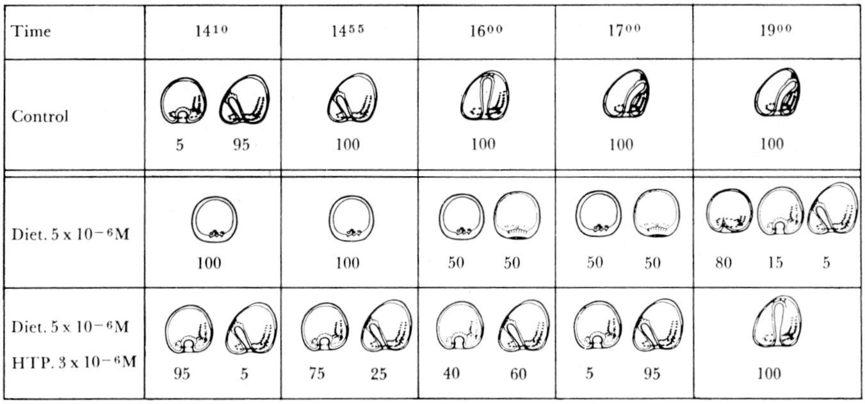

Time	14^{10}	14^{55}	16^{00}	17^{00}	19^{00}
Control	5 95	100	100	100	100
Diet. 5 x 10^{-6}M	100	100	50 50	50 50	80 15 5
Diet. 5 x 10^{-6}M HTP. 3 x 10^{-6}M	95 5	75 25	40 60	5 95	100

FIG. 10. Inhibitory effects of diethyl dithiocarbamate (Diet.) on morphogenesis and the marked alleviation of inhibition by 5-hydroxytryptophan (HTP). Figures below larvae represent percentage of all observed larvae in each culture. [From Gustafson & Toneby (1970).]

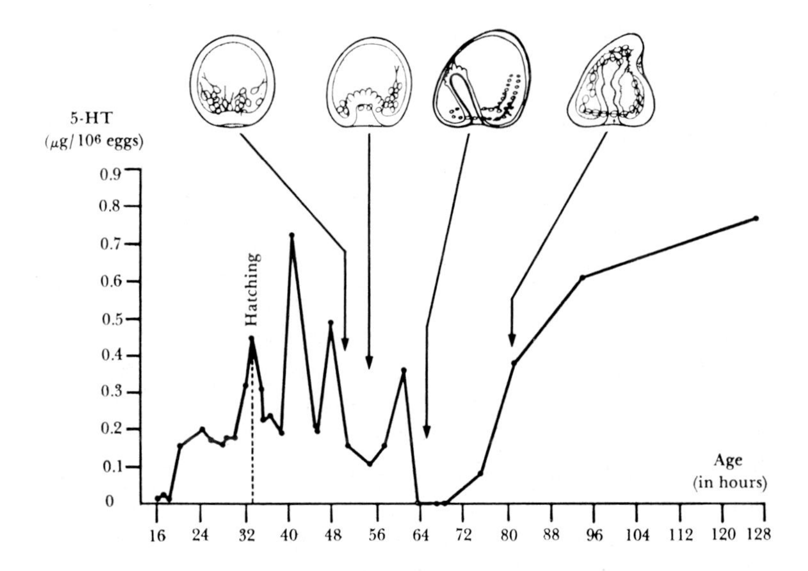

FIG. 11. Variation in the content of serotonin during early development of a sea urchin larva *(Strongylocentrotus)*; curve reproduced from Buznikov *et al.* (1964) (with kind permission) but modified by the insertion of figures to define the progress of morphogenesis.

On the topic of the control of the synthesis and disappearance of serotonin one possible mechanism is that an excess of tryptophan appears when certain cells begin to utilize yolk for manufacturing large amounts of collagen. If so gastrulation might be said to be a direct consequence of the anabolic specialization of the cells.

More important in this context is how the pseudopodal contraction is controlled. We should remember that this activity underlies the migration of the cells from zone 1 and the extension of both the archenteron rudiment and the coelom (cf. Figs. 1 and 2). Time lapse films show that pseudopods extend from pulsatory cell surfaces and it is tempting to think that emission of thin pseudopods is closely related to the formation of pulsatory blebs that are particularly common at the tip of the archenteron rudiment just before the archenteron rudiment starts to elongate into a tube (cf. Fig. 1). The extension of the pseudopods stops as soon as their tips have attached to the wall. We have never seen a pseudopod lengthen once it has attached. On the contrary, their lack of extension is directly followed by shortening as if the pseudopods receive an impulse to change their behaviour when they attach. The nature of this impulse might be related to that of the impulses responsible for contact paralysis. One might for instance consider some mechanism of the type discussed by Gingell (1967). If so any process in the control of the electrophysiological properties of the cell membranes would interfere with gastrulation.

Returning to our pharmacological results, we found that physostigmine (eserine) enormously enhances the contractile activity of the muscular strands directed downward from the coelom (Gustafson *et al.* 1972*b*). Excessive doses (e.g. 10μM) are inhibitory. Quaternary anticholinergics, unexpectedly, displayed little inhibitory effect except at higher doses (e.g. 1mM) when they eventually suppressed the effects of eserine. Their low activity is probably due to the difficulties of strongly polar agents to penetrate the system. The concentrations of tertiary anticholinergics necessary for inhibition are much lower.

The necessity of acetylcholine for the control of contractile activities is underlined by the occurrence of a specific acetylcholinesterase within the larva (Augustinsson & Gustafson 1949). It should be noted that the activity is located in particular at the inner border of the cells, that is, where the pseudopods attach (Fig. 12). Could this mean that the acetylcholine–cholinesterase system contributes to the generation of impulses to the pseudopods by initiating their contractions? Credence to this idea is given by the results of many experiments with anticholinergic agents. Quaternary anticholinergics are very inefficient, which probably reflects their slow penetration. Tertiary anticholinergics, however, are strongly inhibitory with respect to the pseudopod-dependent phase of gastrulation. Even a quaternary lipid-soluble agent, PAD (pyridine-

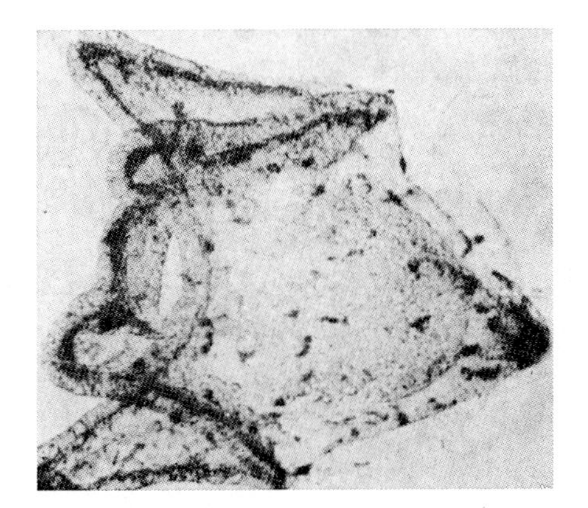

FIG. 12. Distribution of cholinesterase activity in a very early pluteus stage. The activity is particularly intense in the ciliated band of the ectoderm. (The dark spots scattered throughout the body are pigment cells which tend to keep their pigment.)

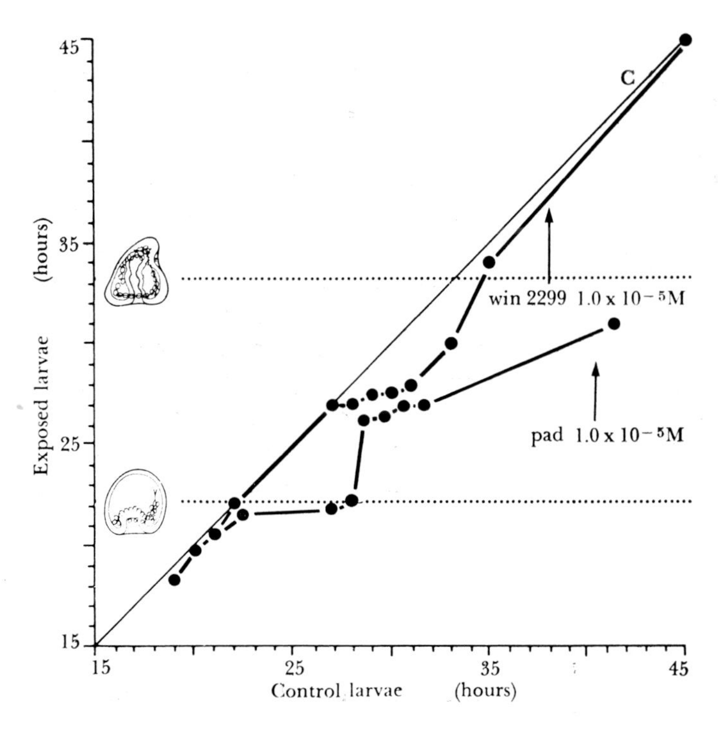

FIG. 13. Morphogenetic effects of two lipid-soluble cholinolytic compounds. Early infolding of the primitive gut has occurred at a rather normal rate, whereas its further extension by pseudopods is clearly affected. For further examples see Gustafson & Toneby (1970).

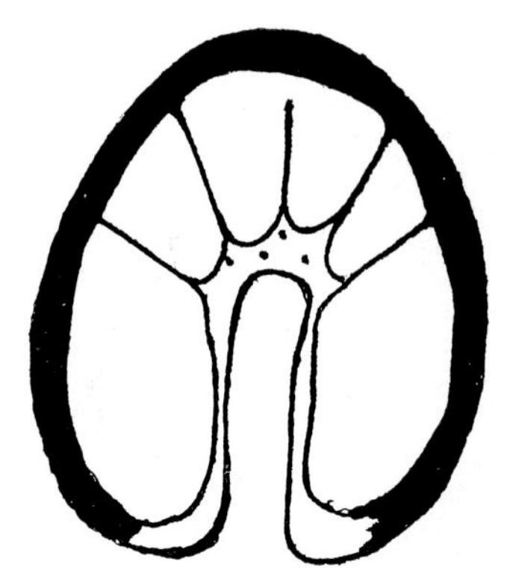

FIG. 14. Diagram of the hypothetical effects of serotonin and the acetylcholine–cholinesterase system. Serotonin in dotted area induces movements related to early invagination of the archenteron rudiment, presumably including the formation of pseudopods. Pseudopods attaching to the ectoderm are stimulated to contract by impulses somehow related to the acetylcholine–cholinesterase system (distributed within the black area).

aldoxime dodecyl iodide), is very efficient (Fig. 13). Furthermore excessive doses of eserine (which blocks most of the muscular activity) efficiently block the second phase of gastrulation after a brief period of intense stimulation. We can thus easily understand why agents that strongly antagonize serotonin as well as acetylcholine (e.g. cyproheptadine) are exceedingly potent inhibitors of gastrulation (Gustafson & Toneby 1970, 1971). Other results bear out the importance of serotonin in the formation of the nervous system of the pluteus larva, a process where there are also intense cellular movements.

In conclusion, I believe at the moment that the relation between serotonin, acetylcholine and morphogenetic movements during gastrulation can be represented by Fig. 14. Whatever the mechanism of action of these agents are, I am confident that morphogenetic movements in the sea urchin embryo are controlled by factors closely related to those in the subsequent muscular contraction.

References

AUGUSTINSSON, K. B. & GUSTAFSON, T. (1949) Cholinesterase in developing sea urchin eggs. *J. Cell Comp. Physiol.* **34**, 311-322

BUZNIKOV, G. A., CHUDAKOVA, I. V. & ZVEZDINA, N. D. (1964) The role of neurohumours in early embryogenesis. I. Serotonin content of developing embryos of sea urchin and loach. *J. Embryol. Exp. Morphol.* **12**, 563-573

GINGELL, D. (1967) Membrane surface potential in relation to a possible mechanism for intercellular interactions and cellular responses: a physical basis. *J. Theor. Biol.* **17**, 451-482

GUSTAFSON, T. & KINNANDER, H. (1956) Microaquaria for time-lapse cinematographic studies of morphogenesis in swimming larvae and observations on sea urchin gastrulation. *Cell Res.* **11**, 36-51

GUSTAFSON, T. & TONEBY, M. (1970) On the role of serotonin and acetylcholine in sea urchin morphogenesis. *Exp. Cell Res.* **62**, 102-117

GUSTAFSON, T. & TONEBY, M. (1971) How genes control morphogenesis. The role of serotonin and acetylcholine in morphogenesis. *Am. Sci.* **59**, 452-462

GUSTAFSON, T. & WOLPERT, L. (1967) Cellular movement and contact in sea urchin morphogenesis. *Biol. Rev. (Camb.)* **42**, 442-498

GUSTAFSON, T., LUNDGREN, B. & TREUFELDT, R. (1972*a*) Serotonin and contractile activity in the Echinopluteus. A study of the cellular basis of larval behaviour. *Exp. Cell Res.* **72**, 115-139

GUSTAFSON, T., RYBERG, E. & TREUFELDT, R. (1972*b*) Acetylcholine and contractile activity in the Echinopluteus. *Acta Embryol. Exp.* **2**, 199-223

Discussion

Wohlfarth-Bottermann: Bauer (1967) has reported cholinesterase activity in the invaginations of the plasmalemma of slime mould plasmodia. The site of this activity seemed to correlate closely with that of the fibrillar system, which is responsible for contractility and thereby for the motive force generation for protoplasmic streaming. To the best of my knowledge nobody has tried to reproduce Bauer's results.

Allison: Do you have any information about melatonin?

Gustafson: We have only carried out one experiment with melatonin so far. The results of that suggest that the effect of melatonin is insignificant. This inefficacy might be due to a slow rate of penetration of the molecule, possibly on account of its acetyl group.

Allison: The reason I ask is that Banerjee *et al.* (1972), working on *Stentor*, have shown that melatonin prevents polymerization of microtubules and is also bound to tubulin *in vitro*.

Gustafson: That is very interesting. In future experiments we shall try higher doses of melatonin.

Abercrombie: Is there any evidence of the involvement of these substances in the movement of the kinds of cells we have been discussing earlier?

Gustafson: Yes. There is some evidence that monoamines are engaged in the motility of neurons and even in the morphogenetic movements of neurons, notably in the sea-urchin larva. According to the analytical data of Buznikov *et al.* (1964), serotonin reappears in the late gastrula. One important source of this serotonin is the mast-cell-like pigment cells (Gustafson & Toneby 1971). These cells, which are derivatives of the pseudopod-forming cells at the archenteron tip which are so important during gastrulation, gradually accumulate a red pigment which is probably not echinochrome but a tryptophan derivative like serotonin itself. The pigment cells first cluster around the tip of the archenteron and probably give off serotonin which induces the bubbling activity of the coelomic rudiment, apparently an important activity in the morphogenesis of the coelom. Later, numerous pigment cells invade the circumventral ciliated band. What is their function? If we treat the larva with agents that degranulate mast cells, that is, agents which make them give off histamine, the cilia begin to beat vigorously. Evidently, one function of the pigment cells is to release serotonin and thus to stimulate the swimming movements of the larva. This conclusion is supported by the observation that exogenous serotonin stimulates the cilia. In the early pluteus larva there is evidence of another role for the pigment cells. It is important in this connection to mention that the red colour of the pigment granules of the pigment cells can turn green and that greenish granules may be released from the cells. In the pigment cells, or where green granules have been deposited, the adjacent ciliated epithelial cells begin to behave strangely: they contract, round up, break their contact with their neighbours and begin to rotate. Their behaviour resembles that of primary mesenchyme cells. One might guess that they have 'picked up something' from the pigmented mesenchymal cells or from the released granules and that this factor induces the high motility, characteristic of mesenchymal cells. Soon the former ciliated cells reveal other characteristics of mesenchyme cells: they enter the body cavity and send out pseudopods which are highly motile. Gradually the movements calm down and the pseudopods begin to look like axons. That they also have an axonal function is supported by the fact that they make up a rather reproducible system and that they make contact with various parts of the contractile mesendodermal system. Synapse-like structures have also been observed at the contact points. Furthermore if we irritate the periphery of the larva, mechanically or chemically, contractions are induced where the axon-like structures make contact with the stomach wall, etc. What we have studied is therefore no doubt the formation of neurons. Under the influence of a pigment cell, the ciliated cells are transformed into neurons. But what do the pigment cells do to the ciliated cells? We think that they release heavy doses of serotonin which induces the vigorous motility of the future neurons.

We have tried to find some support for this hypothesis. If this factor is serotonin, the number of cells affected by each pigment cell should increase if the larvae were treated with precursors of serotonin, for example, tryptophan.

In many such experiments (E. Ryberg, unpublished results; see also Gustafson & Toneby 1971) enormous clusters of 'neuroblasts' appeared in larvae treated with 50μM-L-tryptophan. The neuroblasts entered the body cavity of the larvae and formed axon-like processes as in normal larvae.

The suggested relation between serotonin and cell motility related to the formation of sea-urchin neurons reminds me of some experiments on the effect of serotonin, dopamine, noradrenalin etc., on the behaviour of neurons in cell culture (summarized by Geiger 1963). Axonal flow, the rotation of the nuclei and other types of motility were markedly increased in the presence of these agents which also increased the motility of certain glial cells. In summary, monoamines appear to participate in the control of cell motility in a number of cases.

In order to explore the possible function of monoamines in morphogenetic movements in vertebrates, we have carried out some experiments with eggs of *Xenopus*, which have favourable surface–volume ratios. After treatment of the eggs with certain hydroxylase inhibitors, gastrulation was strongly retarded (Gustafson & Toneby 1971). The closure of the neural plate into a tube was also defective. Pigmentation was affected, as was, apparently, the formation of collagen. The larvae finally became very peculiar. The defective development of collagen and pigment presumably reflects reduced hydroxylation of the proline residues in protocollagen and of tyrosine respectively, but the other defects could be due to reduced formation of monoamines. If monoamines act in the morphogenesis of the vertebrate embryo (including the human embryo) many drugs, for example antihistamines which also have antagonistic action against serotonin and other monoamines, should be used with great care by pregnant women.

References

BANERJEE, S., KERR, V., WINSTON, M., KELLEHER, J. K. & MARGULIS, L. (1972) Melatonin: inhibition of microtubule-based oral morphogenesis in *Stentor coeruleus*. *J. Protozool.* **19**, 108-113

BAUER, L. G. (1967) On the similar orientation of fibrillar structures and rows of esterolytic invaginations in plasmodia of the slime mould *Physarum confertum* Macbr. *J. Exp. Zool.* **164**, 69-80

BUZNIKOV, G. A., CHUDAKOVA, I. V. & ZVEZDINA, N. D. (1964) The role of neurohumours in early embryogenesis. I. Serotonin content of developing embryos of sea urchin and loach. *J. Embryol. Exp. Morphol.* **12**, 563-573

GEIGER, R. S. (1963) The behaviour of adult mammalian brain cells in culture. *Int. Rev. Neurobiol.* **5**, 1-52

GUSTAFSON, T. & TONEBY, M. (1971) How genes control morphogenesis. The role of serotonin and acetylcholine in morphogenesis. *Am. Sci.* **59**, 452-462

Time lapse studies on the the motility of fibroblasts in tissue culture

MITCHELL GAIL

National Cancer Institute, Bethesda, Maryland

Abstract This article reviews our efforts to characterize and measure fibroblast motility and to determine what factors influence motility. Though the distinction does not always hold, we have organized our exposition into extrinsic and intrinsic factors influencing motility. Extrinsic factors include cell–substrate adhesivity, the medium and cell area density. Pharmacological studies give some insight into intrinsic cellular factors, and, at the risk of over-interpreting our data, we venture the conclusions: (1) structural integrity, which can be disrupted by Colcemid, is required for maximal motility; (2) studies with cytochalasin, procaine and ethylenebis(oxyethylenenitrilo)tetraacetic acid (EGTA), and with various extra-cellular cation concentrations, are consistent with the hypothesis that motive power of fibroblasts derives from a calcium-controlled interaction between myosin and actin; (3) motility and proliferation are dissociable physiological functions.

FIBROBLAST MOTILITY MEASUREMENTS

Abercrombie & Heaysman (1953) pioneered the use of time lapse cinematography in their studies on the locomotion of chick heart fibroblasts. They measured motility in terms of a simple estimate of velocity based on the distance travelled in a given unit of time, but they recognized that this measurement was 'biased' since the motion of these cells was not uniform or linear. Ambrose (1961) noted that 'the movements of cells when isolated and spreading on a glass are largely at random'. We recognized the attractiveness of a two-dimensional random walk model which could characterize the motility of fibroblasts in terms of a single constant—the diffusion constant. We therefore sought to test the hypothesis that fibroblasts executed a pure two-dimensional random walk (Gail & Boone 1970).

Trypsinized 3T3 mouse fibroblasts were planted sparsely (about 30 cells/mm^2) in Dulbecco–Vogt medium containing 10% (v/v) foetal calf serum. After

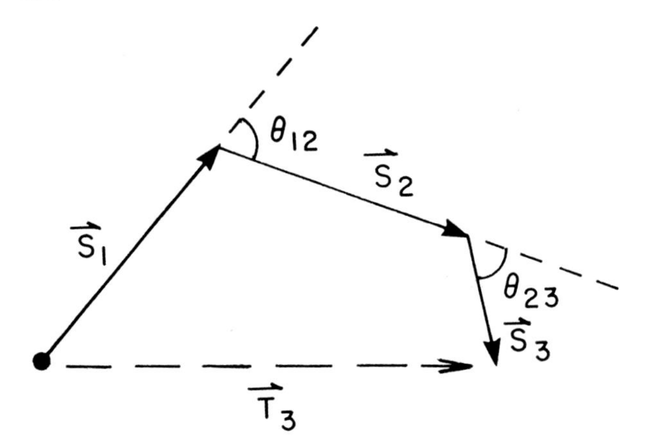

FIG. 1. Diagram of measurements made on cell motion over three time intervals. $\mathbf{T}_3$ represents the total displacement after three time intervals, $\mathbf{S}_i$ the segmental displacement during the ith interval, and θ_{ij} the intersegmental angle between $\mathbf{S}_i$ and $\mathbf{S}_j$.

incubation for 24 h followed by a change of medium, the fibroblasts were filmed at a rate of 0.5 frame/min. We could measure both population growth and individual cell displacements in the observation field (0.69 mm^2). Fig. 1 shows the types of measurements on cell displacement used to investigate the random walk hypothesis. We observe the position of each cell every 2.5 h. We index each successive observation as $i = 1, 2, \dots n$. Fig. 1 depicts three such observations, $n = 3$. We denote the total vector displacement from the origin at the end of the ith observation by $\mathbf{T}_i$. Likewise, $\mathbf{S}_i$ represents the segmental displacement during the ith time interval and θ_{ij} the angle between the ith and jth displacement vectors.

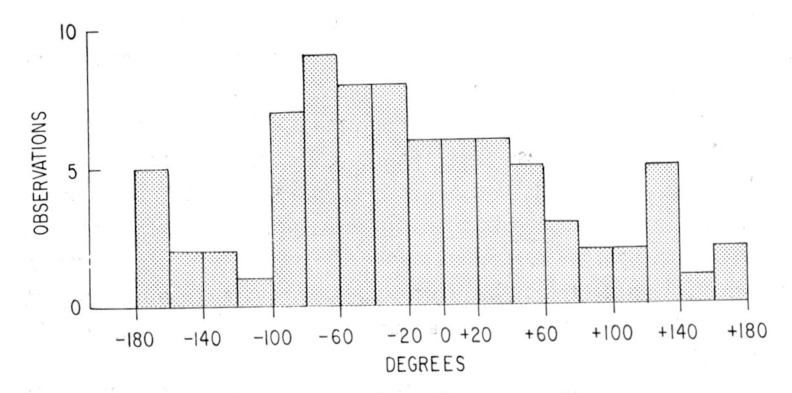

FIG. 2. Histogram of 80 intersegmental angle measurements for 2.5 h intervals. The distribution is non-uniform with concentration about 0°.

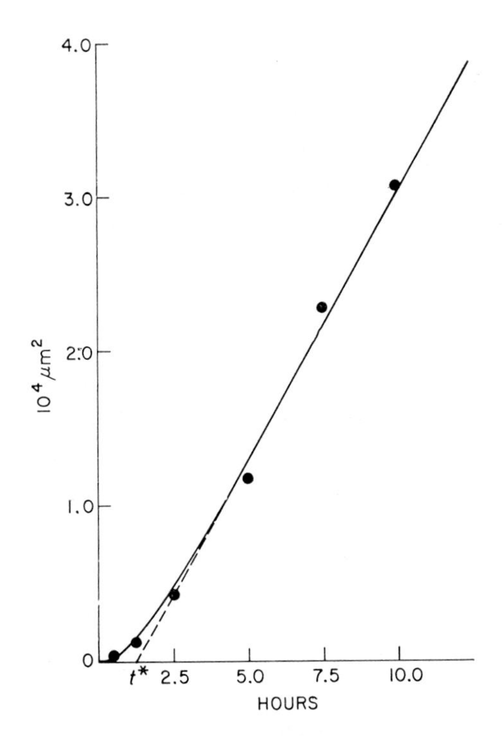

FIG. 3. Mean square displacement, $<T^2>$, is plotted as a function of observation time. Equation (1) ($D^* = 858\ \mu m^2/h$ and $t^* = 1.23$ h) is shown as solid line. The dashed line indicates linear extrapolation to the time intercept, t^*.

Fig. 2 shows that the angles between successive segmental displacements, θ_{ij}, are not uniformly distributed but tend to be concentrated within the —100 to +100° range (the non-uniformity is significant at the 1% level). This means that the cells tend to persist somewhat in their direction of motion from one 2.5 h time interval to the next; they are not pure random walkers.

The plot of mean square displacement $\langle T^2 \rangle$ against time (Fig. 3) confirms the fact that these cells are persistent (rather than 'pure') random walkers. After 2.5 h, $\mathbf{T}_1$ is measured for each cell under study. Then T^2 ($= \mathbf{T}_1 \cdot \mathbf{T}_1$) is computed for each cell and the mean square displacement, $\langle T^2 \rangle$, is calculated and plotted, and similarly at 5 h intervals for T^2 ($= \mathbf{T}_i \cdot \mathbf{T}_i$). If fibroblasts executed a pure random walk, the plot of mean square displacement, $\langle T^2 \rangle$, against time would be a straight line passing through the origin. The effect of persistence is to induce a small non-linearity for small times as shown in Fig. 3.

Drawing on the one-dimensional results of Fürth (1920), we modified the random walk theory to accommodate the persistence effect. The modified theory predicted the mean square displacement of a persistent random walker should follow the locus (1), where D^* and t^* are constants and t is time. For $t \gg t^*$, equation (1) reduces to the simple straight line equation (2), whose slope is $4D^*$. As is evident in Fig. 3, equation (1) fits the observed data closely, and the straight line locus [equation (2)] obtains for $t > 2.5$ h.

$$\langle T^2 \rangle = 4D^* \{t - t^* [1 - \exp(-t/t^*)]\} \quad (1)$$

$$\langle T^2 \rangle = 4D^*(t - t^*) \quad (2)$$

D^* is a good measure of motility. The larger D^* is, the greater is the motility. 3T3 fibroblasts have D^* values ranging from about 200 to 1000 $\mu m^2/h$ under normal conditions. D^* is called the augmented diffusion constant since it can be shown that the effect of persistence is to raise the mean square displacement above that of an otherwise identical pure random walker. However, in the hypothetical case of a pure random walker with diffusion constant D, equation (1) also applies with $t^* = 0$ and $D^* = D$. So these methods are applicable to persistent and pure random walkers alike. We (Gail & Boone 1970) have established the statistical properties of estimates of D^* and have shown how to construct confidence intervals on D^*. Statistical methods for comparing two treatment groups and for pooling data have also been devised (Gail & Boone 1972*a*). One attractive feature of measuring motility in terms of D^* is that the greater the number of cells studied the smaller becomes the confidence interval about D^*. This cannot be said of velocity measurements which exhibit a large coefficient of variation no matter how many cells are studied (Abercrombie & Heaysman 1953). Indeed this is to be anticipated, since the velocities of these cells are inherently variable, both in magnitude and direction, whereas D^* is a constant parameter of a non-deterministic model. Another advantage is that measurements are made at low cell area density, eliminating cell–cell interactions which complicate the interpretation of motility measurements based on migration into a tissue culture 'wound'.

The persistent random walk model has been shown to apply to all varieties of fibroblasts examined including 3T3, their SV40 viral transformants, spontaneous transformants, embryonic fibroblasts and lines carried under various culture regimens (e.g. 3T12). It also applies to macrophages and would be expected to apply to a wide range of cell types on an appropriate time scale. The model is misleading when cells are under a predominating tactic influence.

EXTRINSIC FACTORS WHICH INFLUENCE FIBROBLAST MOTILITY

In an effort to standardize our experimental system for studying fibroblast motility, we found that several extrinsic factors required control, and some of these were of biological interest.

We showed that cell–substrate adhesivity is a determinant of fibroblast motility, the cells moving faster on the less adhesive substrates (Gail & Boone 1972*a*), and used the fraction of cells distracted by a standardized air blast as our measure of this adhesivity. The cells were sparsely planted in a Petri dish and covered to a height of 1.4 mm with culture medium. The air blast was focused on a 4 mm^2 area of the Petri dish and cells were counted before and after the blast to determine the fraction distracted. We showed that 3T3 fibroblasts and their SV40 viral transformants, SV3T3, adhered to Pyrex more firmly than to cellulose acetate. Both these cell types also showed an increase in D^* of about 50% on cellulose acetate over that on Pyrex. Phytohaemagglutinin-M (PHA-M) profoundly increased cell–substrate adhesivity between 3T3 cells and both Pyrex and plastic Petri dishes (Falcon). PHA-M likewise greatly increased the adhesivity of SV3T3 cells to these substrates. Although PHA-M did not alter 3T3 doubling time, it reduced D^* by about 90%. For the SV3T3 fibroblasts, PHA-M did prolong the doubling time somewhat and reduced D^* by more than 95%. We interpreted these results as showing that D^* varies inversely with cell–substrate adhesivity in the range of adhesivities studied. Probably there exists an optimal cell–substrate adhesivity, less than which cell motility decreases, but we have not demonstrated this by experiment.

Cell area density is another extrinsic determinant of fibroblast motility. Abercrombie & Heaysman (1953) demonstrated a 'significant inverse relationship between the speed of movement of a cell (chick heart fibroblast) and the number of other cells with which it was in contact'. We (Gail & Boone 1971*a*) studied this phenomenon by measuring cell area density instead of contact number (Abercrombie & Heaysman had shown these variables to be strongly correlated). Fig. 4 shows that D^* decreases dramatically with increasing 3T3 cell area density, but that SV3T3 motility is only slightly diminished at comparable densities. If we adopt the regression model [equation (3)], where D_0 is the

$$D^* = D_0 - G\psi \tag{3}$$

limiting value of D^* as the cell area density, ψ (in cells/mm^2), tends to zero, then G (the negative slope of a plot of D^* against ψ) is a numerical measure of what we term density inhibition of motility. The respective G values for 3T3 and SV3T3 cells as shown in Fig. 4 were 3.22 and 0.19 (μm^2/h)/(cell/mm^2). Control experiments showed that the slowing of 3T3 cells cannot be ascribed to the

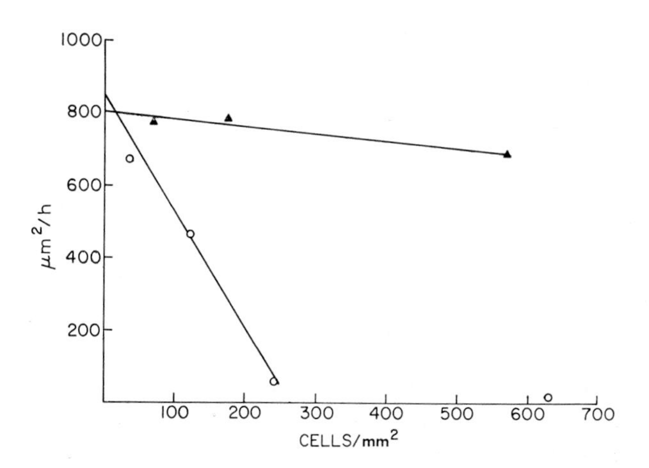

FIG. 4. Estimates of the augmented diffusion constant, D^*, are plotted as a function of cell area density for 3T3 (○) and SV3T3 (▲). The linear plots represent maximum likelihood 'fittings' to the three data points to which they are closest.

depletion of any postulated 'locomotion factors'. The rate of collision of SV3T3 cells (contacts/10 h) was as great as that for 3T3 cells at low densities. Moreover, the SV3T3 collision rate increased indefinitely with increasing cell density, whereas the 3T3 cells appeared to reach a limiting rate of collision (mainly between near neighbours) of about 6–10 (collision/cell)/10 h for all densities exceeding 100 cells/mm². Thus the lack of density inhibition of motility in SV3T3 cells cannot reflect a lower collision rate. We gained the impression that 3T3 fibroblasts are held in a matrix of neighbouring cells at high density, nearly fixed in position. By contrast, SV3T3 fibroblasts seethe above, below and around each other at high density. These observations suggest that density inhibition of motility mainly reflects cell–cell adhesivity, large G values being associated with great cell–cell adhesivity. Consistent with this interpretation are the following facts. (1) Abercrombie & Heaysman (1953) demonstrated that such slowing depended on cell–cell contact and not merely on high cell area density. (2) Our studies show the average contact time of 3T3 is 152 min/contact and that of SV3T3 cells with comparable D^* values only 50 min/contact. (3) 3T3 cells reach a limiting collision rate as cell area density increases, whereas that for SV3T3 cells increases indefinitely with increasing density. (4) Sheets of untransformed fibroblasts can support significant tensions [0.3 N/cm² (3×10^4 dyn/cm²)], reflecting substantial cell–cell adhesivity (James & Taylor 1969). (5) Intercellular 'nexuses', possible morphological concomitants of cell–cell adhesivity, are rare in SV3T3 but common in 3T3 (McNutt & Weinstein, unpublished results). (6) The random walk theory predicts that the formation of transient adhesions between independent random walkers will reduce the augmented diffusion constant D^* of the resulting couplet. Although the hypothesis that density inhibition of motility results from transient adhesive bonds between contacting cells is attractive, direct confirmation, such as the studies of Coman (1944) which showed that the force required to separate normal cervical epithelial cells is greater than that required to separate cervical carcinoma cells, is not available.

Studies now in progress show that many tumorigenic cell lines exhibit little or no density inhibition of motility (C. W. Boone, unpublished findings). This is true for SV40 viral transformants, spontaneous transformants, and methylcholanthrene-induced fibrosarcoma explants. Thus density inhibition of motility resembles contact inhibition of locomotion (Abercrombie & Heaysman 1954). This term describes a 'directional prohibition of movement' toward a contacted cell, and this phenomenon is experimentally distinguishable from the slowing of contacted cells (Abercrombie & Heaysman 1953). Abercrombie *et al.* (1957) found that mouse sarcoma cells showed little contact inhibition of locomotion compared to mouse or chick heart fibroblasts. Exceptional tumorigenic fibroblasts which do exhibit contact inhibition of locomotion have been described (Defendi *et al.* 1963; Macieira-Coelho 1967). It remains to be seen whether exceptional tumorigenic fibroblasts exhibiting substantial density inhibition of motility will also be found.

The culture medium is, of course, an extrinsic determinant of cell motility. It must provide a suitable nutritional environment and maintain physiological temperature and pH. Korohoda (1971) has shown that in one hour migrating chick embryo fibroblasts consume about 19 μl oxygen/10^6 cells whereas cells in suspension consume only about 1/5 that amount and non-migrating, 'contact inhibited' cells about 1/8. (However proliferation is also probably suppressed in the latter two cases.)

Specific non-dialysable serum factor(s) are also required for maximal motility. Lipton *et al.* (1971) claim to have isolated a serum fraction, stable to heat and pH, required for cell migration into tissue culture wounds and distinct from two other fractions required respectively for cell proliferation and cell survival. One might object that this wound migration factor was required only to separate mutually adherent cells at the edge of the wound and not to promote motility itself, but we have confirmed the importance of non-dialysable serum factors in sparsely planted fibroblasts (Gail & Boone, unpublished results). In media containing only 2% (v/v) serum or in 'growth-factor-free medium' prepared by depletion as described by Smith *et al.* (1971), sparsely planted cells remained viable but did not proliferate or move well ($D^* = 20\text{–}50\ \mu m^2/h$). Addition of 10% (v/v) foetal calf serum, which had been exhaustively dialysed against phosphate buffer, increased the motility ($D^* = 300\ \mu m^2/h$) within 20 h, and cell proliferation followed.

In the context of non-dialysable serum factors required for maximal motility, one cannot dismiss the possible importance of protein-bound hormones. Von Hamm & Cappel (1940) showed that thyroxine and insulin promote fibroblast proliferation.

INTRINSIC CELLULAR FACTORS WHICH INFLUENCE MOTILITY

The intrinsic cellular properties important to motility include cell structure, contractile proteins, mechanisms of motility control and the relation between motility and proliferation. We investigated these pharmacologically.

The migrating fibroblast is lenticular and generally moves in the direction of its 'leading edge'. Abercrombie (1961), noting this axial organization, asserted that 'the main locomotory organ of the fibroblast is its ruffled membrane ... at the front end'. Recent studies tend to confirm this view (Abercrombie *et al.* 1970). Ingram's (1969) side views of moving fibroblasts reveal that these cells attach to the substrate anteriorly and posteriorly only. The migrating fibroblast thus resembles a shoe with its unattached central arch. Porter (1966) has suggested that a 'cytoskeleton', composed of microtubules, is required to maintain such asymmetry, and indeed microtubules have repeatedly been identified in fibroblasts (Buckley & Porter 1967; Goldman & Follett 1969).

We (Gail & Boone 1971*b*) have studied the effect of Colcemid on fibroblast motility since it and colchicine are known to cause dissolution of microtubules (Tilney 1968) without binding to myosin, F-actin or G-actin (Borisy & Taylor 1967). Colcemid (0.36 μg/ml) dramatically altered fibroblast motility within one hour. The cells lost their axial organization and became 'rounded or polygonal' (Miszurski 1949), their cell outline area nearly doubled, and the 'ruffling membrane' was no longer localized to the 'leading edge', but instead became fairly uniformly distributed about the cell periphery and eventually occupied about 81% of the cytoplasmic periphery (compared to 22% before treatment with Colcemid). At the same time, cell motility was more than halved compared to control values. These effects were reversed by changing to Colcemid-free medium. During treatment with Colcemid, cell–substrate adhesivity was decreased, as a result of which motility should be enhanced. Other studies (see later) show there is no direct relationship between motility and proliferation, so we could not ascribe the decrease in motility to the anti-mitotic effects of Colcemid. The most appealing hypothesis was that maximal fibroblast motility depends on axial cell structure and Colcemid slows fibroblasts by dissolving the cytoskeleton which sustains this structure.

Buckley & Porter (1967) identified actin-like filaments 7.5 nm in diameter in fibroblasts and suggested that these probably 'provide the motive power'. Ishikawa *et al.* (1969) provided convincing evidence that these filaments were indeed actin by demonstrating characteristic arrowhead complexes in the presence of heavy meromyosin. Recently, Adelstein & Conti (1972) have isolated myosin and actin from fibroblasts. Since the principal contractile proteins of skeletal muscle have been isolated from fibroblasts, we shall invoke the current theory of skeletal muscle contraction to explain subsequent results. According to this theory (Bendall 1969; Ebashi & Endo 1968; Huxley 1969), skeletal muscle contraction begins when membrane-bound calcium ions are released into the cytosol where they depress troponin, a myosin inhibitor. The myosin then hydrolyses ATP and binds to actin, leading to muscular contraction. (Magnesium must be present for contraction.) The relaxation phase begins when the calcium chelating system in the sarcoplasmic reticulum reduces the free calcium concentration from 10μM in the contracted state to below 0.1μM. Troponin then inhibits myosin ATPase activity. If ATP were unavailable at this point, electrostatic bonds between myosin and actin would freeze the muscle in 'rigor'. In the presence of ATP and magnesium ions, these bonds are broken and the muscle relaxes.

Cytochalasin B has been shown to inhibit L-cell fibroblast motility and cytoplasmic cleavage (Carter 1967). We confirmed these results in 3T3 fibroblasts and showed that motility was inhibited at lower doses (10μg/ml) than those required for inhibition of cytoplasmic cleavage (Gail & Boone 1971*c*). Within one hour motility was inhibited, and cells divided and became rounded and refractile under phase contrast. Even though D^* was decreased, ruffling was visible throughout the cytoplasmic periphery. These effects on motility and proliferation were reversed within ten hours after substitution of cytochalasin-free medium. Cell–substrate adhesivity is diminished during treatment; this decrease would be expected to increase motility. Wessells *et al.* (1971) suggested that cytochalasin disrupts the functioning of contractile microfilaments. Central to this argument are Schroeder's (1969) studies showing that the 5–8 nm 'contractile ring' filaments of the cleaving *Arabacia* egg disappear after cytochalasin treatment, which also prevents cleavage. These studies are the more significant since Perry *et al.* (1971) demonstrated that such filaments in the newt egg during cleavage bind heavy meromyosin and are therefore probably actin. Our observations might thus represent an effect of cytochalasin on actin filaments, but other mechanisms are conceivable (Zigmund & Hirsch 1972; Holtzer & Sanger 1972).

Skeletal muscle contraction is initiated by an increase in free calcium in the cytosol. We (Gail & Boone, unpublished results) have shown that fibroblast motility and proliferation are inhibited in the presence of a calcium chelating agent, ethylenebis(oxyethylenenitrilo)tetraacetic acid (EGTA). We added 1.5mM-EGTA to a medium consisting of 10% (v/v) foetal calf serum and 90% (v/v) calcium-free Dulbecco–Vogt salts, which yields a free calcium concentration of about 0.1 μM. Within ten hours exposure to this medium, D^* was reduced (as was ruffling) and about half the cells became round and refractile under phase contrast within 20 h. We interpreted this latter change as characteristic of low cell–substrate adhesivity, and confirmed this by air-blast experiments. These changes in proliferation and motility were reversed within 30–40 h after restoring a calcium-containing medium. Low cell–substrate adhesivity usually enhances motility. Small round cell forms are not necessarily slow movers: we have observed very motile rounded forms which had low cell–substrate adhesivity. Moreover, the rounded cells in EGTA were as motile as their flattened counterparts, both such cell forms moving only a little. We therefore surmise that EGTA medium does not act on cell motility indirectly through an action on cell morphology or on cell–substrate adhesivity but rather exerts a direct effect on the calcium-dependent contractile proteins. Treatment with a medium containing high calcium concentration (10mM) had no appreciable effect on motility or proliferation; neither did exposure to medium containing only 0.1mM-magnesium, though effects due to magnesium might be observed at much lower concentrations.

The fact that procaine inhibits fibroblast motility and proliferation is possibly related to the calcium requirement (Gail & Boone 1972*b*). Pharmacological doses of procaine (2.4mM) inhibit the motility (and ruffling) within one hour and proliferation within three hours. Ten hours after substituting procaine-free medium these effects are reversed. Since we are unable to demonstrate any changes in cell–substrate adhesivity due to procaine, we believe that procaine acts directly on the contractile system and in particular on the elements of calcium control. Feinstein & Paimre (1969) have reviewed the evidence that local anaesthetics (procaine and tetracaine) prevent contraction of the frog sartorius by inhibiting the release of intracellular calcium into the cytosol. Feinstein (1963) had previously shown that 5mM-caffeine caused contractions and increased efflux of intracellular ^{45}Ca from the frog sartorius muscle. The inhibition of these effects of caffeine by both procaine and tetracaine suggested to Feinstein & Paimre (1969) that the anaesthetic prevents release of calcium into the cytosol mainly by maintaining low calcium permeability in membranes bounding intracellular calcium stores. In support of these ideas Hertz & Weber (1965) showed that caffeine causes rapid release of calcium bound to sarcoplasmic reticulum in the frog, which presumably explains why caffeine can induce contractions even in depolarized muscle. Furthermore, Gruener (1967) showed that sarcolemma-free fibres of sartorius muscle in the frog contract in response to calcium even in the presence of 3mM-procaine. Thus procaine probably does not act directly on troponin, actin or myosin.

In one film, 1mM-theophylline and 1.2mM-dibutyryl cyclic AMP together decrease the motility of L-cells within 20 min (Johnson *et al.* 1972), but no such slowing was noted with SV3T3 cells in a single experiment (Johnson, unpublished results). Since L-cells and SV3T3 cells treated thus became more lenticular and flattened (Johnson *et al.* 1971) and also were resistant to detachment by trypsin or EGTA (Johnson & Pastan 1972), it is tempting to suppose that theophylline and dibutyryl cyclic AMP primarily increase cell–substrate adhesivity with secondary decreases in motility. Yet preliminary air-blast experiments show, if anything, that these drugs cause a slight decrease in the adhesivity of 3T3 and SV3T3 cells to Falcon plastic (Gail & Johnson, unpublished results). Just how these drugs decrease L-cell motility, alter morphology and increase resistance to trypsin and EGTA remains obscure, but since these drugs are structurally analogous to caffeine they might affect calcium disposition directly or indirectly by raising the concentration of cyclic AMP (Rasmussen 1970).

Our observations on sodium and potassium suggest these ions are relatively unimportant in controlling fibroblast motility or proliferation (Gail & C. W. Boone, unpublished results). Using sucrose to maintain osmolarity at 350 mosmol, we showed that 3T3 fibroblasts moved and proliferated well in media whose cationic compositions were greatly altered (e.g. 56mM-Na, 5.4mM-K; 154mM-Na, $<$0.5mM-K; and 56mM-Na, 73mM-K). Cells died under extreme hyperkaliaemia (56mM-Na, 103mM-K), but our general impression is that fibroblasts are robust towards changes in extracellular sodium and potassium concentrations. We have also shown that high concentrations of tetrodotoxin (4μg/ml), a specific inhibitor of sodium conductance (Takata *et al.* 1966), are without effect on fibroblast motility or proliferation. Likewise 99mM-tetraethylammonium chloride, which decreases potassium conductance when applied internally in axon preparations (Armstrong & Binstock 1965) and which appears to have a similar effect when applied externally to striated muscle (Hagiwara & Watanabe 1955), has no effect on fibroblast motility, though proliferation is inhibited (Gail & C. W. Boone, unpublished results). We conclude from these experiments that sodium and potassium are less important than calcium in regulating fibroblast motility.

We have stressed the importance of calcium in controlling fibroblast motility. Possibly equally important is its function in the control of fibroblast proliferation. We noted that fibroblasts proliferate neither in the presence of EGTA nor after treatment with procaine. Sheppard (1971) reported that spontaneous transformants (3T6) and PyV3T3 viral transformants exhibit contact inhibition of proliferation (6×10^4 cells/cm^2) in the presence of 1mM-theophylline and and 0.1mM-dibutyryl cyclic AMP and subsequently resume superconfluent proliferation when these agents are removed. These facts suggest that calcium helps to control proliferation, especially since the results of the experiments with sodium and potassium (see before) indicate that proliferation is relatively insensitive to large changes in the extracellular concentration of these ions and to tetrodotoxin.

Fibroblast motility and proliferation are often inhibited under the same circumstances. For example, a depletion of the amount of calcium and treatment with procaine inhibit both proliferation and motility. Likewise contact inhibition of proliferation and density inhibition of motility are both evident at confluence, and both are 'released' by making a 'wound' in the confluent monolayer (Todaro *et al.* 1965). It is tempting to suppose that these two physiological cell functions are under common control. However, we found to the contrary that proliferation can be greatly suppressed without much effect on motility (Gail *et al.* 1972). After removal of gammaglobulin by Cohn precipitation, newborn calf serum was heated at 70 °C for 30 min and added to Dulbecco–Vogt salts to make a 'factor-free medium' containing 10% (v/v) heat-treated serum. Sparse populations of BALB/c-3T3 fibroblasts exhibited substantial motility in this medium (D^* was 270–545 μm^2/h), but DNA synthesis (measured by [^{3}H]TdR autoradiography) was almost completely suppressed. Three days later the addition of dialysed calf serum increased the motility (within 10 h), increased [^{3}H]TdR uptake (within 23 h), and cell division (within 30 h). These experiments showed that non-dialysable, heat-labile serum factors are required for BALB/c-3T3 proliferation, and that these cells survive and move well (though not maximally) in the factor-free medium. These results support the contention of Lipton *et al.* (1971) that serum can be fractionated into relatively heat-labile proliferation factor(s) and relatively heat-stable motility and survival factor(s). 10mM-Cytosine arabinoside profoundly inhibits incorporation of [^{3}H]TdR by 3T3 cells within four hours of treatment, yet motility is undiminished within 20 h of such treatment (Gail & C. W. Boone, unpublished results). These experiments again show that 3T3 proliferation and motility are dissociable functions and need not rise and fall together. Our observation that sparse populations (37–70 cells/mm^2) of BALB/c-3T3 in factor-free medium exhibit substantial motility ($D^* > 270$ μm^2/h) and negligible proliferation suggests that density inhibition of motility observed at confluence (500–1200 cells/mm^2) depends on cell–cell adhesions and is not merely a secondary or concomitant effect of proliferative arrest (Njeuma 1971).

SUMMARY

We have shown that 3T3 mouse fibroblasts execute a persistent random walk and that their motility can be measured in terms of D^*, the augmented diffusion constant. Statistical methods for comparing the motilities of two treated populations and for computing confidence limits on D^* have also been developed. The persistent random walk model applies to all fibroblast lines yet examined, and we expect it to be useful in measuring the motility of many other types of cells in tissue culture.

Several extrinsic factors affect motility and must be controlled (or at least measured) to permit meaningful interpretation of motility changes. Cell area density is such a factor, motility generally decreasing with increasing cell area density. This phenomenon, which we call density inhibition of motility, is of direct biological interest, since many tumorigenic cell lines show little or no such effect while the non-tumorigenic 3T3 cell slows dramatically with increasing density. Cell–substrate adhesivity is another extrinsic determinant of motility, D^* generally increasing as the adhesivity decreases. The medium must provide nutritional and serum factors required for motility and a congenial ionic and thermal environment.

We pursued a variety of pharmacological investigations in an effort to elucidate important intrinsic cellular determinants of motility. Aware of the danger of over-interpreting such results, we nonetheless propose that: (1) maximal fibroblast motility requires specific axial structure, and by dissolving the microtubular cytoskeleton Colcemid inhibits this motility; (2) actin and myosin have been found in fibroblasts, and our studies are consistent with the supposition that these contractile proteins provide the motive power in fibroblasts as they do in skeletal muscle and that their action is likewise under calcium control. Cytochalasin B, which might disrupt actin filaments, has been shown to inhibit fibroblast motility and proliferation reversibly. Procaine, which might prevent the release of sequestered calcium into the cytosol, inhibits fibroblast motility and proliferation reversibly. EGTA, which lowers calcium concentration to 0.1 μM, inhibits motility and proliferation reversibly. These functions are relatively unaffected by alterations in external sodium and potassium concentrations, and motility is unaffected by tetrodotoxin or tetraethylammonium chloride, at least within 24 h. All these experiments support the contention that fibroblast motive power is derived from a calcium-regulated interaction between actin and myosin, as in skeletal muscle; (3) fibroblast motility and proliferation are dissociable functions, and density inhibition of motility may be quite unrelated to contact inhibiton of proliferation.

ACKNOWLEDGEMENTS

I thank Dr C. W. Boone, co-author on most of this work, and Mr C. S. Thompson for his reliable laboratory assistance. I acknowledge the suggestions of Dr J. J. Gart, Dr T. Pollard, Mrs Dorothy B. Gail and others at the National Institutes of Health. I am grateful to Rockefeller University Press and Academic Press for permission to include previously published figures.

References

ABERCROMBIE, M. (1961) The bases of the locomotory behavior of fibroblasts. *Exp. Cell Res.* (Suppl.) **8**, 188-198

ABERCROMBIE, M. & HEAYSMAN, J. E. M. (1953) Observations on the social behavior of cells in tissue culture. I. Speed of movement of chick heart fibroblasts in relation to their mutual contacts. *Exp. Cell Res.* **5**, 111-131

ABERCROMBIE, M. & HEAYSMAN, J. E. M. (1954) Observations on the social behavior of cells in tissue culture. II. Monolayering of fibroblasts. *Exp. Cell Res.* **6**, 293-306

ABERCROMBIE, M., HEAYSMAN, J. E. M. & KARTHAUSER, H. M. (1957) Social behavior of cells in tissue culture. III. Mutual influence of sarcoma cells and fibroblasts. *Exp. Cell Res.* **13**, 276-291

ABERCROMBIE, M., HEAYSMAN, J. E. M. & PEGRUM, S. M. (1970) The locomotion of fibroblasts in culture. I. Movements of the leading edge. *Exp. Cell Res.* **59**, 393-398

ADELSTEIN, R. & CONTI, M. A. (1972) The characterization of contractile proteins from platelets and fibroblasts. *Cold Spring Harbor Sym. Quant. Biol.* **37**, 599-605

AMBROSE, E. J. (1961) The movements of fibrocytes. *Exp. Cell Res.* (Suppl.) **8**, 54-73

ARMSTRONG, C. M. & BINSTOCK, L. (1965) Anomalous rectification in the squid giant axon injected with tetraethylammonium chloride. *J. Gen. Physiol.* **48**, 859-872

BENDALL, J. R. (1969) in *Muscles, Molecules and Movement*, pp. 67-69, American Elsevier Publishing Company, New York

BORISY, G. G. & TAYLOR, E. W. (1967) The mechanism of action of colchicine. Binding of colchicine-^{3}H to cellular protein. *J. Cell Biol.* **34**, 525-533

BUCKLEY, T. K. & PORTER, K. R. (1967) Cytoplasmic fibrils in living cultured cells. A light and electron microscope study. *Protoplasma* **64**, 349-380

CARTER, S. B. (1967) Effects of cytochalasins on mammalian cells. *Nature (Lond.)* **213**, 261-264

COMAN, D. R. (1944) Decreased mutual adhesiveness, a property of cells from squamous cell carcinoma. *Cancer Res.* **4**, 625-629

DEFENDI, V., LEHMAN, J. & KRAEMER, P. (1963) Morphologically normal hamster cells with malignant proprieties. *Virology* **19**, 592-598

EBASHI, S. & ENDO, M. (1968) Calcium ion and muscle contraction. *Prog. Biophys. Mol. Biol.* **18**, 125-183

FEINSTEIN, M. B. (1963) Inhibition of caffeine vigor and radiocalcium movements by local anesthetics in frog sartorius muscle. *J. Gen. Physiol.* **47**, 151-172

FEINSTEIN, M. B. & PAIMRE, M. (1969) Pharmacological action of local anesthetics on excitation-contraction coupling in striated and smooth muscle. *Fed. Proc.* **28**, 1643-1648

FÜRTH, R. (1920) Die brownsche Bewegung bei Berücksichtigung einer Persistenz der Bewegungsrichtung. *Z. Phys.* **12**, 244-256

GAIL, M. H. & BOONE, C. W. (1970) The locomotion of mouse fibroblasts in tissue culture. *Biophys. J.* **10**, 980-993

GAIL, M. H. & BOONE, C. W. (1971*a*) Density inhibition of motility in 3T3 fibroblasts and their SV40 transformants. *Exp. Cell Res.* **64**, 156-162

GAIL, M. H. & BOONE, C. W. (1971*b*) Effect of Colcemid on fibroblast motility. *Exp. Cell Res.* **65**, 221-227

GAIL, M. H. & BOONE, C. W. (1971*c*) Cytochalasin effects on BALB/3T3 fibroblasts: dose-dependent, reversible alteration of motility and cytoplasmic cleavage. *Exp. Cell Res.* **68**, 226-228

GAIL, M. H. & BOONE, C. W. (1972*a*) Cell–substrate adhesivity: a determinant of cell motility. *Exp. Cell Res.* **70**, 33-40

GAIL, M. H. & BOONE, C. W. (1972*b*) Procaine inhibition of fibroblast motility and proliferation. *Exp. Cell Res.* **73**, 252-255

GAIL, M. H., SCHER, C. D. & BOONE, C. W. (1972) Dissociation of cell motility from cell proliferation in BALB/c-3T3 fibroblasts. *Exp. Cell Res.* **70**, 439-443

GOLDMAN, R. D. & FOLLETT, E. A. C. (1969) The structure of the major cell processes of isolated BHK21 fibroblasts. *Exp. Cell Res.* **57**, 263-276

GRUENER, R. (1967) Caffeine contractures in sarcolemma-free muscle fibres. *J. Physiol. (Lond.)* **191**, 106P-108P

HAGIWARA, S. & WATANABE, A. (1955) The effect of tetraethylammonium chloride on muscle membrane examined with an intracellular electrode. *J. Physiol. (Lond.)* **129**, 513-527

HERTZ, R. & WEBER, A. (1965) Caffeine inhibition of Ca uptake by muscle reticulum. *Fed. Proc.* **24**, 208

HOLTZER, H. & SANGER, J. W. (1972) Cytochalasin B: problems in interpreting its effect on cells. *Dev. Biol.* **27**, 443-446

HUXLEY, H. E. (1969) The mechanism of muscular contraction. *Science (Wash. D.C.)* **164**, 1356-1366

ISHIKAWA, H., BISCHOFF, R. & HOLTZER, H. (1969) Formation of arrowhead complexes with heavy meromyosin in a variety of cell types. *J. Cell Biol.* **43**, 312-328

INGRAM, V. M. (1969) A side view of moving fibroblasts. *Nature (Lond.)* **222**, 641-644

JAMES, D. W. & TAYLOR, J. F. (1969) The stress developed by sheets of chick fibroblasts *in vitro*. *Exp. Cell Res.* **54**, 107-110

JOHNSON, G. S. & PASTAN, I. (1972) Cyclic AMP increases the adhesion of fibroblasts to substratum. *Nat. New Biol.* **236**, 247-249

JOHNSON, G. S., FRIEDMAN, R. M. & PASTAN, I. (1971) Restoration of several morphological characteristics of normal fibroblasts in sarcoma cells treated with adenosine-3′ : 5′-cyclic monophosphate and its derivatives. *Proc. Natl. Acad. Sci. U.S.A.* **68**, 425-429

JOHNSON, G. S., MORGAN, W. D. & PASTAN. I. (1972) Regulation of cell motility by cyclic AMP. *Nature (Lond.)* **235**, 54-56

KOROHODA, W. (1971) Interrelations of motile and metabolic activities in tissue culture cells. I. Respiration in chicken embryo fibroblasts actively locomoting, contact inhibited, and suspended in a fluid medium. *Folia Biol. (Kraków)* **19**, 41-51

LIPTON, A., LINGER, I., PAUL, D. & HOLLEY, R. W. (1971) Migration of mouse fibroblasts in response to a serum factor. *Proc. Natl. Acad. Sci. U.S.A.* **68**, 2799-2801

MACIEIRA-COELHO, A. (1967) Dissociation between inhibition of movement and inhibition of division in RSV transformed human fibroblasts. *Exp. Cell Res.* **47**, 193-200

MISZURSKI, B. (1949) The effects of colchicine on resting cells in tissue cultures. *Exp. Cell Res.* (Suppl.) **1**, 450-451

NJEUMA, D. L. (1971) Non-reciprocal density-dependent mitotic inhibiton in mixed cultures of embryonic chick and mouse fibroblasts. *Exp. Cell Res.* **66**, 244-250

PERRY, M. M., JOHN, H. A. & THOMAS, N. S. T. (1971) Actin-like filaments in the cleavage furrow of newt egg. *Exp. Cell Res.* **65**, 249-253

PORTER, K. (1966) in *Principles of Biomolecular Organization (Ciba Found. Symp.)*, pp. 308-345, Little Brown, Boston

RASMUSSEN, H. (1970) Cell communication, calcium ion, and cyclic adenosine monophosphate. *Science (Wash. D.C.)* **170**, 404-412

SCHROEDER, T. E. (1969) The role of contractile ring filaments in dividing *Arabacia* egg. *Biol. Bull. (Woods Hole)* **137**, 413-414

SHEPPARD, J. R. (1971) Restoration of contact-inhibited growth to transformed cells by dibutyryl adenosine 3′ : 5′-cyclic monophosphate. *Proc. Natl. Scad. Sci. U.S.A.* **68**, 1316-1320

SMITH, H. S., SCHER, C. D. & TODARO, G. J. (1971) Induction of cell division in medium lacking serum growth factors by SV 40. *Virology* **44**, 359-370

TAKATA, M., MOORE, J. W., KAO, C. Y. & FURMAN, F. A. (1966) Blockage of sodium conductance increase in lobster giant axon by tarichatoxin (tetrodotoxin). *J. Gen. Physiol.* **49**, 977-988

TILNEY, L. G. (1968) Studies on the microtubules in heliozoa. IV. The effect of colchicine on the formation and maintenance of the axopodia and the redevelopment of pattern in *Actinosphaerium nucleofilum* (Barrett). *J. Cell Sci.* **3**, 549-562

TODARO, G. J., MATSUYA, Y., BLOOM, S., ROBBINS, A. & GREEN, H. (1965) in *Growth Regulating Substances for Animal Cells in Culture* (Defendi, V. & Stoker, M., eds.), pp. 87-101, Wistar Institute Press, Philadelphia

VON HAMM, E. & CAPPEL, L. (1940) The effect of hormones upon cell growth in vitro. *Am. J. Cancer* **39**, 354-359

WESSELLS, N. K., SPOONER, B. S., ASH, J. F., BRADLEY, M. O., LUDUEÑA, M. A., TAYLOR, E. L., WRENN, J. T. & YAMADA, K. M. (1971) Microfilaments in cellular and developmental process. Contractile microfilament machinery of many cell types is reversibly inhibited by cytochalasin B. *Science (Wash. D.C.)* **171**, 135-143

ZIGMUND, S. H. & HIRSCH, J. G. (1972) Cytochalasin B: inhibition of D-2-deoxyglucose transport into leukocytes and fibroblasts. *Science (Wash. D.C.)* **176**, 1432-1434

Discussion

Wolpert: Although your conclusions may be right, I feel that you cannot rely too much on the data obtained by applying chemical agents outside the cell. Physiologists say that it is very hard to change the internal concentration of calcium. Thus, the external addition of EGTA, which allows a lot of material to permeate through the whole membrane and which interferes in other processes, surely cannot be used to substantiate the proposed function of calcium internally.

Gail: I agree it is difficult to interpret these pharmacological data, but I was encouraged because the effects of EGTA were reversible. Within 40 h of EGTA-treated cells being placed in EGTA-free complete medium, the cells

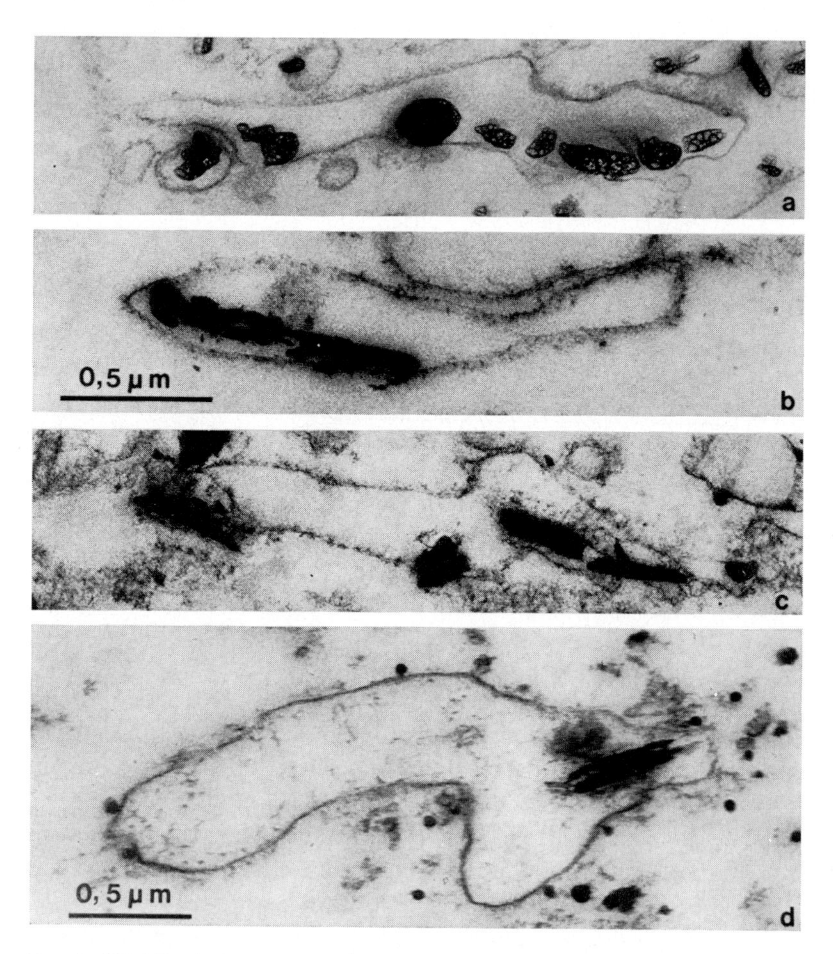

FIG. 1 (Wohlfarth-Bottermann). (*a*)—(*c*) Calcium oxalate precipitations in the endoplasmic reticulum of *Chaos chaos*. (*d*) Precipitations of calcium in the endoplasmic reticulum of *Amoeba proteus* (from Reinold & Stockem 1972).

resumed normal motility and proliferation rates; only about 20% of the cells died.

Wohlfarth-Bottermann: Braatz & Komnick (1970) and Reinold & Stockem (1972) discovered a calcium-pumping system in slime mould plasmodia and in amoebae by cytochemical methods, after a short pre-fixation or after glycerination. Fig. 1 shows crystals of calcium oxalate in calcium-accumulating vacuoles of *Amoeba proteus* and *Chaos chaos*. The transport of added calcium acetate through the membrane is ATP-dependent, just as in the vesicles of the sar-

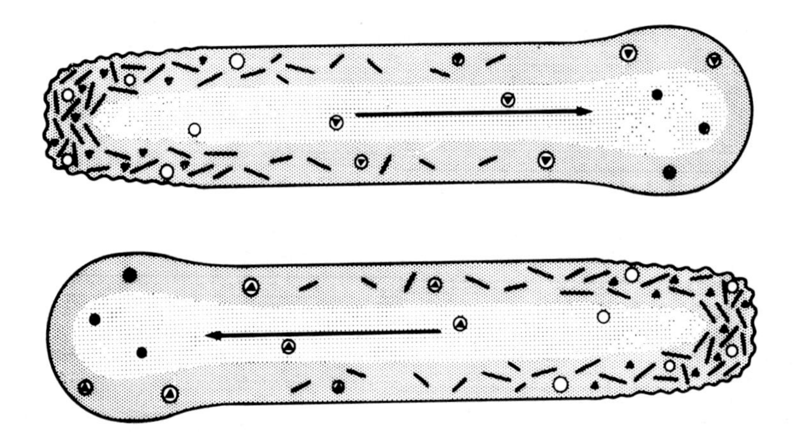

FIG. 2 (Wohlfarth-Bottermann). Schematic representation of protoplasmic shuttle streaming in a plasmodial strand of *Physarum*. The rods indicate fibrils, which contract after assumed calcium release at the tail region. At the front region calcium (▲) is presumably accumulated within vacuoles (from Komnick *et al.* 1972).

coplasmic reticulum (L-system) of the cross-striated muscle. These vacuoles are able to accumulate calcium ions from a concentration of 10^{-6}M. Calcium ions are probably important for the regulation of contraction of the actomyosin system in amoebae and slime moulds.

If this is so, we should be able to predict for the shuttle streaming in *Physarum*, where we know the exact location of the contracted and the relaxed areas, that the vacuoles will be empty on the contracted pole, whereas at the relaxed pole, the vacuoles will be filled with calcium crystals. Analogously with the shuttle streaming, this situation should change within every one or two minutes (Fig. 2). Our preliminary results are in accordance with this prediction.

Porter: Is there any evidence of this kind of calcium chelator in cultured tissue cells, such as 3T3 cells?

Wohlfarth-Bottermann: Dr Allison has told me (personal communication) that a similar system of vacuoles might exist in tissue cells or cultured cells.

Allison: Two situations have been found: in one, the vesicles in the mitotic spindle may have a similar function, and in the other, as found by Woodin, if polymorphs are exposed to leucocidin, vesicles in the cytoplasm accumulate large amounts of calcium phosphate. It is not yet shown formally that these are the same as the sarcoplasmic reticulum, but it is likely that they represent similar systems able to concentrate calcium. Also, in many cells such as macrophages, there are smooth vesicles in the cytoplasm the function of which is totally unknown, but some of which might be able to store calcium and release it when appropriately stimulated.

Gingell: Would not a more direct way of investigating the action of calcium, magnesium and EDTA have been iontophoresis into the fibroblast? Surely that would give you rapid results.

Gail: That is a good idea. Nelson *et al.* (1972) have impaled fibroblasts with electrodes and measured trans-membrane potentials of 10–16 mV. So the experiments you propose seem technically feasible and of great interest.

Gingell: Coming back to the adhesion measurements, if there is any case where one could be sure that adhesion is not being mediated by a specific chemical bond, it would between the cell and Pyrex or plastic. However, the degree of absorption of protein for these surfaces is not known. This increased adhesivity, as you called it, to Pyrex as opposed to plastic is precisely what one would predict from purely physical analysis of the attractive forces (Parsegian & Gingell 1972).

Gail: My experiments are in accord with theory in that respect.

Curtis: Have you tested whether low concentrations of calcium affect adhesiveness?

Gail: At 0.2mM-calcium I can detect no effect on cell–substrate adhesiveness. This puzzled me until I found a report by Weiss (1960) who noticed the same thing at these calcium concentrations. However, EGTA, which drastically reduces calcium concentration, does facilitate air-blast distraction of the cells, and this is not unexpected since similar chelating agents are commonly used to distract cells.

Curtis: But isn't the effect on motility the exact opposite of the one you would expect in view of the effects on adhesiveness?

Gail: Yes, and that is why measurement of cell–substrate adhesiveness was an appropriate control. In spite of the fact that cell–substrate adhesiveness was decreased during EGTA treatment, which would be expected to promote motility, the cells exhibited dramatically reduced motility. This shows EGTA does not slow the cells by increasing cell–substrate adhesiveness and suggests that EGTA affects the contractile mechanism directly.

Albrecht-Bühler: Did you ever find any influence of conditioning of the medium on cell motility?

Gail: Normally we let the cells migrate for ten hours and then watch their motility over a further 12 h. I have never noticed systematic slowing over the first two days of filming under our conditions. Over longer periods slowing can be demonstrated. Concerning early motility, I have not noticed that conditioning is required to attain maximal D^* values.

Albrecht-Bühler: How long after plating do you begin your measurements?

Gail: We plate the cells, incubate them for 24 h, change the medium and start time lapse studies at that point. Only ten hours after the initiation of time lapse cinematography do we begin the motility measurements.

Huxley: Over how long does EGTA act? What motion of the cells do you see?

Gail: EGTA acts over about 5–10 h. After 20 h about half the cells become spherical and refractile, the rest remaining flattened, but the motility of both the flattened and spherical forms is negligible. Ruffling is diminished in the presence of EGTA. During recovery in complete medium, normal motility returns before normal proliferation rate. The cells take about 40 h to recover.

The effect of procaine is different. Ruffling, motility and, as far as I can tell, proliferation all stop almost immediately (within 1–3 h), and the recovery is correspondingly faster (about 10 h).

Curtis: Couldn't the decreased motility simply be due to an almost total loss of adhesion of the cells to the substrate, so that the cells are rounded up?

Gail: With procaine, cell–substrate adhesiveness does not change significantly. With EGTA, the cells adopt two forms: flattened and round. I suggest that the rounded forms are less adhesive than the flattened forms, but both exhibit very little motility.

Trinkaus: How does procaine act?

Gail: Feinstein & Paimre (1969) have reviewed the evidence that procaine prevents the release of calcium into the cytosol by maintaining low calcium permeability in the membranes bounding intracellular calcium stores. Gruener (1967) showed that sarcolemma-free fibres from the frog sartorius muscle contract in response to calcium, even in the presence of 3mM-procaine. So procaine probably acts by preventing calcium release and not by a direct effect on troponin, actin or myosin (see pp. 296 and 297).

Gustafson: In the sea-urchin larva, many local anaesthetics such as procaine and tetracaine, and also a number of spasmolytics such as papaverine, cause retardation of the second phase of gastrulation (Gustafson & Toneby 1970). However, a large increase in the potassium content of the sea water has little effect in this respect.

de Petris: Replacement of up to 75% of sodium by potassium had no effect at all on cap formation. However, EGTA, at a concentration which reduced that of free extracellular calcium to less than 10^{-8}M, was without effect on cap formation. In our experiments, however, the preincubation with EGTA was not prolonged as in yours and therefore we probably did not deplete appreciably the intracellular calcium.

Porter: In the fish scale, the action of potassium is probably on the nerve; it doubtless depolarizes the membrane and blocks any further impulses to the chromatophore. With calcium, we recorded similar effects to those you observed; a low concentration of calcium stopped pigment motion in the aggregated state and blocked the reassembly of microtubules.

Gail: You have intrigued me, Dr Porter, because you seem to have demonstrated that the microtubules are providing the motive force for the motion of these pigment granules. The mechanism I was considering was that energy is used to establish the highly organized radial microtubular configuration in which the granules are dispersed. This organized structure should be in a low entropy state, and it is natural to imagine that a trigger, such as potassium, could initiate spontaneous reversion to the high entropy state of microtubular disarray in which the pigment granules have moved in and are contracted.

You have previously stressed the importance of microtubules in guiding fibroblast motion and in maintaining cell shape (Porter 1966). We have considered how colchicine can decrease fibroblast motility, but I still hesitate to assume that microtubules provide the motive force for the displacement of whole fibroblasts in the same way they might provide such forces for pigment granules or other intracellular organelles.

Abercrombie: Many years ago, we suggested there were two explanations for the density-dependent decline of motility (Abercrombie & Heaysman 1953). One was mutual interference by sticking together, and the other was contact inhibition. You seem now to be concentrating on mutual interference. From what one knows about contact inhibition it would seem to slow cells, simply because after collision they take time to change direction. Would you agree?

Gail: Yes. You have described how when two cells collide they are momentarily immobilized thus tending to reduce the average square displacement. They also change direction, making them less persistent and, hence, slower random walkers. Both these effects of collision would be expected to diminish D^*, my measure of motility. The formation of transient cell–cell adhesions must also reduce D^*. Further, 3T3 cells might exhibit greater density inhibition of motility than tumorigenic lines because of their greater mutual adhesiveness.

Curtis: Surely, Coman's method (Coman 1961), which you use, is one of the more inaccurate methods of measuring cell–substrate adhesiveness, since most of the energy measured must be the energy expended in deforming the cells as they peel apart.

Gail: Coman's method (1944) has the virtue of intuitive simplicity. Two adhering cells are impaled, one on a rigid needle and the other on a flexible needle. As the cells are pulled apart, the flexible needle is deflected to an extent proportional to the tension between the cells. I agree that this is no measure of the energy of cell–cell adhesive bonds, but it gives an idea of the maximum tensions required to pull cells apart.

Gingell: If the shape of the contact area is constant, it does not matter what happens in the cell with regard to stretching, tension and distortion when you pull the cells apart. If you alter the shape of the contact area when you are pulling the surfaces apart you get a difference in the force; the energy is, of course, constant but the force is dependent on geometry, so if part of the work is tending to slide the surfaces tangentially as opposed to pulling them apart normally you will measure a smaller force (Gingell 1971). Can you distinguish between bending the cells and adhesion *per se* by the air blast?

Gail: Certainly the cell profile is important. The air-blast method generates shearing forces by the rapid displacement of overlying medium and these shearing forces depend on cell profile. As with other shearing methods used to measure cell–substrate adhesiveness, one cannot measure the energy required to distract the cell, but the air-blast method at least gives one a qualitative assessment of shearing forces required to remove the cells.

Curtis: I think that Dr Gingell has put his finger on some of the difficulties inherent in Dr Gail's technique. Bikerman (1957) described the theory of peeling of adhesive surfaces. Peeling appears to happen with Coman's method, and so simple measurements of the force applied give results which cannot simply be interpreted as measurements of the adhesive force. Additionally 'drainage' of the medium into the opening gap might produce transient forces far larger than the adhesive force to be overcome.

Wolpert: That would make no difference whatsoever; the total force will simply be transmitted by the deformed cell.

Trinkaus: Another objection is that raised by Weiss & Coombs (1963), namely that the cell surface might fracture upon separation of the cells.

Porter: In defence of Coman's studies, I am sure he never felt they were better than crude. McNutt *et al.* (1971) have studied the number of nexuses ('gap junctions') between cervical carcinoma cells; apparently there are substantially fewer between the carcinoma cells than between cells of the normal epithelium.

Gail: McNutt (unpublished findings, 1972) has found fewer nexuses between SV3T3 cells than between 3T3 controls. More work must be done before we can come to any firm conclusions, especially in view of Dr Heaysman's comments (p. 187).

Wessells: We have been using drugs like papaverine both on single cells and on the salivary epithelial system (Ash *et al.* 1973). The results with papaverine might be pertinent in attempting to distinguish between intra- and extra-cellular functions of calcium. In smooth muscle papaverine is known to be a relaxant and an agent that will prevent contraction if one stimulates the muscle cell electrically. It is thought to prevent the entry of calcium from outside the cells. When one applies papaverine to the salivary system at times when filaments rather like those in moving cells are present, one observes the same relaxation phenomenon that one sees with cytochalasin, although there is no visible effect on the morphology of the filament system. Now, if one exposes a papaverine-treated epithelium to 3% trypsin–pancreatin, there is no sign of any loss of integrity in the epithelium. Contrast this result with effects of culturing in the absence of calcium, where there is also a relaxation and the epithelium looks exactly like an epithelium which has been treated with papaverine or cytochalasin. But attempts to separate the epithelium from mesoderm with trypsin and pancreatin result in disintegration of the epithelium into single cells. The conclusion is that one may be able to use a drug like papaverine to alter intracellular calcium concentrations and thereby change processes which might be dependent on filaments. But, as that occurs, adhesive relations between the cells are apparently *not* altered.

Allison: I am still puzzled by the fact that you observe a near-random walk, because in the macrophages we have studied, we see obvious persistence. Is this because your experiments are carried out over a rather long time? If the time span is shorter, do you find a greater departure from a random walk?

Gail: Yes. We find that cells persist in their direction of motion over 2.5-h but not 5-h intervals (Gail & Boone 1970). If I could observe these cells over a long time interval, say 100 h, I would happily use the pure random walk theory since these cells would in effect be executing a pure random walk. Note that for very large times equation (1) in my paper reduces to $\langle T^2 \rangle = 4D^*t$, which is characteristic of the pure random walk. The more elaborate persistent random walk theory [equations (1) and (2) in text] is required precisely because we must work within intermitotic time intervals, which are only about 20 h.

Albrecht-Bühler: In this case, doesn't the value of the augmented diffusion constant, D^*, depend on time?

Gail: No, it does not.

Allison: The time factor should determine the shoulder but not the slope.

Albrecht-Bühler: But D^* is dependent on ρ, the constant of persistence tendency [see equation (1)]. If the walk of the cells becomes more random at larger time intervals, ρ has to decrease and hence so does D^*.

Gail: If I observe a group of cells over a long time, I will have no evidence that the motion is persistent over short time intervals. That is, the plot of mean square displacement against time will appear to be a straight line passing through the origin with slope $4D^*$. However, the value D^*, determined from this slope on a large time scale, will be given by $D^* = 4D(1 + \rho)$, where D is the diffusion constant which would have been observed had there been no persistence even over short time intervals. Thus the diffusion constant observed on the long time scale is augmented by persistence over short time intervals, even though one has no direct evidence for this short interval persistence from studies over long time intervals. Equations (1) and (2) (in text) enable us to estimate the same D^* which would have been observed over long time intervals from data obtained over short time intervals where the effects of persistence are obvious.

Albrecht-Bühler: If you change the interval between two observations, do you find any influence on the constants D^* or ρ?

Gail: No; this is a continuous time scale and I have used 2.5-h intervals merely for convenience. Other intervals give the same D^* values.

References

ABERCROMBIE, M. & HEAYSMAN, J. E. M. (1953) Observations on the social behaviour of cells in tissue culture. I. Speed of movement of chick heart fibroblasts in relation to their mutual contacts. *Exp. Cell Res.* **5**, 111-131

ASH, J. F., SPOONER, B. S. & WESSELLS, N. K. (1973) Effects of papaverine and calcium-free medium on salivary gland morphogenesis. *Dev. Biol.*, in press

BIKERMAN, J. J. (1957) Formation and rupture of adhesive joints. *Proc. 2nd Int. Congr. Surf. Sci.* **3**, 427-432

BRAATZ, R. & KOMNICK, H. (1970) Histochemischer Nachweis eines Calcium-pumpenden Systems in Plasmodien von Schleimpilzen. *Cytobiologie* **2**, 457-463

COMAN, D. R. (1944) Decreased mutual adhesiveness, a property of cells from squamous cell carcinoma. *Cancer Res.* **4**, 625-629

COMAN, D. R. (1961) Adhesiveness and stickiness: two independent properties of the cell surface. *Cancer Res.* **21**, 1436-1438

FEINSTEIN, M. B. & PAIMRE, M. (1969) Pharmacological action of local anesthetics on excitation-contraction coupling in striated and smooth muscle. *Fed. Proc.* **28**, 1643-1648

GAIL, M. H. & BOONE, C. W. (1970) The locomotion of mouse fibroblasts in tissue culture. *Biophys. J.* **10**, 980-993

GINGELL, D. (1971) Computed force and energy of membrane interactions. *J. Theor. Biol.* **30**, 121-149

GRUENER, R. (1967) Caffeine contractures in sarcolemma-free muscle fibres. *J. Physiol. (Lond.)* **191**, 106P-108P

GUSTAFSON, T. & TONEBY, M. I. (1970) On the role of serotonin and acetylcholine in sea urchin morphogenesis. *Exp. Cell Res.* **62**, 102-117

KOMNICK, H., STOCKEM, W. & WOHLFARTH-BOTTERMANN, K. E. (1972) Ursachen, Begleitphänomene und Steuerung zellulärer Bewegungserscheinungen. *Fortschr. Zool.* **21**, 1-74

MCNUTT, N. S., HERSHBERG, R. A. & WEINSTEIN, R. S. (1971) Further observations on the occurrence of nexuses in benign and malignant human cervical epithelium. *J. Cell Biol.* **51**, 805-825

NELSON, P. G., PEACOCK, J. & MINNA, J. (1972) An active electrical response in fibroblasts. *J. Gen. Physiol.* **60**, 58-71

PARSEGIAN, V. A. & GINGELL, D. (1972) *J. Adhes.*, in press

PORTER, K. (1966) in *Principles of Biomolecular Organization (Ciba Found. Symp.)*, pp. 308-345, Little, Brown, Boston and Churchill, London

REINOLD, M. & STOCKEM, W. (1972) Darstellung eines ATP-sensitiven Membransystems mit Ca^{+}-transportierender Funktion bei Amöben. *Cytobiologie* **4**, 182-194

WEISS, L. (1960) The adhesion of cells. *Int. Rev. Cytol.* **9**, 167-220

WEISS, L. & COOMBS, R. R. A. (1963) The demonstration of the rupture of cell surfaces by an immunological technique. *Exp. Cell Res.* **30**, 331-338

Interactions of normal and neoplastic fibroblasts with the substratum*

JU. M. VASILIEV and I. M. GELFAND

Institute of Experimental and Clinical Oncology, Academy of Medical Sciences of USSR and Laboratory of Mathematical Biology, Moscow State University

Abstract The transition of cultured fibroblasts from an unattached to an attached state proceeds in two main stages: radial attachment and polar attachment. In the first stage, the ring of flattened cytoplasm spreads on the substratum (lamellar cytoplasm). In the second stage, anterio-posterior polarity is developed; lamellar cytoplasm and sites of attachment to the substratum become preferentially localized in the front part of the cell. The morphology of normal embryo-fibroblasts in these two stages is described and possible mechanisms of the formation of lamellar cytoplasm and of the polarization are briefly discussed. The morphology of the regions of lamellar cytoplasm was found to be abnormal in the cultures of several lines of neoplastic fibroblasts. Abnormal formation of this structure was observed at the stage of radial attachment and became even more apparent during polar attachment, and possibly is the basis of deficient attachment of these neoplastic cells to the substratum. Microcinematographic analysis suggests that deficiency in attachment to the substrate is chiefly responsible for modifying the locomotory behaviour of neoplastic cells and, in particular, the results of cell–cell collisions. These modifications lead to formation of diverse morphological patterns in isolated cultures of neoplastic fibroblasts and in mixed cultures of neoplastic and normal fibroblasts; examples of these patterns are described. It is suggested that deficient attachment of the neoplastic cells to the intercellular substrata is important in the invasive growth of these cells *in vivo*.

The contact with solid substratum induces the transition of cultured fibroblasts from an unattached into an attached state. We shall discuss here one aspect of this transition, namely, the changes of cell morphology caused by contact with the substratum. The morphological characteristics related to the attached state were abnormal in several lines of neoplastic fibroblasts. We shall describe these abnormalities and discuss their possible effect on the locomotory behaviour of neoplastic cells.

The main methods for examination of cultures were scanning and trans-

* Contributed *in absentia*.

mission electron microscopy as well as time lapse microcinematography. Trypsinized cells of mouse embryos in their second passage were used and are referred to as normal cells. The hamster embryo cells used in several experiments behaved similarly to mouse cells. The lines of neoplastic fibroblasts included: (a) L line of mouse fibroblasts; (b) CIM and (c) S-40 lines, both established in this laboratory in 1971 from the cells of two primary mouse sarcomas induced by the implantation of plastic films; and (d) HEC-40 line, established from a fibroblast culture treated with green monkey cytomegalovirus (by courtesy of Dr. G. I. Deitchman, Laboratory of Immunology, Institute of Experimental and Clinical Oncology). The L cells were weakly oncogenic: 10^6 cells induced tumours in pre-irradiated C3H mice. All other lines were highly oncogenic: transplantation of 10–100 cells to syngeneic hosts was sufficient to produce rapidly growing tumours.

Unless otherwise stated, the substratum was plane glass. Eagle or 199 medium with 10% serum was used. The cells for scanning electron microscopy were fixed in 2.5% glutaraldehyde, dehydrated in alcohol and shadowed with gold (for details see Vasiliev *et al.* 1969, 1970; Domnina *et al.* 1972).

MORPHOLOGICAL ALTERATIONS OF NORMAL CELLS ACCOMPANYING ATTACHMENT

These alterations have two main stages (Witkowski & Brighton 1971; Domnina *et al.* 1972). In the early phases (i.e. during the radial attachment) the cell has radial symmetry; later the anterior–posterior polarity is developed (the stage of polar attachment). The transition from the radial to polar stage has yet to be studied in detail. Before we compare the morphology of normal and neoplastic cells at these two stages of attachment, we shall define some of the terms used in the text.

By lamellar cytoplasm or lamelloplasm we refer to the region of cytoplasm at the cell periphery which is well spread on the substratum. The thickness of lamellar cytoplasm varies from 0.1 to 1.0 μm. Its upper surface appears relatively flat by scanning electron micrography, although the degree of this flatness depends considerably on the method of fixation and dehydration (Boyde & Veselý 1972). Lamellipodia (Abercrombie *et al.* 1970) are formed at certain edges of lamellar cytoplasm. A cortical layer up to 0.1–0.2 μm thick is seen between the upper and lower membranes of lamellar cytoplasm (Fig. 1). This is the layer from which ribosomes and other organelles are excluded; it is filled with a moderately electron-dense substance that looks amorphous or contains 5–7 nm microfilaments arranged parallel to each other or in the form of a network. Regions where the membrane is close to the substratum are seen

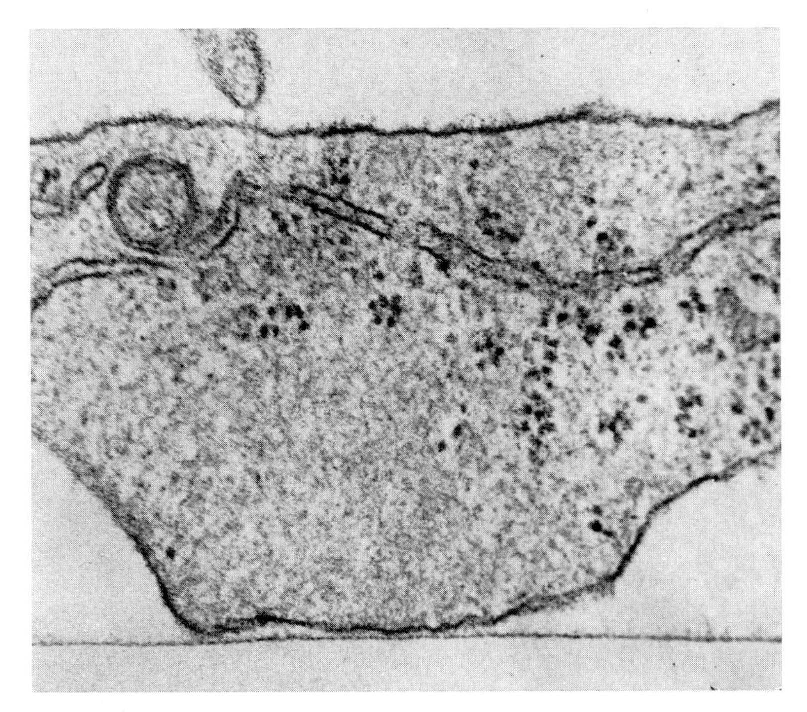

FIG. 1. Section through the lamellar cytoplasm of the front part of a normal fibroblast. The plane of section is perpendicular to the cell axis. The region of close approach to the substratum can be seen. × 72 000.

at the lower cell surface. These structures have been described by Abercrombie *et al.* (1971) and Brunk *et al.* (1971).

Fibrillar bundles consist of parallel closely packed 5–7 nm microfilaments (Fig. 2). One or two microtubules are usually seen in sections near these bundles and in parallel with them. These bundles have been observed in fibroblasts by several authors (Buckley & Porter 1967; Goldman & Follett 1969; Spooner *et al.* 1971).

All cytoplasmic outgrowths which have a diameter of about 0.1–0.2 μm are described by the term 'microvilli'.

The endoplasm is the region of cytoplasm around the nucleus, which contains numerous vesicular organelles (sacs of endoplasmic reticulum, lysosomes, inclusions etc.). Peripheral cytoplasm that does not contain these inclusions will be referred to as ectoplasm. Lamellar cytoplasm is part of the ectoplasm; cytoplasmic processes which are not spread on the substratum are also regarded as parts of the ectoplasm.

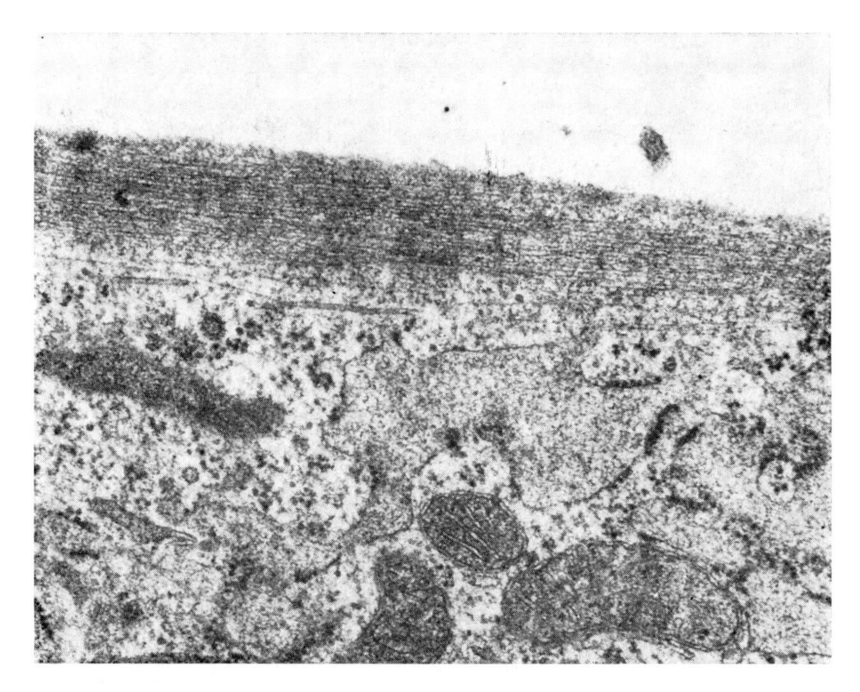

FIG. 2. Fibrillar bundle accompanied by a microtubule near the lateral edge of the polarized cell. The plane of section is parallel to the substratum. × 30 000.

Radial attachment

Thirty minutes after seeding, some cells had assumed a spherical shape with several microvilli or lamellipodia spreading upon the substratum from their lower pole. Most cells (80–90 %) at that time had already a flatter central part and a ring of lamellar cytoplasm about 8–12 μm wide (Fig. 3a). The edge of this ring was usually smooth, except occasionally for a few short microvilli. The preferential orientation of filaments in lamellar cytoplasm was radial; these filaments were not packed into discrete fibrillar bundles.

Areas of close approach to the substratum were seen at the lower surface of lamellar cytoplasm, especially near its external edge (Fig. 4), although they were also seen at the lower surface of the central part of the cell.

Lamellar cytoplasm is formed only upon the surface of an appropriate ('adhesive') substratum. When glass was covered with a non-adhesive film of phospholipids and cuts were made (by a blade) in this film, the cells attached themselves only to the narrow strips of glass revealed through the cuts (Ivanova

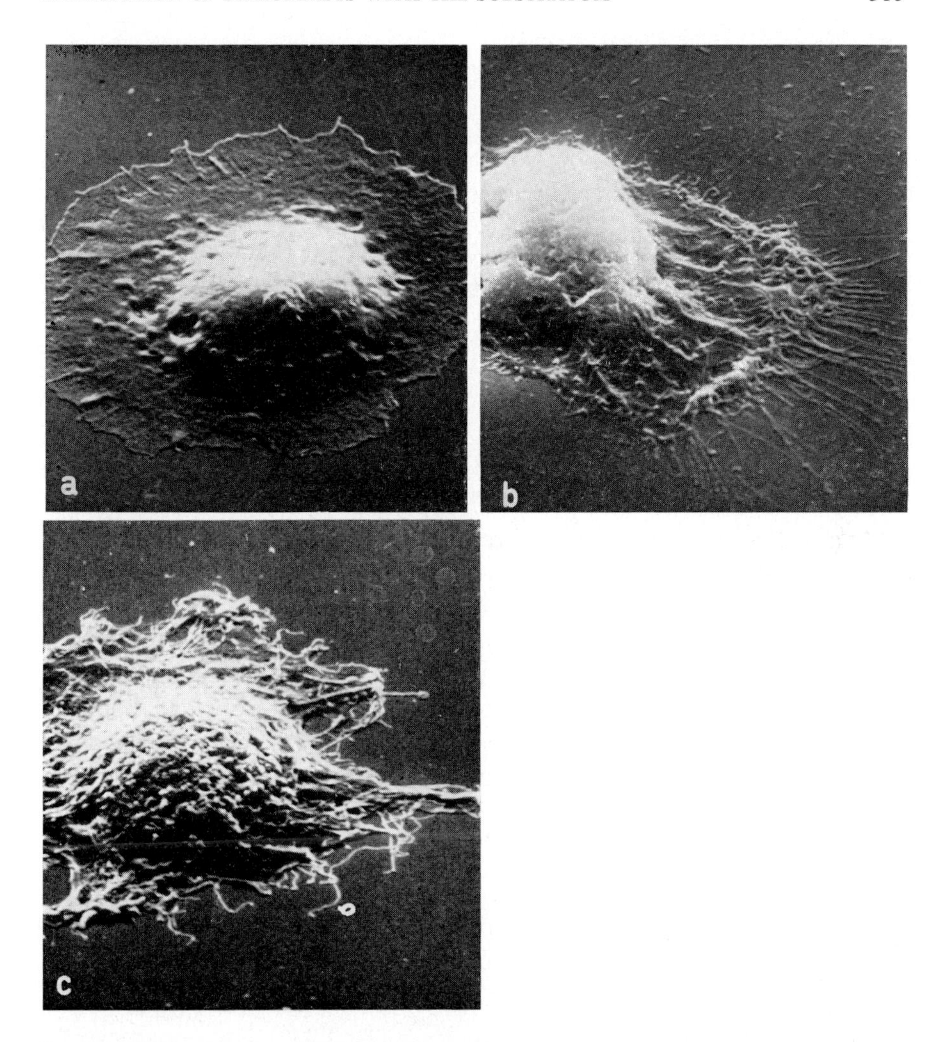

FIG. 3. Scanning electron micrographs of normal and neoplastic cells 30 min after seeding: (a) normal fibroblast showing the wide ring of lamellar cytoplasm (× 2 800); (b) a cell of the L line with its uneven lamellar part with long microvilli at the edge (× 2 800); (c) a cell of the S-40 line showing the lamellar cytoplasm with jagged edge. × 1 900.
The tilt angle in all the scanning electron micrographs was 45°.

& Margolis 1973). In this event, the cells had an elongated not circular form 30 min after seeding, because lamellar cytoplasm was developed only upon the adhesive areas of the substratum (Fig. 5a).

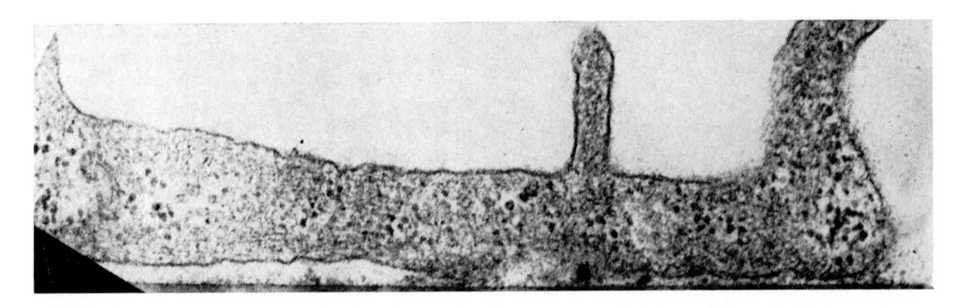

FIG. 4. Section of lamellar cytoplasm of a normal fibroblast 30 min after seeding upon the substratum. × 82 800.

Polar attachment

As an example of a well-polarized cell it is convenient to describe a fibroblast moving into a wound made in dense culture. The wide plate of lamellar cytoplasm (anterior lamella) forms the front part of such a cell; numerous regions of close approach are seen at its lower surface. The endoplasm is divided into two parts: the anterior part is well attached to the substratum (its lower surface is almost flat and contains many regions of close approach) and the posterior part is unattached (its lower surface has curved cylindrical shape without any regions of close approach). The transition from an attached to an unattached part of the endoplasm varies from gradual in some cells to

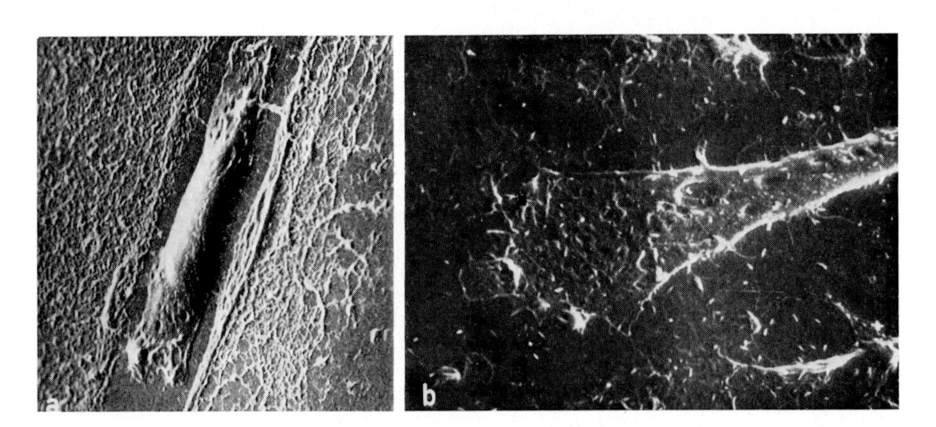

FIG. 5. Scanning electron micrographs of normal cells seeded upon the strip of glass surface surrounded by the phospholipid film: (a) 30 min after seeding, the lamellar cytoplasm is polarized (× 1 620); (b) 24 h after seeding, showing the region of lamellar cytoplasm at the cell pole. × 3 600.

abrupt in others. The tail process is unattached except for few microvilli and small zones of lamellar cytoplasm near its end.

The presence of longitudinal fibrillar bundles accompanied by microtubules is a distinctive characteristic of polarized cells. These regularly spaced bundles are seen in the lamellar cytoplasm (Fig. 6a), and often pass inside the cortical layers. These bundles are also present in the endoplasm, especially near its lateral edges and its lower surface. Similar structures are seen in the tail process. The polarized form of fibroblast can be modified experimentally in various ways. The cells may be further elongated by attaching them to the narrow strip of an adhesive substratum. The body of these cells is long and cylindrical with small plates of lamellar cytoplasm on one or both poles (see Fig. 5b).

Colcemid and other drugs which disrupt the microtubules inhibit and reverse the polarization (Vasiliev *et al.* 1969, 1970; Gail & Boone 1971; Goldman 1971). Colcemid-treated cells have a ring of lamellar cytoplasm surrounding the well-spread endoplasm. In contrast to the non-treated cells which are at the stage of radial attachment (see above), Colcemid-treated fibroblasts in the wound do not have the smooth rounded external edge; numerous lamellipodia are observed there. The parallel orientation and close packing of filaments in the fibrillar bundles are gradually diminished until these bundles disappear completely in Colcemid-treated fibroblasts. Many 5 nm microfilaments radiating in various directions are seen in the lamellar cytoplasm.

The shape of a polarized normal cell varies considerably with the density of the culture. It is not clear what the exact shape of these cells is in very dense cultures. Examination of sections of these cultures reveals 4–5 layers of cells (Fig. 7a), each cell forming lamellae of considerable length upon the surface of other fibroblasts and/or upon the intercellular substance. The ultrastructure of these lamellae is similar, although possibly not identical, to those formed by cells upon the glass. As yet we do not know the width of these lamellae, or whether they are formed preferentially at one pole of the cell or at both poles.

ATTACHMENT OF NEOPLASTIC CELLS

Radial attachment

Thirty minutes after seeding, most L fibroblasts did not develop the ring of lamellar cytoplasm but only had the microvilli radiating from the lower pole. The shape of the more spread L cells (about 10–25% of the whole population) was abnormal: the peripheral part of the cytoplasm was thicker and less smooth than normal lamellar cytoplasm (Fig. 3b). Numerous microvilli up to 10 μm

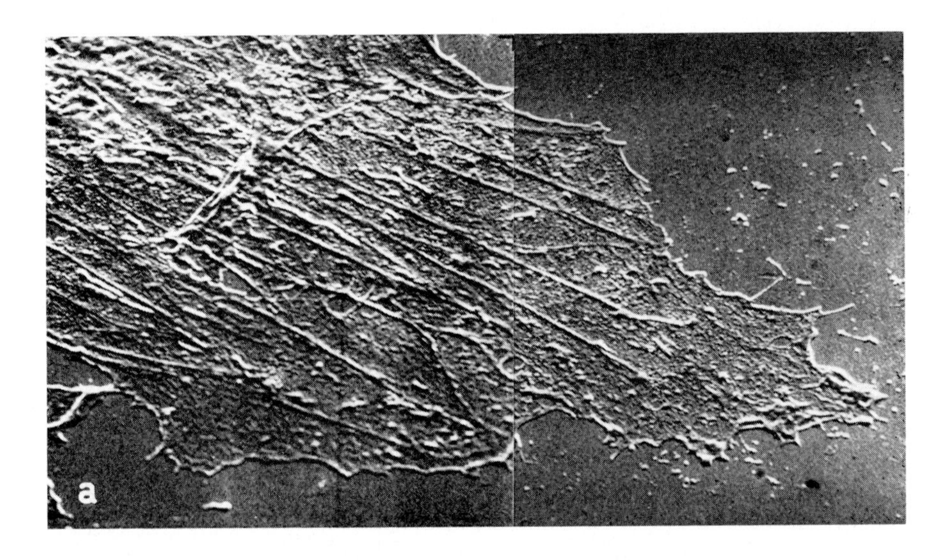
a

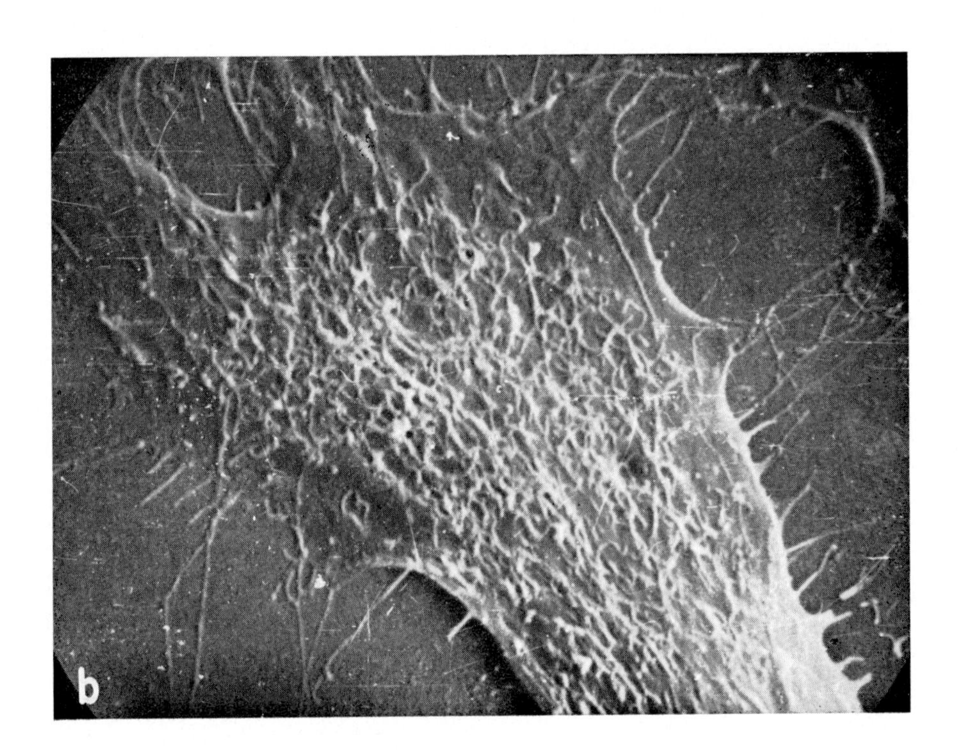
b

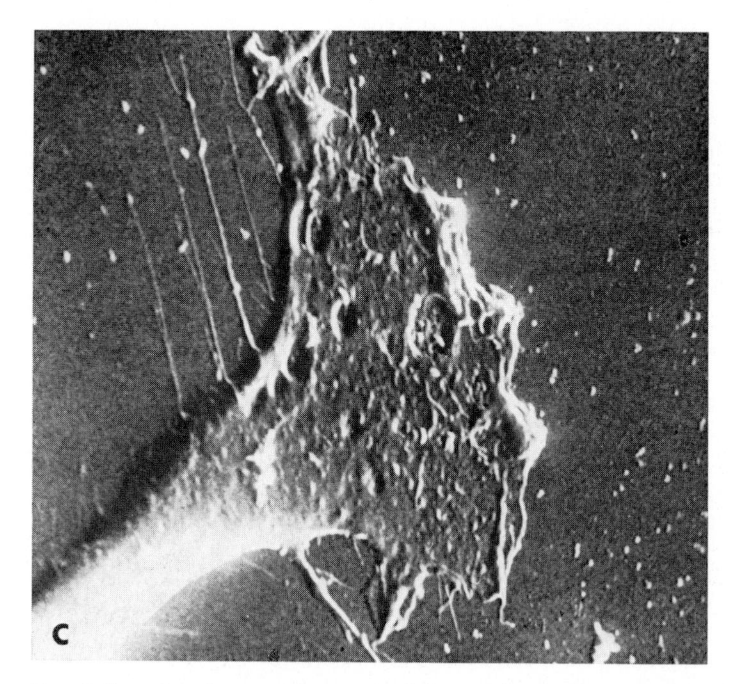

FIG. 6. Scanning electron micrograph of the anterior lamella of normal and neoplastic cells moving into the wounds: (a) normal fibroblast, showing wide anterior lamella with fibrillar bundles (× 3 200); (b) a cell of the L line with small lamellar region and numerous microvilli on the lateral edges and on the upper surface (× 6 000); (c) a cell of the S-40 line with a small lamellar region at the end of unspread cytoplasmic process (× 4 000).

long extend from their edges. About 1–2 hours after seeding, the proportion of these more-spread cells in the population increased but their shape remained abnormal.

In the experiments with other neoplastic fibroblasts (CIM, S-40 and HEK-40), most cells did form the ring of lamellar cytoplasm. However the shape of this ring often differed from that of normal cells: width and thickness were more variable, the external edge was less smooth (Fig. 3c), and microvilli 3–5 μm in length were often seen at the edge. Only a few cells in each population had the completely normal morphology.

Polar attachment

The cells of all the lines examined migrated into a wound, although the rate of migration of some lines (L and HEK-40) was lower than that of normal fibro-

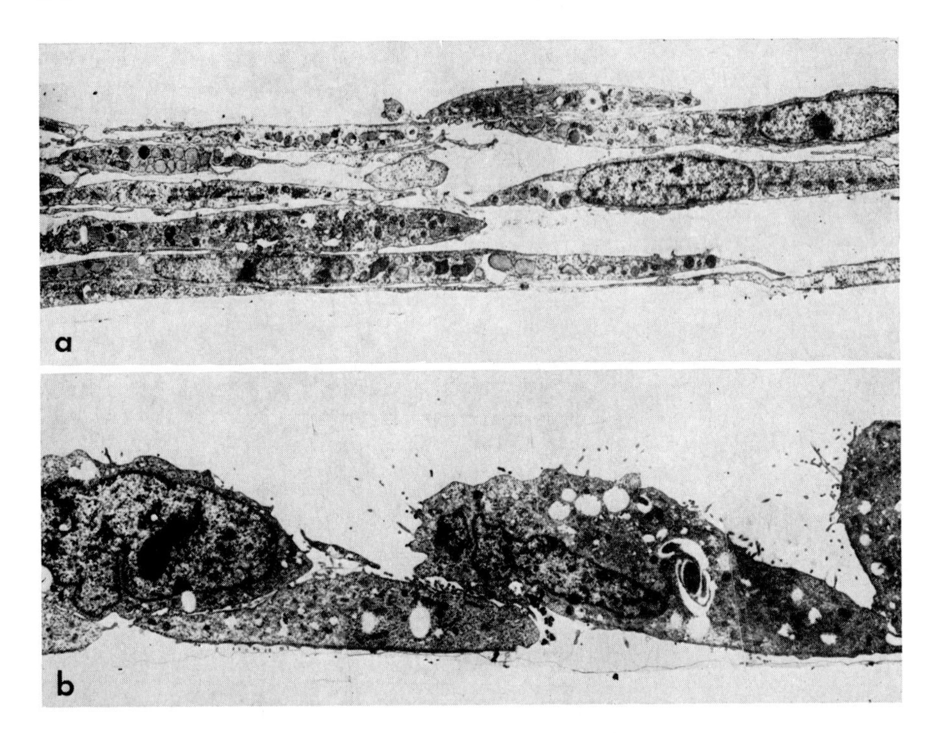

FIG. 7. Sections of dense isolated cultures of (a) normal mouse fibroblasts and (b) L cells. The plane of sections is perpendicular to the substratum. Several layers of overlapping cells are visible in normal culture (a), while only a monolayer is formed in neoplastic culture (b). × 4 000.

blasts. We used these migrating cells to examine the morphology of polarized neoplastic cells.

The shape of neoplastic cells varied considerably not only from one line to another but within the same line. Three main morphological classes of the shapes of neoplastic cells could be distinguished.

The first class consisted of cells that had no areas of lamellar cytoplasm. Their body was almost hemispherical but more often these cells had several ectoplasmic processes about 2–5 μm long. Microtubules and one fibrillar bundle were often seen in sections of these processes. The surface of these processes was almost cylindrical or conical so that the area of their contact with the substratum was small. Long microvilli or very small areas of flattened cytoplasm were visible at the ends. The shape of the anterior and posterior processes was almost identical.

The cells that had appreciable regions of lamellar cytoplasm composed the second class (Fig. 6b, c). Each of these regions was usually smaller in area than the anterior lamella of a normal fibroblast. Some neoplastic fibroblasts had only one lamellar region, usually in the front part of the cell. Other cells had several (2–4) separate lamellar regions, situated in various parts of the cell, for example in the ends of cytoplasmic processes.

The structure of lamellar regions in these cells often showed various abnormalities: (*a*) the edge of these regions could be very jagged; (*b*) numerous microvilli could be seen radiating from this edge; (*c*) the lamellar regions could be thicker and their upper surface less smooth than in normal cells; and (*d*) fibrillar bundles could be absent, or decreased in number, and less regular in distribution. This less-ordered distribution of fibrillar structures on the periphery of neoplastic cells has been observed by Ambrose *et al.* (1970) and McNutt *et al.* (1971). Characteristics of the cortical layer and of the regions of close approach to the substratum have now been studied and will be reported later.

Lamellar regions of each particular cell only showed some of these abnormalities. Even in certain cells which had lamellar cytoplasm of almost normal size and shape, however, detailed examination often revealed certain abnormalities. These 'quasi-normal' fibroblasts may be regarded as the third main morphological class of neoplastic cells. The shape of one cell could change considerably during microcinematographic observation, so that the cell shifted from one morphological class to another.

The cultures of different neoplastic lines had different proportions of cells belonging to various classes and also differed in the details of cell morphology within each class. For instance, most cells of the L line belonged to the first and second classes; there were hardly any 'quasi-normal' cells. Within the first and second class, cells with long cytoplasmic processes were rare; numerous microvilli were often seen at the upper cell surface.

In the cultures of the CIM and S-40 lines many quasi-normal cells were seen but cells of the second class with 1–3 long processes predominated. Very elongated bipolar cells of the first class were also present.

In summary, populations of all the lines of neoplastic fibroblasts examined contained many cells whose ability to form lamellar regions of cytoplasm was defective. In certain lines (CIM, S-40 and HEK-40) this deficiency was more obvious at the stage of polar attachment than at the stage of radial attachment.

At present these abnormalities are apparent only by qualitative morphological examination, especially with scanning electron microscopy. We do not have any good quantitative tests for them. Even the area of lamellar cytoplasm cannot be measured precisely: in electron micrographs of neoplastic cells it is difficult to differentiate abnormal lamellar cytoplasm from other parts of the cell. In phase-contrast microphotographs it is easy to measure the area of ectoplasm relative to that of the whole cell. In normal cells the area of lamellar cytoplasm is almost equal to that of ectoplasm since the area of unspread cytoplasmic processes is small. The relative area of ectoplasm was 60–62% of the total area for mouse and hamster fibroblasts. This area was significantly lower for the neoplastic cells of L line (30–35%) and CIM line (40–45%). However in other cell lines unspread ectoplasmic processes were prominent and this increased the relative area of ectoplasm to near normal values. More precise methods to measure the formation of lamellar cytoplasm have yet to be developed. Needless to say, crude physical methods based on the measurements of forces detaching the cells from the substratum are not adequate for this purpose.

INTERACTIONS OF NEOPLASTIC CELLS WITH THE SUBSTRATUM AND THEIR LOCOMOTORY BEHAVIOUR

Comparison of microcinematographic and electron microscopic data suggests that deficiencies in the formation of lamellar cytoplasm are the structural basis of many abnormal traits in the locomotory behaviour of neoplastic cells. We shall consider one aspect of this behaviour, cell–cell collisions.

The consequences of such collisions obviously depend on the attachment of the contacting cell regions to the substratum. Collision of the attached leading edge of one cell with the leading edge or another attached part of the second cell may lead to contact inhibition, that is, to local inhibition of the formation of lamellipodia and to retraction of the leading edge, sometimes accompanied by formation of cell–cell adhesion (Abercrombie 1970). Alternatively the surface of the second cell may be progressively overlapped by the lamellar region of the first. Another possibility arises when the leading edge of one cell collides with an unattached part of another cell; this may result in underlapping, that is the leading edge passes under the unattached part (Abercrombie 1970; Boyde *et al.* 1969; Trinkaus *et al.* 1971; Weston & Roth 1969).

We found that contact inhibition with a slight retraction of the leading edge was the most common type of cell–cell collisions in sparse cultures of normal cells. Two main modifications of collisions were apparent in sparse cultures of neoplastic cells (L and CIM) examined microcinematographically. (*a*) Retraction of the leading edge accompanying contact inhibition was more pronounced in neoplastic cells. This often led to the detachment and disappearance of the whole cellular process carrying the leading edge. Possibly this increased retraction was due to a defective attachment of these processes to the substratum. (*b*) Neoplastic cells often had unattached lateral edges, and therefore were

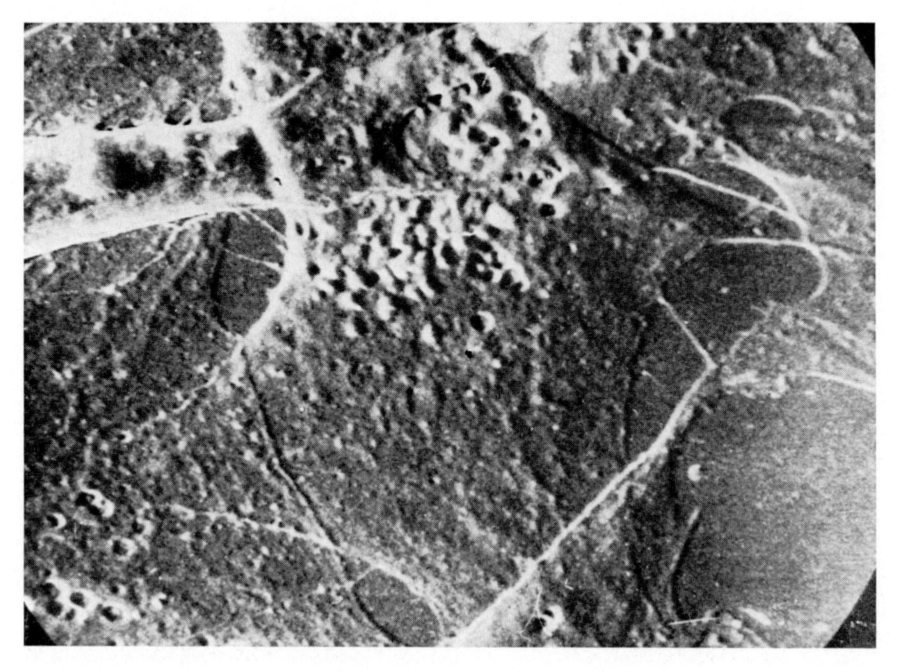

FIG. 8. Scanning electron micrograph of a region in normal culture with numerous overlaps of lamellar regions of various cells. × 1 800.

easily underlapped by the leading edges of normal cells in mixed cultures. Such underlapping was often observed in the isolated cultures of CIM cells but not in those of L cells. Possibly this difference was due to the fact that attachment of L cells was more defective than that of CIM cells. Many CIM cells had firmly attached lamellar regions carrying their leading edges. In most L cells even the area near the leading edge, possibly, was not well attached and could not pass under the body of other cells.

Overlappings were rare in sparse cultures of all cell types. When cell density in culture increased, analysis of individual cell–cell collisions became impossible. Scanning electron microscopy revealed numerous overlaps of lamellar regions even in moderately dense cultures of normal cells (Fig. 8). As mentioned before, these overlappings eventually resulted in the formation of multilayered cell sheets in dense cultures of normal fibroblasts. It is not clear how these overlappings developed and why locomotion of cells in dense culture was diminished although these cells apparently continued to form lamellar cytoplasm over the surfaces of other cells. Overlapping was rare even in dense cultures of L cells; these cells formed a monolayer (Fig. 7b). In dense cultures of CIM cells,

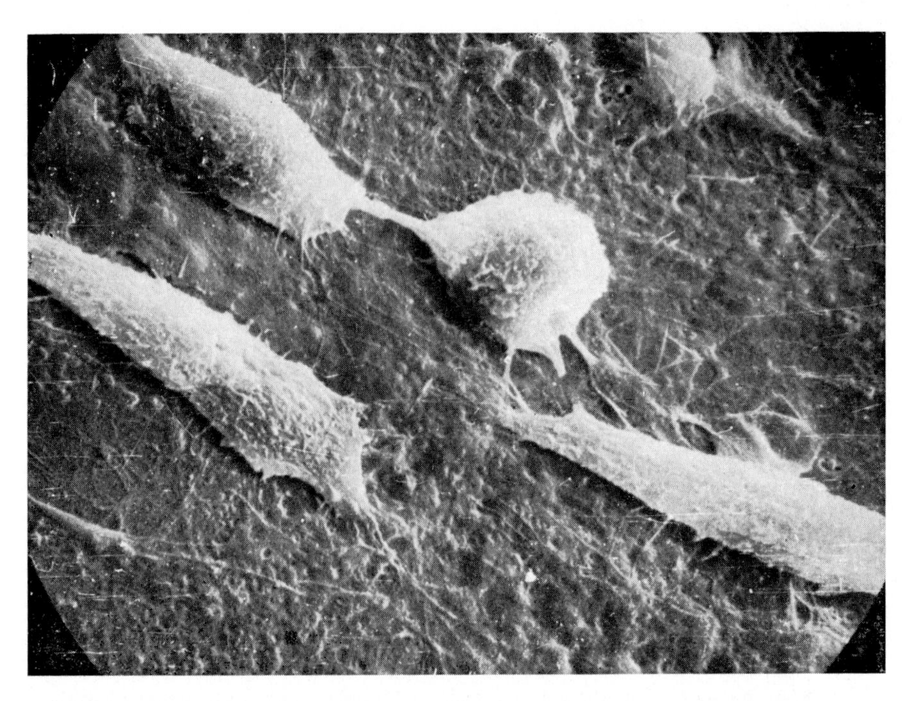

FIG. 9. Scanning electron micrograph of mixed culture of normal and L cells showing poorly spread L cells on the surface of normal cell sheet. × 2 000.

we noticed multilayered cell sheets, while in sparse cultures of CIM cells we saw criss-cross patterns. In contrast, the parallel orientation of cells was clear in dense cultures of this line. These neoplastic fibroblasts had narrow leading edges and, possibly, only in dense cultures was the frequency of collisions of these edges with surfaces of other cells high enough to produce an oriented cell pattern. However, this point deserves further study.

The behaviour of L cells and normal fibroblasts in mixed cultures deserves special comment. In these cultures the more firmly attached normal fibroblasts gradually detached L cells from the glass. Microcinematographic analysis had shown that this detachment was due to underlapping of L cells by normal fibroblasts and to more pronounced retraction of the leading edges of L cells after contact (see Domnina *et al.* 1972 for details). Once dislodged from the glass, L cells attached themselves to the upper surfaces of normal fibroblasts, although this attachment was weak; bipolar cells without lamellar cytoplasm were usually seen on this surface (Fig. 9). The rate of translocation of L cells on the surface of normal fibroblasts was usually small. L cells multiplied actively on the surface of normal fibroblasts and eventually formed a monolayer over the sheet of normal cells.

Thus, examination of only two types of neoplastic cells suggests that defective attachment to the substratum is sufficient to produce considerable diversity of morphological patterns in culture. Depending on the population density and on the degree of deficiency characteristic for the cell line these cultures may be monolayered or multilayered and consist of oriented or unoriented cells, etc. It is significant that of all the cells examined the best monolayers were formed by the L fibroblasts having the most deficient attachment. The same L fibroblasts in mixed culture formed colonies on the top of the normal cell sheet, that is, they displayed the characteristic property of transformed cells. Obviously it is dangerous to regard monolayering or mutual orientation as distinctive characteristics of normal compared to neoplastic cells without further analysis.

Growth characteristics of neoplastic cells also seem to be affected by their deficient attachment. L cells stop multiplication in dense isolated cultures but grow actively on the top of a normal cell sheet (Domnina *et al.* 1972). However, detailed analysis of growth properties of neoplastic cells is beyond the scope of this paper.

DISCUSSION AND CONCLUSION

Formation of lamellar cytoplasm

After the initial contact with the substratum, the cell first forms a lamellar cytoplasm and then becomes polarized. Let us look at the characteristics of these two processes.

One of the preliminary consequences of cell contact with an appropriate ('adhesive') substratum is the formation of attachment sites. These sites might be identical to the regions of close approach to the substratum but this remains to be proven. Attachment of some site of the cell surface seems to induce a reaction which leads to formation of an attached region of lamellar cytoplasm around this site. The nature of this surface-induced reaction might be similar to that of phagocytosis or secretion. We shall avoid the temptation to speculate in detail about the possible components of this reaction and, in particular, about the nature of signals produced by an attached cell surface, about the role of microfilaments in the intracellular transport of materials to the site of formation of new surface, etc. The formation of lamellar cytoplasm seems to be self-perpetuating: formation of one attachment induces formation of another, and so on. The total area of lamellar cytoplasm continues to increase until an upper limit is reached. At any point of the advancing cell edge formation of new lamellar cytoplasm is induced locally: if lamellar cytoplasm is not attached to a certain point of the substratum, it cannot spread further around this point. This might be the case at the edge of non-adhesive substratum. Attachment to the substratum is essential not only for the formation of lamellar cytoplasm but also for maintenance of its structure. Detachment of the cell causes immediate disappearance of lamellar regions. Possibly the presence of attachment sites has some continuous influence upon the state of cytoplasm around them.

Polarization

Direction of cell polarization might be determined by various extracellular factors, such as contacts with other cells, structure of the substratum etc., which could possibly modify the rate of formation of lamellar cytoplasm in certain directions. Initial differences in the rate of formation of lamellar cytoplasm are then stabilized by an intracellular Colcemid-sensitive process (Vasiliev *et al.* 1970). A good example of stabilization is shown by cells seeded upon a narrow strip of adhesive substratum. First, formation of lamellar cytoplasm is stopped in certain directions by the non-adhesive substratum. Then, those cell edges where the formation of lamellar cytoplasm has been hindered are transformed into lateral stable edges of elongating cell. Formation of new surface continues at the leading edge of lamellar cytoplasm on polarized cell (Abercrombie *et al.* 1970). In other words, polarization limits the possible directions of the formation of lamellar cytoplasm.

Polarization is correlated with the formation of fibrillar bundles accompanied by the microtubules. Destruction of microtubules by Colcemid results in the eventual disintegration of fibrillar bundles. This suggests an organizing role for microtubules in the formation of bundles, although its nature remains unknown. Neither do we know how the orientation of microtubules can be determined by the uneven formation of lamellar cytoplasm in various directions.

Deficient formation of lamellar cytoplasm by neoplastic fibroblasts

We have seen that the degree of alteration of the lamellar region varies considerably from one neoplastic line to another and from cell to cell within each population. The attachment of one neoplastic cell at various times can be either very abnormal or almost normal. Nevertheless, at any given time the presence of numerous cells with definitely abnormal lamellar regions was characteristic for all the cultures of neoplastic fibroblasts which we examined.

We conclude that some deficiency of the formation of lamellar cytoplasm is characteristic for these lines. The ability to adopt the polarized shape is retained by neoplastic fibroblasts; the process of polarization remains Colcemid-sensitive. The direction of polarization of neoplastic cells seems to be determined by extracellular factors in the same way as that of normal cells, although the effectiveness of these factors might be modified by the deficient formation of lamellar cytoplasm.

How regularly is a deficiency of this type developed in the course of neoplastic evolution of fibroblasts? Obviously many more cell lines have to be studied and the development of quantitative methods for evaluation of these deficiencies is essential before this question can be answered. Some observations on deficient attachment of neoplastic cells to the substratum have been made (see Abercrombie & Ambrose 1962). In this connection, some workers (Barker & Sanford 1970; Sanford *et al.* 1970) have compared the cytological properties of a large number of lines of mouse embryo cells at various stages of 'spontaneous' neoplastic conversion *in vitro*, and found that 'retracted cytoplasm' was one of the morphological changes of cells *in vitro* usually consistent with the ability to grow as tumours *in vivo*; this change was due to decreased spreading of these cells upon the substratum. These authors suggest that growth pattern changes associated with neoplastic conversion result from a loss of surface–substrate dependency.

However, it would be an oversimplification to expect a strict correlation between the deficient attachment of cells *in vitro* and their ability to produce tumours *in vivo*. The inability of cells to grow as tumours might be a result of secondary changes, such as antigenic alterations, inability to induce stroma formation, etc.

Cells cultivated *in vitro* for a long time can reach an advanced stage of 'neoplastic development' (Foulds 1969), as shown by the changes in many characteristics. Nevertheless these cells need not exhibit high oncogenic properties *in vivo*. This is probably the case with the weakly oncogenic cells of L line that demonstrate strikingly abnormal attachment *in vitro*.

We have seen that defective formation of lamellar cytoplasm could be the basis of many abnormal traits of the behaviour of neoplastic cells in cultures. Deficient attachment to the substrata might also be important for invasive growth *in vivo*. Normal fibroblasts *in vivo* are spread upon the network of connective-tissue fibres, which often have the wide thin ectoplasm, possibly identical to lamellar regions formed *in vitro*.

Cells and intercellular fibres of connective tissue *in vivo* often form sheet-like structures. Possibly, poorly attached neoplastic cells are excluded from these normal structures but can attach themselves to their surface and grow there in the same way as they grow at the surfaces of the normal fibroblasts in mixed cultures. This suggestion is in agreement with the fact well known to pathologists: tumour cells often infiltrate into the interstices between normal tissue structures (Willis 1967). More detailed studies of the distribution of invading tumour cells with regard to the normal connective tissue structures would be interesting in this connection.

ACKNOWLEDGEMENTS

The new experimental data in this paper were obtained in collaboration with A. P. Cherny, L. V. Domnina, V. I. Guelstein, O. Y. Ivanova, L. B. Margolis, S. G. Komm, E. E. Krivitzkaja and L. V. Olshevskaja.

We wish to thank Mr Abercrombie for summarizing our paper so accurately and well at the symposium.

References

ABERCROMBIE, M. (1970) Contact inhibition in tissue culture. *In Vitro* **6**, 128-142

ABERCROMBIE, M. & AMBROSE, E. J. (1962) The surface properties of cancer cells: a review. *Cancer Res.* **22**, 525-548

ABERCROMBIE, M., HEAYSMAN, J. E. M. & PEGRUM, S. M. (1970) The locomotion of fibroblasts in culture. III. Movements of particles on the dorsal surface of the leading lamella. *Exp. Cell Res.* **62**, 389-398

ABERCROMBIE, M., HEAYSMAN, J. E. M. & PEGRUM, S. M. (1971) The locomotion of fibroblasts in culture. IV. Electron microscopy of the leading lamella. *Exp. Cell Res.* **67**, 359-367

AMBROSE, E. J., BATZDORF, U., OSBORN, J. S. & STUART, P. R. (1970) Sub-surface structures in normal and malignant cells. *Nature (Lond.)* **227**, 397-398

BARKER, B. E. & SANFORD, K. K. (1970) Cytologic manifestations of neoplastic transformation *in vitro*. *J. Natl. Cancer Inst.* **44**, 39-64

BOYDE, A., GRAINGER, F. & JAMES, D. W. (1969) Scanning electron microscopic observations of chick embryo fibroblasts *in vitro* with particular reference to the movement of cells under others. *Z. Zellforsch. Mikrosk. Anat.* **94**, 46-55

BOYDE, A. & VESELÝ, P. (1972) Comparison of fixation and drying procedures for preparation of some cultured cell lines for examination in the SEM in *Scanning Electron Microscopy* 1972 (Part II). Proceedings of the Workshop on Biological Specimen Preparation for Scanning Electron Microscopy III Research Institute Chicago.

BRUNK, U., ERICSSON, J. L. E., PONTÉN, J. & WESTERMARK, B. (1971) Specialization of cell surfaces in contact inhibited human glia-like cells *in vitro*. *Exp. Cell Res.* **67**, 407-415

BUCKLEY, J. K. & PORTER, K. R. (1967) Cytoplasmic fibrils in living cultured cells. *Protoplasma* **64**, 349-360

DOMNINA, L. V., IVANOVA, O. Y., MARGOLIS, L. B., OLSHEVSKAJA, L. V., ROVENSKY, Y. A., VASILIEV, JU. M. & GELFAND, I. M. (1972). Defective formation of the lamellar cytoplasm by neoplastic fibroblasts. *Proc. Natl. Acad. Sci. U.S.A.* **69**, 248-252

FOULDS, L. (1969) *Neoplastic Development*, vol. I, Academic Press, London & New York

GAIL, M. H. & BOONE, C. W. (1971) Effect of Colcemid on fibroblast motility. *Exp. Cell Res.* **65**, 221-227

GOLDMAN, R. D. (1971) The role of three cytoplasmic fibers in BHK-21 cell motility. I. Microtubules and the effects of colchicine. *J. Cell Biol.* **51**, 752-762

GOLDMAN, R. D. & FOLLETT, E. A. C. (1969) The structure of major cell processes of isolated BHK-21 fibroblasts. *Exp. Cell Res.* **57**, 263-276

IVANOVA, O. Y. & MARGOLIS, L. B. (1973) The use of phospholipid membranes for preparation of cell cultures of given shape. *Nature (Lond.)* **242**, 200

MCNUTT, N. S., CULP, L. A. & BLACK, P. H. (1971) Contact-inhibited revertant cell lines isolated from SV40–transformed cells. II. Ultrastructural study. *J. Cell Biol.* **50**, 691-708

SANFORD, K. K., BARKER, B. E., PARSHAD, R., WESTFALL, B. B., WOODS, M. W., JACKSON, J. L., KING, D. R. & PEPPERS, E. V. (1970) Neoplastic conversion *in vitro* of mouse cells: cytologic, chromosomal, enzymatic, glycolytic and growth properties. *J. Natl. Cancer Inst.* **45**, 1071-1096

SPOONER, B. S., YAMADA, K. M. & WESSELLS, H. K. (1971) Microfilaments and cell locomotion. *J. Cell Biol.* **49**, 595-613

TRINKAUS, J. P., BETCHAKU, T. & KRULIKOWSKI, L. S. (1971) Local inhibition of ruffling during contact inhibition of movement. *Exp. Cell Res.* **64**, 437-444

VASILIEV, JU. M., GELFAND, I. M., DOMNINA, L. V. & RAPPOPORT, R. I. (1969) Wound healing processes in cell cultures. *Exp. Cell Res.* **54**, 83-93

VASILIEV, JU. M., GELFAND, I. M., DOMNINA, L. V., IVANOVA, O. Y., KOMM, S. G. & OLSHEVSKAJA, L. V. (1970) Effect of Colcemid on the locomotory behaviour of fibroblasts. *J. Embryol. Exp. Morphol.* **24**, 625-640

WESTON, J. A. & ROTH, S. A. (1969) Contact inhibition behavioural manifestions of cellular adhesive properties *in vitro* in *Cellular Recognition* (Smith, R. T. & Good, R. A., eds.), pp. 29-37, Meredith Corporation, New York

WILLIS, R. A. (1967) *Pathology of Tumours*, 4th edn., Butterworth, London

WITKOWSKI, J. A. & BRIGHTON, W. D. (1971) Stages of spreading of human diploid cells on glass surfaces. *Exp. Cell Res.* **68**, 372-380

Discussion

Gail: Using the air-blast method (Gail & Boone 1972), we have confirmed that SV3T3 fibroblasts are less adherent to Pyrex or cellulose acetate substrates than are 3T3 cells. These observations agree with the morphological data presented here by Vasiliev and Gelfand.

Abercrombie: There is the difficulty of the difference in profile, I suppose.

Gail: Yes. The data only measure distractibility.

Steinberg: The difference in profile might be due to a difference in the intrinsic roundness of the cell, such as could be caused by the tension below its surface. In other words, I don't think you can unquestioningly accept the flatness of the cell as a sure way of judging its adhesiveness to the substrate.

Curtis: I agree. I wonder, both in general cases and in the experiments of Vasiliev and Gelfand, how much the flattening is more a measure of internal structures—microfilaments and microtubules—than just a general rheological behaviour of the cell. I do not think we can answer this yet.

Vasiliev and Gelfand:* We think one should distinguish between adhesion and attachment. Both terms are somewhat vague and seem to differ in meaning. Adhesion may be assessed by measuring the force of attraction between two contacting surfaces (per unit area? per cell?). Perhaps attachment might be assessed by measuring the number and area of the regions of close approach to the substrate (attachment sites). There are reasons for supposing that certain neoplastic cells have reduced attachment to the substrate. The number and area of their regions of close approach may be abnormal but this does not imply that cell surface in these regions has decreased adhesion to the substrate.

Flatness is a characteristic of upper cell surface. Often it seems to be correlated with the attachment: the more flattened cells and parts of cells have better attachment than the more rounded ones. However, it is not clear whether the flattening is a direct consequence of attachment. The changes of flatness and of attachment in neoplastic cells may both result from the alteration of some third parameter. We favour this last suggestion as we regard both flattening and formation of the attachment sites as various manifestations of the structural reorganization accompanying cell transition from an unattached into an attached state.

Curtis: Possibly the important difference between a tumour cell and a normal cell lies internally rather than on the surface.

Vasiliev and Gelfand:* Yes, this is possible because formation of lamellar cytoplasm and of the attachment sites may involve many intracellular events as well as surface changes.

Trinkaus: When we studied the deformability of *Fundulus* deep cells, we found that during early gastrula, when they are engaged in locomotion, they are significantly more deformable than in early blastula (Tickle & Trinkaus 1973). This suggests that they are less rigid. I think this is consistent with what you were describing.

Bell (1972), working with us, has been studying the locomotory activity and contact behaviour of 3T3 cells and polyoma-transformed 3T3 cells. The latter have long, spindly processes which tend to adhere only at the end. What he has found, in complete agreement with the results of Vasiliev and Gelfand, is that the criss-crossing which everyone has taken as an indication of a reduction of contact inhibition is probably not that at all. He has shown that it is due entirely to underlapping, rather than overlapping. This underlapping comes about as the result of two factors. First, Py3T3 cells have relatively few points of ad-

* These comments were added after the Symposium as a result of postal exchanges.

hesion to the substratum and therefore many regions where other cells can move underneath. Second, they advance with narrow processes whose ruffling is localized at the tips. These narrow processes are easily able to slip under other cells. 3T3 cells, which tend to form monolayers, have broad lamellipodia and thus less adhesion-free edge. Therefore, it is more difficult for these cells to underlap one another to any great degree.

There are two important conclusions to be drawn from these observations. (1) One cannot draw conclusions about contact inhibition or lack of it by looking at end results; one must watch the process. (2) On the criterion of lack of overlapping, Py3T3 cells do not show reduced contact inhitibion of movement. Significantly, Bell has also shown that when Py3T3 cells meet head-on, there is an inhibition of ruffling, just as in normal 3T3 cells.

*Vasiliev and Gelfand**:* The classical description of contact inhibition by Abercrombie includes inhibition of the formation of lamellipodia and retraction of the leading edge. According to these criteria, the neoplastic fibroblasts which we examined demonstrated obvious contact inhibition. Retraction was often even more pronounced in these cells than in normal fibroblasts. We are glad to hear that Professor Trinkaus has made similar observations with other neoplastic strains.

Albrecht-Bühler: Several years ago, Professor Curtis proposed a relation between shearing forces on the membrane and adhesiveness (Curtis 1960). We found (pp. 44–47) that polyoma-transformed 3T3 cells actually move six times less than normal cells, which means that they would have less adhesiveness in terms of Professor Curtis' ideas.

Curtis: Unfortunately, I now feel that my ideas (Curtis 1960) which Dr Albrecht-Bühler favours are probably incorrect. There is no experimental evidence supporting them.

References

BELL, P. B., JR. (1972) Criss-crossing, contact inhibition, and cell movement in cultures of normal and transformed 3T3 cells. *J. Cell Biol.* **55**, 16A

CURTIS, A. S. G. (1960) Cell contacts: some physical considerations. *Am. Nat.* **94**, 37-56

GAIL, M. H. & BOONE, C. W. (1972) Cell–substrate adhesivity: a determinant of cell motility. *Exp. Cell Res.* **70**, 33-40

TICKLE, C. A. & TRINKAUS, J. P. (1973) The deformability of *Fundulus* deep cells, in press

Cell movement in confluent monolayers: a re-evaluation of the causes of 'contact inhibition'

MALCOLM S. STEINBERG

Department of Biology, Princeton University

Abstract Abercrombie and Heaysman proposed two explanations for 'contact inhibition' (of 'cell movement') which we here restate in the following way: (1) there exists a restraint to locomotion in the direction of a cell contact or (2) there exists a restraint to overlapping with another cell. Either mechanism would produce both monolayering and the observed temporary stalling when two cells collide in an open area. We wished to decide between these alternative explanations and sought a situation in which they would produce different consequences. We reasoned that if translocation in the direction of a cell contact is directly restrained, then a cell in contact on all sides with other cells should be restrained from moving in all directions. But if the restraint is directly against overlapping, then a surrounded cell, while restrained from crossing over any of its neighbours, should nevertheless be free to exchange places on the substratum with any of them. Accordingly, I filmed confluent monolayers of both 3T3 fibroblasts (with Dr E. Martz) and chick embryonic liver cells (with Dr D. Garrod). In both cases, analysis of the films showed that although the formation of monolayers was extensive, cells in mutual contact on all sides continued to mill about amongst one another. We conclude that avoidance of overlapping rather than inhibition of locomotion causes these cells to form monolayers, and suggest that the differences in the strengths of cell–cell compared to cell–substratum adhesion explain the observed 'contact inhibition of overlapping'.

The hypothesis of 'contact inhibition' (Abercrombie & Heaysman 1954) was formulated in order to explain the tendency of populations of cultured fibroblasts to remain in a monolayer on the substratum. Several phenomena have subsequently been broadly construed as expressions of contact inhibition [see Martz & Steinberg (1973) for review]. Besides the disinclination of the cells to form multilayers, these include the slowing of a cell and local paralysis of ruffling by contact with another cell, the tendency of the elongated cells of certain lines to align themselves, and the inhibition of cell replication observed in confluent cultures of most normal cells. However, it has by no means been established that all these phenomena are pleiotropic expressions of a single, central mechanism.

My concern here will be the same as that of Abercrombie and Heaysman: to explain why certain cell populations tend to remain in a monolayer. The original hypothesis of contact inhibition consisted of two parts. The first proposition was that when a moving cell collides with another cell, its continued locomotion in the direction of travel before collision will be inhibited. Such behaviour would in itself, without further elaboration, be sufficient to explain the formation of monolayers. The second proposal was that when the inhibited cell's locomotion is resumed, the direction of movement will usually be away from the axis of collision. The second part of the hypothesis is irrelevant to the question of monolayering itself, and bears instead upon other features of cell movement in monolayer cultures. I shall therefore focus attention upon the first proposal. This inhibition of locomotion on collision has been directly observed (Abercrombie & Ambrose 1958) *inter alia* in time lapse films of cell cultures. Proof that this response, repeated over and over again, statistically accounts for the degree of monolayer formation observed has not been presented in any case, and would require an enthusiasm for the filming and scrutiny of pairing encounters not yet evinced even in Sweden. Nevertheless, the existing observations do constitute an empirical foundation for this proposal.

If monolayering is due to this cellular response to contact with another cell, then to what is this cellular response itself due?

When one cell is brought to a standstill by collision with another cell in an open area, two things are happening at once. The cell ceases translocation in the direction of the collision, and it neither overlaps nor underlaps the obstructing cell. Either behaviour might be primary with regard to the other. That is to say, the injunction might be 'Thou shalt not translocate in the direction of thy contacted neighbour'. As a consequence of obeying this injunction, a cell would have to refrain from moving above or below its neighbour. On the other hand, if the injunction were 'Thou and thy neighbour shalt not superimpose the one on the other', a cell would have to stop translocating in the direction of the obstructing cell, once contact was complete.

How can one distinguish cause from effect? When two cells collide in the open, cessation of forward motion and failure to overlap go hand in hand. But might there be some other circumstance that would permit cause and effect to be separated? Consider the confluent monolayer itself. What would be the motile behaviour of a cell deep within it? If spontaneous formation of a monolayer results from a prohibition against movement in the direction of a cell contact, then a cell within such a monolayer, contacted on all sides by other cells, should be prevented from moving in any direction: it should be immobilized.

But if spontaneous monolayering results from a prohibition against superimposition of cells, then a cell within such a monolayer, while prevented from crossing over or under any of its neighbours, could nonetheless remain free to exchange places on the substratum with any other: the cells could mill about like sheep in a pen. Thus the primacy of either prohibition could be established by observing whether totally surrounded cells within a population spontaneously forming a monolayer stay put or move about.

We have conducted two such investigations. Dr D. Garrod* and I have recorded the behaviour of liver parenchyma cells from seven-day-old chick embryos in a confluent monolayer on a Falcon plastic surface by time lapse cinematography. About 8×10^6 trypsin-dissociated primary cells in 2 ml of Eagle's minimal essential medium containing 10% horse serum, penicillin (100 i.u./ml), streptomycin (100 μg/ml) and amphotericin B (Fungizone) (0.25 μg/ml) were plated in a 35 mm tissue culture dish and maintained at 37 °C in an atmosphere of air with 5% carbon dioxide. The parenchyma cells sorted out from the fibroblast-like cells to form a group of islands. The behaviour of the cells in one such island was filmed at six-second intervals for six hours with low-power phase contrast optics. The resulting film was analysed frame-by-frame. Each cell nucleus appearing in the film was assigned a number and a detailed comparison was made of the positions of all cells on the first and last frames of the film.

The rare (if ever observed) moving of a cell over others in the compact island showed that a stricture against superimposition was widely obeyed. Nevertheless, movements of individual cells from one position to another in the island were clearly apparent. In order to quantitate these movements, cell speed and neighbour exchanges were measured.

NEIGHBOUR EXCHANGES

While cell boundaries could sometimes be observed in these low-power phase contrast films, often the exact position of the boundary between two cells could not be determined. Consequently the neighbours of a cell could not usually be identified with precision. Accordingly, assignments of neighbours were made in the most satisfactory way we could devise. On tracings of the positions of all nuclei in the first and last frames of the film, certain cells appeared most likely to be in mutual contact, either because their boundaries could be discerned or because of the proximity of their nuclei. Such cells were considered to be *probable neighbours*. When four nuclei lay close to one another at the corners of

* *Present address:* Department of Zoology, University of Southampton, England.

a quadrilateral, the diagonals were compared and only the shorter one was taken to connect a pair of probable neighbours. The pair of cells joined by the longer diagonal were considered to be non-neighbours. When both diagonals were equal, all four cells were considered *possible neighbours*. All possible and probable neighbours in the first and last frames were then listed.

Because we wished to consider only the behaviour of surrounded cells, we eliminated a number of cells from further consideration. These included (1) cells at the edge of the island in either frame in the film, (2) cells adjacent to a region in which nuclei could not be clearly made out, (3) cells entering or leaving the field during the film, and (4) cells at the edge of the photographic field. Because of their different reaction to contact with other cells and their post-mitotic separation movements (Abercrombie & Gitlin 1965; Martz 1973), the eleven cells that divided during the film (and their daughters) were also eliminated. This left for analysis 85 of the total of 186 different cells present in the two frames.

To each of these 85 cells an average of 3.9 probable neighbours (and 2.2 possible neighbours) was assigned on the first frame and 3.7 probable and 2.4 possible neighbours on the last frame. A cell was deemed to have lost an original neighbour only when the latter had been designated as a probable neighbour in the first frame but not even as a possible neighbour in the last frame. Similarly, a cell was counted as having gained a neighbour only when the latter had been assigned as a probable neighbour on the last frame, but not even as a possible one on the first frame. We felt that these criteria were sufficiently stringent to underestimate the true exchange of neighbours. In this way, the average cell was found to have lost 1.0 (25.6 %) out of its original 3.9 neighbours, and to have replaced these with 0.8 (20.5 %) new neighbours.

CELL SPEED

Speeds were calculated as the straight-line distance traversed by a cell between the first and last frames over the six hours of filming. The mean distance travelled by all cells present in both frames was 20.9 μm, that is a speed of 3.5 μm/h. This value is an underestimate because the paths of the cells were often observed to be crooked. However, this calculation refers to mean speed *over the substratum*, while the speed relative to neighbouring cells is, in a sense, more pertinent to our interest. For example, suppose that an island of liver parenchyma cells moved across the substratum as a coherent body, with all its component cells fixed in position. Here, mean speed over the substratum, though not lacking in interest, would not measure the mean speed of *mutual* cell displacement, which would be zero. In fact, the cell island we filmed did drift within the field. We allowed for this movement in the following way. A calibrated grid was superimposed first on one frame and then on the other, and the positions of all cells on both the x and y axes at both times were recorded. By summing the individual cell displacements along the two axes so that equal and opposite displacements cancelled each other, we determined the average displacement in each direction (i.e. the displacement of the island as a whole). The tracings of the two frames were then superimposed with the second frame offset to allow for the displacement of the island as a whole, and the 'corrected' straight-line displacements of all cells present in both frames were determined. The average dislocation of a cell relative to its neighbours in six hours was thus calculated to be 7.8 μm, that is, a mean *speed of mutual displacement* of 1.3 μm/h. Owing to the use of straight-line distances, this value also is an underestimate (D. Garrod & Steinberg, unpublished results).

The purpose of this investigation was to decide whether the formation of monolayers is due to an inhibition of locomotion in the direction of a contacted neighbour or to an inhibition of cell superimposition *per se*. The results so far presented might appear, at first glance, to resolve this question in favour of the latter. However, another interpretation is possible. Might it be, for example, that gaps between cells in the monolayer release a local contact inhibition of *locomotion*, thereby initiating local cell shifts, which cause new gaps to open, and so on? If this were the case, the cell movements we have described would be due to the lack of all-round contact rather than in spite of it. In fact, gaps were occasionally seen to open up between cells in the island of liver cells. When this occurred, the bordering cells always extended ruffling pseudopods into the gap, soon closing it again. While it seemed that these occasional gaps could not account for all the cell movements we observed, we could not exclude the possibility that these movements resulted from the opening up of many 'cracks' too small to be noticed. Further evidence on this, as well as information relating to the generality of cell locomotion within confluent monolayers, was obtained from an analysis of the movements of 3T3 cells in a confluent culture.

MOVEMENTS OF 3T3 CELLS IN A CONFLUENT MONOLAYER

Dr E. Martz* and I filmed the growth of an originally sparse 3T3 culture through confluence until after its cell density was constant. The cells were plated on coverslips in Dulbecco and Vogt's modification of Eagle's basal

* *Present address:* Department of Pathology, Harvard Medical School, Boston, USA.

medium containing 3.3 % calf serum, and equilibrated with 10 % carbon dioxide in air. During the filming, medium was added by continuous perfusion to avoid abrupt changes, and was constantly stirred to promote homogeneity. The low serum concentration counteracted an increase in stationary density caused by the stirring. The use of Nomarski (10× objective) rather than phase contrast optics allowed better definition of cell boundaries. The interval between exposures was 150 s before saturation and 240 s afterward. Fuller details are given in Martz & Steinberg (1972).

The following four measurements were made on individual cells at hourly intervals for 7.6 days: (1) the amount of a cell's perimeter in contact with other cells; (2) the number of cells contacting the cell being followed; (3) local cell density; and (4) the speed of nuclear translocation. In addition, the mitotic histories of all cells were recorded, and data on neighbour exchange were collected after confluence was achieved.

Todaro *et al.* (1964) have shown the pronounced inhibition of nuclear overlapping by 3T3 cells. Under the conditions used in our film, the ratio of nuclear overlaps observed to those expected (had the cells been distributed randomly) was less than 0.02. This is below the previously recorded minimum value of 0.08 for chick cells (see Curtis 1961; Curtis & Varde 1964; Abercrombie 1965; Abercrombie *et al.* 1968). 3T3 cells (including the population we observed) also show marked 'postconfluence inhibition of cell division' (Martz & Steinberg 1972) and have been used in many investigations of this phenomenon [see Martz & Steinberg (1972) for references]. 3T3 cells show '*density* inhibition of motility' (i.e. of translocation) (Gail & Boone 1971), and, more particularly, the cells we filmed demonstrated *contact* inhibition of cell speed at subconfluent densities (Martz 1973).

Having established that the filmed cells were displaying the various forms of 'contact inhibition', we were interested to note whether they were therefore obligated to remain immobilized after confluence was achieved. It was quickly apparent that they were not. Again there arose the question of whether the observed neighbour exchanges could be due to local, invisible gap formation. This time, however, observations relevant to the question could be made.

The filmed cell population grew from 3.6×10^3 cells/cm^2 at the start of filming to 2×10^4 cells/cm^2 at saturation five days later. Confluence, however, was reached after three days, when the last visible gap on the area of coverslip filmed was covered by cells. During the 53 h between initial confluence and saturation, growth continued (at a decreasing rate) as more and more cells crowded together on the fixed area of coverslip until their number there almost doubled. One would expect the opening of gaps—visible or invisible—to become less and less frequent as the crowding of the still monolayered cells increased to its maximum at saturation. Consequently, if the cell locomotion causing neighbour exchange is due to the temporary release of an inhibition at gaps, neighbour exchange should be markedly greater between confluence and saturation than it is after saturation is achieved.

To examine this postulate, stringent criteria for probable neighbours were again devised [different from those of Garrod (see before) but as stringent], and neighbour exchanges were tabulated for the 53 h between confluence and saturation and the 53 h after saturation. It was found that the average cell exchanged 47% of its probable neighbours during the first interval and 36% of its probable neighbours during the second. (These differences were not statistically significant [$P \approx 0.31$].) It does not seem, therefore, that the cell translocations observed can be attributed to the opening of even invisible gaps between the monolayered cells (Martz and Steinberg, unpublished observations).

CONCLUSIONS

These results do *not* prove that a cell's motile machinery is unaffected by contact with another cell. That remains to be resolved. However, the results described here do clearly establish that whatever restrains these totally surrounded cells from moving over one another's surfaces does not accomplish this by immobilizing them. For that reason we believe the expression 'contact inhibition of movement' to be a potentially misleading one when applied to the inhibition of cell superimposition. We prefer 'contact inhibition of overlapping' as more descriptive of the facts as they are presently known. In terms of the alternatives set forth near the outset, it appears that when one cell stops short and avoids superimposition upon colliding with another, it is most likely that the stopping is the effect, and avoidance of superimposition the cause, rather than *vice versa*.

Two cautions against misinterpretation seem called for. First, there is no reason to think that the injunction against overlapping is a flat prohibition; it is usually rather a discouragement of overlapping that might vary in intensity with circumstances. Second, the overlapping that has been shown to be discouraged is that of major portions of cells, and not of minor portions. Because superimposition to the extent of nuclear overlap is discouraged, it does not necessarily follow that the overlapping of small projections, filopodia etc., is also discouraged to the same extent. Consequently, the common observation of appreciable 'microoverlapping' does not cast doubt upon the existence of a 'contact inhibition of overlapping' phenomenon, since the latter belongs to a more macroscopic province.

Finally, if monolayering is due to a restraint upon cell overlapping, then to what might the restraint upon overlapping itself be due? Our results provide no answer to this question, but Abercrombie has previously suggested one. He pointed out (Abercrombie 1961) that overlapping would be discouraged if the colliding cell adhered to the substratum more strongly than to the upper surface of the cell with which it collides. It would then have difficulty in exchanging its stronger adhesions to the substratum for weaker adhesions to the upper surface of the obstructing cell. The difficulty experienced by cells in spreading upon the upper surfaces of other cells in an established monolayer, brought out during this conference [cf. Middleton (pp. 251-262) and Trinkaus (pp. 233-244)], lends experimental support to this possibility.

To the extent that differential adhesion governs the configurations adopted by cell populations on substrata, the correspondences between particular configurations and particular sets of cell–cell and cell–substratum relative adhesive strengths can be deduced from the application of thermodynamic principles embodied in the differential adhesion hypothesis (reviewed in Steinberg 1963, 1964, 1970). These correspondences have recently been explored (Martz and Steinberg, unpublished data) [for a preliminary report see Martz & Steinberg (1973)] with the result that a surprising variety of unique cell population configurations, some of them very familiar to the tissue culturist, prove capable of being determined in this manner.

ACKNOWLEDGEMENTS

The original research upon which this paper is based was supported by Grants GB-2315 and GB-5759X from the National Science Foundation, and by a special grant for research (P-532) from the New Jersey Division of the American Cancer Society. Eric Martz held a National Science Foundation Graduate Fellowship and a Damon Runyon Memorial Fund Postdoctoral Fellowship.

References

ABERCROMBIE, M. (1961) The bases of the locomotory behaviour of fibroblasts. *Exp. Cell Res.* (Suppl.) **8**, 188-198

ABERCROMBIE, M. (1965) in *Cells and Tissues in Culture* (Willmer, E. N., ed.), pp. 177-202, vol. 1, Academic Press, New York

ABERCROMBIE, M. & AMBROSE, E. J. (1958) Interference microscope studies of cell contacts in tissue culture. *Exp. Cell Res.* **15**, 332-345

ABERCROMBIE, M. & GITLIN, G. (1965) The locomotory behaviour of small groups of fibroblasts. *Proc. R. Soc. Lond. B Biol. Sci.* **162**, 289-302

ABERCROMBIE, M. & HEAYSMAN, J. E. M. (1954) Observations on the social behavior of cells in tissue culture. II. 'Monolayering' of fibroblasts. *Exp. Cell Res.* **6**, 293-306

ABERCROMBIE, M., LAMONT, D. M. & STEPHENSON, E. M. (1968) The monolayering in tissue culture of fibroblasts from different sources. *Proc. R. Soc. Lond. B Biol. Sci.* **170**, 349-360

CURTIS, A. S. G. (1961) Control of some cell–contact relations in tissue culture. *J. Natl. Cancer Inst.* **26**, 253-268

CURTIS, A. S. G. & VARDE, M. (1964) Control of cell behaviour: topological factors. *J. Natl. Cancer Inst.* **33**, 15-26

GAIL, M. H. & BOONE, C. W. (1971) Density inhibition of motility in 3T3 fibroblasts and their SV40 transformants. *Exp. Cell Res.* **64**, 156-162

MARTZ, E. (1973) Contact inhibition of speed in 3T3 and its independence from postconfluence inhibition of cell division. *J. Cell Physiol.* **81**, 39-48

MARTZ, E. & STEINBERG, M. S. (1972) The role of cell–cell contact in 'contact' inhibition of cell division: a review and new evidence. *J. Cell Physiol.* **79**, 189-210

MARTZ, E. & STEINBERG, M. S. (1973) Contact inhibition of what? An analytical review. *J. Cell Physiol.* **81**, 25-38

STEINBERG, M. S. (1963) Reconstruction of tissues by dissociated cells. *Science (Wash. D.C.)* **141**, 401-408

STEINBERG, M. S. (1964) in *Cellular Membranes in Development (22nd Symp. Soc. Study Dev. Growth)* (Locke, M., ed.), pp. 321-366, Academic Press, New York

STEINBERG, M. S. (1970) Does differential adhesion govern the self-assembly of tissue structure? Equilibrium configurations and the emergence of a hierarchy among populations of embryonic cells. *J. Exp. Zool.* **173**, 395-434

TODARO, G. J., GREEN, H. & GOLDBERG, B. D. (1964). Transformation of properties of an established cell line by SV40 and polyoma virus. *Proc. Natl. Acad. Sci. U.S.A.* **51**, 66-73

Discussion

Goldman: With regard to mapping the position of nuclei and not whole cells, is it possible that you are looking at the intracellular displacement of nuclei in cells which are not translocating?

Steinberg: Some of the movements could be due to that in certain cells. But I gave you the mean values, not the extremes. There are nuclei which lose half their neighbours, for example, and are found quite far away.

Curtis: When you spoke of overlapping, did you mean nuclear overlapping or cell overlapping?

Steinberg: We observe nuclear overlapping.

Trinkaus: Is this true for both the liver parenchymal and 3T3 cells?

Steinberg: Yes.

Wessells: Though I think your technique is important, one is obliged to prove that cells are really moving. Recall the case of the embryonic neural tube and neural retina; nuclei synthesize DNA in one position and then move to a different position when they enter into mitosis (Sidman 1961). If you subjected that system to your analysis, I suppose you would observe a 100% change every 12 h.

Trinkaus: What about distances?

Wessells: The distances moved are of the order of several cell diameters. The

Steinberg technique should be used for epithelial populations undergoing morphogenetic changes. Burnside & Jacobson (1968) studied the way in which cells in the amphibian neural plate change shape markedly as the neural tube forms, and distinguished the actual outlines of the cell periphery. Could one go to that situation and test out the technique?

Steinberg: Because we were unable to observe the liver cell boundaries in every case, we had to use nuclear positions for our analysis. But when cell boundaries could be observed, it was clear that the nuclear movements did reflect actual cell translocations. Using Nomarski optics, we were able to see the cell boundaries of the 3T3 cells more clearly, and again there was no doubt that the cells themselves were changing position. Nevertheless, it still is most convenient to use nuclear positions for the analysis.

The other point about the Burnside–Jacobson work is interesting because it reveals an absence of relative movement of cells in the neural plate. All the neighbours are conserved. I am not trying to generalize beyond the evidence that I have presented. In certain morphogenetic systems cells do change neighbours, and in others they evidently do not. I suspect that certain kinds of cell junctions are basically anchors around which the cell cannot move.

The folding of the urodele neural plate is an example of a morphogenetic movement in which there is no slippage of cells relative to one another within the plate. When the configuration of a cell population changes without cell slippage, microfilamentar or perhaps microtubular mechanisms seem likely to be involved. Then the shape change of the tissue is likely to be the sum and consequence of all the individual cell shape changes. The studies of Gustafson & Wolpert (1963; 1967) on morphogenesis in the sea urchin suggest other examples of the action of such a mechanism.

Cells do slip during sorting out and in various examples of spreading, as for example in the splitting of mesoderm from the endoderm during gastrulation in any vertebrate embryo. Here the mesoderm moves ventrally, penetrating between the ectoderm and endoderm, and the endoderm moves dorsally on the inner surfaces of the mesoderm. Then and only then, a different kind of morphogenetic mechanism, based on differences in intercellular adhesive strengths, is likely to be guiding the movements.

Wolpert: Do you have any electron micrographic information about the junctions in your system? Light microscopy can be very misleading about the nature of the gaps.

Steinberg: No. I would suppose that the junctions between the 3T3 cells are the same as those seen in other 3T3 cultures.

Middleton: In an epithelial sheet migrating in culture from an explant of pigmented retina the individual cells within the sheet do move relative to one another. For example, as the sheet expands more cells are introduced into the margin of the sheet to accommodate the increasing circumference of the sheet (Middleton, unpublished results).

Steinberg: Don't they just stretch?

Middleton: No, definitely not. Ultrastructural observations (Middleton & S. Pegrum, unpublished results) show that the cells within such a sheet have contact specializations similar to those that we have demonstrated between pigmented retina cells in dispersed culture (see Middleton Fig. 2, p. 258). However, I suspect that the movement of individual cells within the sheet is made possible by the fact that these contacts can be broken down and re-formed very rapidly. Certainly Heaysman & Pegrum (1973) have shown that rather similar contact specializations can be formed rapidly between colliding chick heart fibroblasts.

Trinkaus: Although you say that you can sometimes see boundaries, I am not sure. When there are that many cells in the field at once the magnification is too low to see the detailed behaviour of a membrane less than 1 μm thick.

Steinberg: The cells can become vanishingly thin and we are dealing with low magnifications so we are limited in resolution. But with Nomarski optics, the contrast gets greater with lower magnification, so that one can see the edge area of the cell, although one cannot see many details.

Trinkaus: But with underlapping it becomes even more difficult to see margins.

Steinberg: Our conclusion that monolayering results from a direct discouragement of cell overlapping would be in no way invalidated if each cell in the whole population sent out thin threads with which it touched every other cell, and all the cells were thus connected. The hypothesis of contact inhibition was intended to explain monolayering, defined as the observation (with ordinary optics) that the nuclear overlap or the cell overlap visible was far less than expected if the cells had been randomly distributed on the substratum. Even if the cells in such a monolayer overlap or reach out and touch one another with thin processes, the non-random distribution of the main cell bodies requires explanation.

Trinkaus: P. Bell (unpublished results) has been studying the detailed activity at the margins of 3T3 cells in culture with high resolution phase contrast optics. Like you, he finds a lot of movement: a cell will pull away from its neighbours and then send out another lamellipodium that will either adhere to the margins of neighbouring cells or, if it encounters a non-adherent region, underlap it. Thus within the sheet individual 3T3 cells might move partially, or occasionally entirely, under others. This might look superficially like overlapping, but it is not—it is always underlapping; the moving cells never leave the substratum. Possibly, they are not aware that the other cell is there.

Up until now, insofar as we know, no one has looked at the detailed movements of a sheet like this at high magnifications with time lapse cinemicrography.

Gail: I like your term 'milling about' to describe the motion of confluent cells because it suggests the random walk model is applicable. We have measured the augmented diffusion constant, D^*, and found values of about 40 $\mu m^2/h$ at confluence compared to 400 $\mu m^2/h$ at low density (Gail & Boone 1971). We can compare these data with your results on speed by considering the root mean square displacements at one hour. Crudely, the square displacement at one hour is given by $4D^*$; thus the root mean square displacement at one hour is given by $\sqrt{4D^*}$ μm (since $t = 1$ h in $4D^*t$). So the ratio of root mean square displacements at one hour for cells at low density compared to cells at confluence is about $\sqrt{400/40} \approx 3.2$ μm, which is in fair agreement with your results derived from speeds.

Goldman: What is meant by the term locomotion in all the cases discussed so far? As I understand it, locomotion is the net displacement of a cell from point A to point B. When the cell reaches point B, no part of that cell touches point A. Therefore the entire cell moves. In studying confluent monolayers, especially with low magnification optics, it would be extremely difficult to determine whether cells move in this way or whether they are just changing shape around an anchorage point.

I also feel that the terms cell locomotion and cell extension get confused. For example, nerve fibre outgrowth has been referred to as cell locomotion; I believe that this phenomenon should be termed outgrowth, not locomotion. We must decide whether we are talking about shape changes or true translocations of cells in any discussion of cell locomotion.

Trinkaus: I agree. We must state whether we mean the advance of the nucleus, of the leading edge, of the trailing edge, or whatever.

Steinberg: I agree with you completely, but suppose for a moment that each of these cells has a root, by which it is fixed to a point on the substratum and from which it can extend itself in various directions. If it extended itself one would get the macroscopic impression of much overlapping. But one definitely does not. This does not mean that no such root can exist, but it does mean that the process of overlapping is prevented. The question is, what prevents it?

Curtis: I find it difficult to see how the experiments distinguish clearly between the two explanations: restriction of movement or restriction of overlap. How do you know that large or small gaps open up between the cells?

Steinberg: All I can tell you is that we did observe such temporary gaps in the liver culture. Although we tried to detect such gaps wherever cell movement was seen, usually we were not able to perceive them. Of course, the resolution is limited, so we cannot eliminate the possibility that the gaps are smaller than can be seen. However, the point is that in these circumstances overlapping *was* prevented, but cell translocation was *not*. Thus it is difficult to account for the failure of overlap on the basis of a supposed prevention of movement which did not occur.

Curtis: These gaps are hard to see even by high-power light microscopy. We certainly cannot see the cells that underlap by light microscopy, although temporarily we do see enormous gaps opening up in some cases.

Trinkaus: That is not so. One can readily see gaps with appropriate microscopy, even with 3T3 cells, which are exceptionally well aligned.

Wolpert: I have a rather simple-minded view of how cells move and I don't think Professor Steinberg can draw any conclusions about the mechanism from his studies here. I believe the fundamental mechanism for cell locomotion is illustrated by the mesenchyme of the sea-urchin embryo. The cell puts out processes which can exert a pulling force. If this is so, then when one considers the interactions of a cell with a particular substratum or with another cell, and whether it will move over it, one must consider all the factors, not just, for example, adhesiveness.

Abercrombie (1961) has described contact inhibition as a cellular phenomenon which prevented overlapping. We have borrowed his term 'contact paralysis' (Gustafson & Wolpert 1967) to describe a local phenomenon in the cell. In the sea urchin, for example, the mesenchyme cells have many pseudopods and I think the consequence of one of these tips making contact is the same as what happens when a ruffled membrane meets another ruffled membrane: the pseudopod usually contracts after extension ceases. We know that these mesenchyme cells can move over the substratum, but not over each other. They are not contact inhibited. Here the reason is obvious; there are many other pseudopods and when one is stopped from moving, others might be extending. So one must always consider the number, size and shape of pseudopods. These determine how the cell behaves. This was clear in Dr Dunn's work (pp. 211-223) where a signal from the point of contact caused local inhibition of all the other pseudopods. One must also consider the adhesiveness and the mechanical properties of the membrane. When cells make contact with each other they will make point contacts if the membranes are stiff. In that case interaction might be restricted, but with other sorts of membranes, because of the mechanical nature of the membrane and adhesiveness, the membrane can spread all the way along and extend the contact and thus the interaction. What sort of tensile forces are generated? How hard can the cell pull? Cells which can pull hard when they make contact will be different from cells that can pull more weakly.

The overall phenomenon of contact inhibition is a multifactorial process implicating all such factors. The key factor from contact inhibition seems to be local contact paralysis, when cells make contact and which in general leads to some local cessation of movement at that particular point. What your data suggest, Professor Steinberg, is that when cells are in contact, there is a great deal of inhibition of local motility. However, when a gap opens up, the cells move into it.

Steinberg: An isolated normal cell stops moving forward when it runs into another cell. Why? Is it because its motor activity is inhibited? The evidence indicates otherwise; it is restrained from overlapping.

Wolpert: You might not get overlapping simply because the pseudopods cannot reach out. You cannot infer anything about the absence of overlapping from your data.

Steinberg: That surrounded cells continue to locomote is a fact. You seem to be suggesting that despite this locomotion, the cells could not extend pseudopods, and that failure to extend pseudopods resulted in their inability to overlap, which is simply one possible mechanism by which contact inhibition of overlapping could be brought about. Fine. I am not championing any particular mechanism to explain contact inhibition of overlapping. I suggested that one possible mechanism for it is differential adhesion but stated that we have no evidence for that. I have not tried to interpret the inhibition of overlapping, but have just said that a cell, although contacted on all sides as fully as cells seem ever to be, in two dimensions, can still move.

Harris: You are trying to decide whether contact inhibition of movement or contact inhibition of overlap is the primary effect. Since movement is not completely inhibited, you conclude that it is the secondary effect and inhibition of overlap is the primary effect. However, overlap is not completely inhibited—at least cytoplasmic overlap isn't—so how can you say that either one or the other is primary?

Steinberg: Because the observed translocations were great enough to have caused much overlapping, but none was observed.

Curtis: There is really considerable cytoplasmic overlap, at least in some cell lines, and therefore you are not going to see the gaps at all. One has the impression that, whatever the mechanism of contact inhibition of movement is and though it starts when edge-to-edge contact of the cell is first established, it increases in effect the further one cell overlaps another.

Abercrombie: We first used contact inhibition to explain the marked change of behaviour and not the total cessation of movement in a completed monolayer (Abercrombie & Heaysman 1954). This change of behaviour still permitted a certain amount of movement, which we found to be very slow. We also found that there was a certain amount of interchange of cells. I think the reason for the movement and interchange might be that there exists what one can regard as a second kind of cell movement. It is closely related to the free movement of cells seen in a sparse culture but it has a difference, namely it is not contact inhibited. This second kind of movement occurs because, when cells make contact, they tend to set into operation their contractile mechanisms, as Professor Wolpert has just said. Consequently, cells completely surrounded by other cells and thus incapable of free locomotion because of contact inhibitions are still capable of moving. I called this second, non-contact-inhibited, kind of movement *associative* movement (Abercrombie 1967).

Trinkaus: Do cells overlap in this associative movement?

Abercrombie: Yes, they do.

Albrecht-Bühler: In support of this, membrane movement, as judged by the transport of gold particles, is not affected at all by confluency in 3T3 cells. In sparse as well as in dense cultures I found almost the same value of the motion constant (in $\mu m^2/min$) even 90 h after plating the cultures (Albrecht-Bühler 1973) (cf. p. 47).

Harris: On the contrary, I find that the centripetal transport of particles on the surface of 3T3 cells is strongly inhibited at confluence.

Trinkaus: The same is true of epithelial cells.

Albrecht-Bühler: The directed movement might be inhibited but the milling around is not.

Harris: It is the directed transport which is related to locomotion.

Trinkaus: This could be evidence that the milling around that you discovered is not related to locomotion.

Albrecht-Bühler: Yes, this is possible.

Goldman: Professor Steinberg, are there still some dividing cells in your cultures? If so, it is possible that space would be created as a cell rounds up to divide. Other surrounding cells would then tend to move into this space and perhaps account for some of the movements which you have described.

Steinberg: In Martz's film of 3T3 cells, which I partially described, the amount of neighbour exchange after saturation when there were very few mitoses was not significantly different from the amount of neighbour exchange at an earlier time, when the cell population was already confluent but mitosis was still continuing. So it would seem that you cannot account for the movements observed on the basis of rounding-up movements associated with mitosis.

Abercrombie: Professor Steinberg's interpretation of contact inhibition as a simple restriction of superimposition seems to me probably right in many instances, and it should certainly not be neglected. However, we do not know of any clear examples of it between cell and cell although we do know that it happens between a cell and an area of non-adhesive substrate, as Carter first described (1967). Carter (1965, 1967) has given two explanations of contact inhibition in these terms. In one, the non-adhesive substrate is so non-adhesive that the cell cannot stick to it at all. In the other, the substrate that confronts the moving cell is merely less adhesive than the one on which it is moving, so that there is competition between the two. We have made crude observations of the interactions between macrophages or sarcoma cells (S180), that show zero heterologous contact inhibition on the one hand and what we believe to be endothelial cells on the other. The macrophages or sarcoma cells can be stopped by running into the endothelial cells. This is probably an example of the kind of inhibition to which Steinberg was referring.

The contact inhibition one observes in sparse cultures of fibroblasts possibly has a different explanation. We have indications, but not conclusive evidence, that it is due to inhibition of the locomotory mechanism within the cell.

Consider a head-on collision between two cells in a sparse culture of free-moving fibroblasts. Apart from the inhibition of movement, three things happen: first, an adhesion rapidly develops between the cells [cf. Dr Heaysman's pictures (pp. 187-194)]. This rules out complete non-adhesion as a cause of contact inhibition, but it still allows the upper surface of cells to be non-adhesive. Secondly, ruffling stops. Ruffling is not essential for cell movement, but its cessation (when it is occurring) is a striking consequence of the collision. The ruffling does not usually cease in the artificial situation, when the cell runs against a patch of non-adhesive substrate; the cell then stops at the edge but continues ruffling. There are, however, exceptions to this; the cell sometimes turns away. Since isolated cells change direction on a supposedly uniform substrate (Abercrombie & Heaysman 1966) a statistical comparison is required between the behaviour at a non-adhesive border and on a uniform substrate. This has not been done.

The third thing that happens during contact inhibition is the retraction that Professor Wolpert has referred to. I have previously assumed (1967) that a spasm of the contractile machinery of the cell is initiated by contact with another cell. Evidence that reinforces this interpretation is the sudden acceleration of cell movement on contact (Abercrombie & Heaysman 1953). If I am correct, the spasm does not seem to be explicable on the hypothesis that contact inhibition is the result of differential adhesion or of absolute non-adhesivity.

A last point is the matter of underlapping. One sees what looks exactly like contact inhibition when one cell goes underneath another. It can push its front end right under another cell until it comes out at the other side, and then undergo the characteristic contact inhibition with retraction. Ruffling ceases as it goes underneath, because there is no room for that activity.

Porter: Why does the top cell not continue to ruffle since it is free to?

Abercrombie: The top cell does continue to ruffle. So this inhibition during

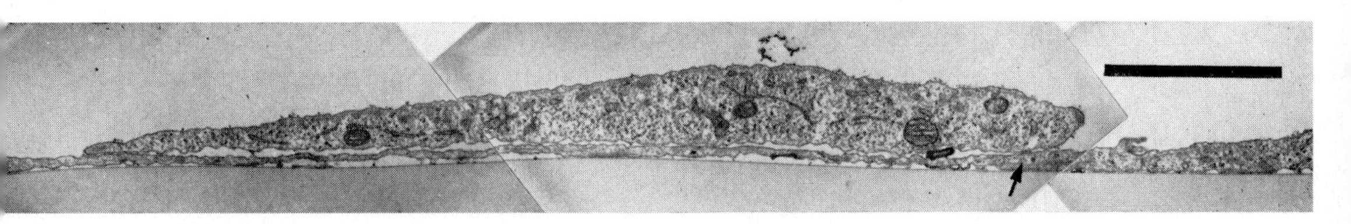

FIG. 1 (Heaysman). Vertical section of a lamella–side collision. The bar represents 2.5 μm.

underlapping is apparently contact inhibition but involves no change of substrate on which the cell is moving.

Heaysman: We have watched the leading edge of a cell that hits the side of another cell advance the whole way under the cell (Fig. 1). No specializations have occurred along the region of underlap, so by the same assumption we used before (pp. 187-194), it has not become contact inhibited. At higher power, one can just see the beginning of a specialization at the region indicated by the arrow. When we watch such a collision in a film, we can see the thin front end coming out the other side of the underlapped cell, and only then do we observe typical contact inhibition. The leading lamella of the underlapping cell is drawn right back and the cells separate.

Abercrombie: But one needs to demonstrate that this is not just a random change of direction of the cell; and this has not been done.

Steinberg: I agree. However, is the direct inhibition of locomotion, which I agree most likely does occur, sufficient to account quantitatively for the failure of overlapping? I don't think it is. I agree that contact inhibition of overlapping does not account for the post-contact behaviour of the cells, such as retraction. Our aim has been to find the reason for the stopping that accompanies the failure to overlap, rather than for the subsequent retraction.

Trinkaus: It is possible that the cell surface in culture varies strikingly in its adhesiveness: that the upper surface is only lowly adhesive, and parts of the under surface highly adhesive, where they adhere to the substratum (see Di-Pasquale & Bell 1972). At the side, the cell could also be adhesive and this would account for the inhibition of ruffling when a cell contacts the side of another. We know adhesions do take place here because of the retraction fibres seen on withdrawal. Perhaps further up the side, above this adhesive zone, the cell surface becomes very lowly adhesive and thus prevents overlapping.

Huxley: Could the mechanism for this be one which would prevent overlapping but still allow strong adhesions? The adhesive units on the membrane of the colliding cell could attach to adhesive units on the membrane of the neighbouring cell. Suppose that the cytoplasm is being pushed forward towards the advancing edge of the cell, at which new membrane is being laid down, and, further, suppose that passing through the lipid part of the membrane are protein molecules which can pick up an actin-like filament in the appropriate orientation inside the cell. As these filaments become attached at the advancing edge of the cell they might normally pull themselves back by the interaction of the actin with assemblies of myosin-like protein in the cytoplasm. While the front of the cell moves forward, these attached actin units are constantly flowing back along the top surface of the cell. On collision of two cells, the part of the actin attachment site which passes through the membrane could join onto the adhesion sites on the other cell membrane, which need not be fixed, but could move about in the plane of the membrane, so that, if one cell tried to pull itself over the top of a neighbouring one (using an active shearing mechanism between cytoplasmic myosin and membrane-attached actin), it would not have any fixed support to pull on, and would pull the attachment sites to the edge of the other cell, leaving the cells attached at their edges.

Steinberg: They might increase in local adhesiveness if they accumulated more anchors.

Huxley: More of these actin filaments attached at the front edge of the cell could accumulate, possibly at the expense of adhesion units underneath the cell, until there were enough of them to pull the cells apart.

Curtis: How would sufficient filaments do that?

Huxley: Maybe initially short actin filaments grow in length when they are immobilized on one place, until eventually they have a good chance of rupturing the external attachment to the neighbouring cell.

Goldman: McNutt *et al.* (1971) demonstrated that there are more sub-membranous microfilaments in regions of contact with other cells. We have made serial sections from the bottom (attachment side) to the top of fully spread BHK-21 and BSC-1 cells (Miller & Goldman, unpublished observations) and we find that there are many bundles of microfilaments on the side of the cell attached to the substrate, whereas there are very few if any on the upper surface of these cells. Thus it appears as if the formation of contacts or adhesions might signal the formation of microfilaments.

Steinberg: More evidence in support of Dr Huxley's idea comes from Jones *et al.* (1970), who proposed that adhesion sites are myosin. These authors prepared antibodies against smooth muscle myosin and found that they inhibited the initiation of intercellular adhesions.

Curtis: But the inhibition was minimal.

Allison: We have been unable to confirm their other claim that with fluorescent antibodies myosin could be detected on the outer surface of the plasma membrane.

Steinberg: It was clear from Dr Heaysman's evidence that promptly after a contact is made, new structures appear both in and below the membrane. Couldn't the membrane be spatially reorganized as a consequence of contact, and that spatial reorganization then triggers further association of material on the inside? I recall Weiss's 'molecular ecology' proposal (Weiss 1947), in which he explained cell 'modulation' as possibly arising through selective, serial sequestration of specific macromolecules, triggered by an environmentally induced change in the cell surface.

Middleton: Contact inhibition between fibroblasts and possibly also between epithelial cells seems to be associated with the formation of a specialized and therefore possibly strong adhesion between the colliding cells (Heaysman & Pegrum 1973; Middleton & S. Pegrum, unpublished results). Possibly therefore we have been wrong in the past in trying to reproduce contact inhibition artificially by confronting the cells with non-adhesive substrates such as agar (Abercrombie 1967) or cellulose acetate (Carter 1967). Perhaps we ought to be investigating instead what happens when cells are confronted with a 'super-adhesive' substrate.

Gail: We have found that phytohaemagglutinin dramatically increases adhesivity of 3T3 and SV3T3 fibroblasts to Pyrex and cellulose acetate substrates (Gail & Boone 1972; and see p. 291).

Harris: If the phytohaemagglutinin could be confined to discrete areas of the glass surface, one could see if the contact with these areas inhibited the movement of cells.

Gail: A possible difficulty is that phytohaemagglutinin located on one edge of a substrate might diffuse into solution. Perhaps we could modify the phytohaemagglutinin to bind it covalently to some portion of a specialized glass substrate.

Porter: Can you be sure that there are sites? Is the cell polarized in a vertical direction? These are endothelial cells, which normally have one face on a basal membrane and the other face free. Are the adhesive sites confined to the zones on the top surface and is the distribution fixed in all these cells? They will develop hemidesmosomes on the basal membrane and desmosomes at the margin.

Curtis: Cells in suspension, which may not be a fair comparison (with the exception of some of the very heavily trypsinized cells), appear to be equally adhesive all over the surfaces. But this might not be comparable with your situation.

Porter: How can you say that? You cannot watch the surfaces of the cells.

Curtis: The kinetics of aggregation give strong evidence for the equal adhesiveness over the surfaces of the component cells. If there are a limited number of adhesive sites the aggregates will very soon be limited in size because all sites on the component cells will already be used in adhesions so that extra cells cannot join on.

How much of the energy of adhesion of a cell is due to specialized areas of adhesion and how much is due to the rest? Roughly, the ratio between the energy of adhesion of the specialized sites and that of the rest could be of the order of 10^2–10^4. Thus if the specialized areas of adhesion occupy about 1 % of the total area or less, they might be unimportant in the overall adhesion of a cell.

Allison: I should like to present a possible explanation for forward movement and contact inhibition which is somewhat different from that put forward by Huxley. Extension requires a fluid cytoplasm. The change from a fluid cytoplasm to a nearly solid cytoplasm—the so-called sol–gel transition—probably represents a polymerization of actin molecules. I suggest that when contact is made either with the substrate or another cell, polymerization of actin filaments is triggered in the subjacent cytoplasm. The membrane could not then move forward in a region of contact but would tend to be retracted by microfilament action.

de Petris: The idea that membrane is fluid could also explain ruffling. The leading edge of a cell moving over a substrate, in particular the part of the cell surface which is not in contact with the substrate, is basically in a fluid (sol) state, that is, it is not rigidly connected with cytoplasmic structures. When this touches another cell, according to the pictures presented by Drs Heaysman (pp. 187-194) and Middleton (pp. 251-262), a zone of contact is established, where the membrane is probably no longer a fluid. There is considerable evidence that in a moving cell the upper part of the membrane is moving backwards, and furthermore, as shown by electron microscopy, that layers of filaments are present beneath the membrane in this region, which probably interact with the membrane, constraining it into a gel state. I imagine streams of free membrane coming from the sides and, if the membrane adheres to the substratum only in limited areas, also from below, from the ruffles, are gelled by interaction with the cytoplasmic filaments and then pulled back as a fairly rigid structure. When a contact is made, a patch of gelled membrane, rigidly connected to the adjacent cell, is established in the area of contact which prevents any further flow of membrane upwards, denying access to more material for ruffles. The ruffle might project from the side or also from below the area of contact, but not from above the area of contact, where a fixed and impermeable layer of membrane has now formed. In the case of epithelial cells, for example, this layer is established along the entire circumference of the cell, forming practically a ring. The total amount of membrane on the upper face is limited. This is why the cells form a monolayer in the epithelial sheet.

Harris: But as the sheet becomes more crowded, shouldn't the surface area of individual cells shrink? Shouldn't they then be more able to overlap?

de Petris: Yes, but only to some extent. The cells might be extending more pseudopods. They cannot really move in any direction, unless there is a gap in the monolayer and the flow of membrane can resume in that direction.

Middleton: In cultures of both fibroblasts (Abercrombie *et al.* 1968) and epithelial cells (Middleton 1972) the overlap index does increase as the cells become less extensively spread with increasing population density.

Curtis: The cells do not become much smaller. More of their surfaces is obscured by overlap (see Curtis & Varde 1964).

Middleton: This is not true of pigmented retina epithelial cells. Ultrastructural observations show that as the population density increases the cells gradually assume a more columnar morphology and thus individual cells occupy less substrate than at low population densities (Middleton & S. Pegrum, unpublished results). In addition Castor (1968) has demonstrated a shift in the distribution of cell volumes towards lower values in dense cultures of an epithelioid cell line.

Wohlfarth-Bottermann: Dr de Petris, what do you think provides the motive force for ruffling?

de Petris: I do not know, although it must be somehow connected with the activity of the cytoplasmic actomyosin system. It is unlikely to reside in the membrane flow alone.

Wolpert: It is not unreasonable to think of the cell surface membrane as having more or less constant surface area during cell movement and there thus being a limit to the amount of membrane available for forming pseudopods (Wolpert & Gingell 1968). For example, when a cell becomes stretched out by pseudopods, it is rare to find pseudopods coming from the middle of the spindle-like shape. When contact between two cells is made a local contraction might be induced. The most convincing evidence for this possibility comes from the observations of Wohlman & Allen (1968) on the giant amoeba *Difflugia.* The pseudopod which it extended has no obvious filamental or axial structure. On contact with the substratum, the pseudopod becomes birefringent, microfilaments appear and contraction takes place.

Harris: In fibroblasts, however, the contraction is continuous, judging by the forces they exert on plasma clot substrata.

Trinkaus: Often one finds that when *Fundulus* deep cells extrude a lobopodium and there is an apparent contact with another cell, there is a 'squaring off' of the end. With this, the rounded contour of the body of the lobopodium is rapidly straightened out and it becomes taut, as if it is under tension. I believe this is the same sort of phenomenon Professor Wolpert reports seeing in the sea urchin.

Porter: Do these lobopodia have adhesion sites on their tips?

Trinkaus: Yes. They certainly adhere often at their tips. But they may adhere along the sides as well.

Gingell: While we are in the realm of speculation I have a few suggestions regarding membrane subunits and membrane permeability. It is becoming increasingly clear that much of the protein and glycoprotein of cell membranes is present as large globular molecules which span the cell membrane (Bretscher 1972) and seem free to diffuse within the plane of the lipid bilayer (Singer & Nicolson 1972). Bretscher (1972) has shown that such mobile glycoproteins bear most of the cell surface negative charge. Since there must be attractive van der Waals forces acting between similar glycoprotein subunits, anything which reduces the electrostatic repulsion between them will cause them to aggregate. Pinto da Silva (1972) has shown that at low pH the subunits aggregate. Alternatively pulling the subunits together by cytoplasmic filaments, like balloons on strings or bridging with suitable molecules (Allison 1972), could cause aggregation.

It seems to me that one can make a case for subunit aggregation leading to increased ionic permeability of the membrane. Aggregated subunit-like particles are seen at gap junctions and where exocytosis vesicles approach the surface membrane (Satir *et al.* 1972). Both situations have been related to a local increase in permeability. In another case, the induction of pinocytosis in amoebae, it is clear that inducers which characteristically cause an increase in permeability are salts and polycations whose action is to lower the electrostatic surface potential (Gingell 1972). They would consequently be expected to cause subunits to aggregate.

References

ABERCROMBIE, M. (1961) The basis of locomotory behaviour of fibroblasts. *Exp. Cell Res.* (Suppl.) **8**, 188-198

ABERCROMBIE, M. (1967) Contact inhibition: the phenomenon and its biological implications. *Natl. Cancer Inst. Monogr.* **26**, 249-277

ABERCROMBIE, M. & HEAYSMAN, J. E. M. (1953) Observations on the social behaviour of cells in tissue culture. I. Speed of movement of chick heart fibroblasts in relation to their mutual contacts. *Exp. Cell Res.* **5**, 111-131

ABERCROMBIE, M. & HEAYSMAN, J. E. M. (1954) Observations on the social behaviour of cells in tissue culture. II. 'Monolayering' of fibroblasts. *Exp. Cell Res.* **6**, 293-306

ABERCROMBIE, M. & HEAYSMAN, J. E. M. (1966) The directional movement of fibroblasts emigrating from cultured explants. *Ann Med. Exp. Biol. Fenn.* **44**, 161-165

ABERCROMBIE, M., LAMONT, D. M. & STEPHENSON, E. M. (1968) The monolayering in tissue culture of fibroblasts from different sources. *Proc. R. Soc. Lond. B Biol. Sci.* **170**, 349-360

ALBRECHT-BÜHLER, G. (1973) A quantitative difference in the movement of marker particles in the plasma membrane of 3T3 mouse fibroblasts and their polyoma transformants. *Exp. Cell Res.* **78**, 67-70

ALLISON, A. C. (1972) in *Cell Interactions (Third Lepetit Colloq.)* (Silvestri, L. G., ed.), p. 156, North Holland, Amsterdam

BRETSCHER, M. (1972) Major human erythrocyte glycoprotein spans the cell membrane. *Nat. New Biol.* **231**, 229-232

BURNSIDE, M. B. & JACOBSON, A. G. (1968) Analysis of morphogenetic movements in the neural plate of the newt. *Dev. Biol.* **18**, 537-552

CARTER, S. B. (1965) Principles of cell motility: the direction of cell movement and cancer invasion. *Nature (Lond.)* **208**, 1183-1187

CARTER, S. B. (1967) Haptotaxis and the mechanism of cell motility. *Nature (Lond.)* **213**, 256-260

CASTOR, L. N. (1968) Contact regulation of cell division in an epithelial-like cell line. *J. Cell Phys.* **72**, 161-173

CURTIS, A. S. G. & VARDE, M. (1964) Control of cell behaviour: topological factors. *J. Natl. Cancer Inst.* **33**, 15-26

DIPASQUALE, A. & BELL, P. B., JR. (1972) The cell surface and contact inhibition of movement. *J. Cell Biol.* **55**, 60A

GAIL, M. H. & BOONE, C. W. (1971) Density inhibition of motility in 3T3 fibroblasts and their SV40 transformants. *Exp. Cell Res.* **64**, 156-162

GAIL, M. H. & BOONE, C. W. (1972) Cell–substrate adhesivity: a determinant of cell motility. *Exp. Cell Res.* **70**, 33-40

GINGELL, D. (1972) in *Membrane Metabolism and Ion Transport* (Bittar, E. E., ed.), p. 317, vol. 3, Wiley, London

GUSTAFSON, T. & WOLPERT, L. (1963) The cellular basis of morphogenesis and sea urchin development. *Int. Rev. Cytol.* **15**, 139-214

GUSTAFSON, T. & WOLPERT, L. (1967) Cellular movement and contact in sea urchin morphogenesis. *Biol. Rev.* **42**, 442-498

HEAYSMAN, J. E. M. & PEGRUM, S. M. (1973) Early contact between fibroblasts—an ultrastructural study. *Exp. Cell Res.* **78**, 71-78

JONES, B. M., KEMP, R. B. & GROSCHEL-STEWART, A. (1970) Inhibition of cell aggregation by antibodies directed against actomyosin. *Nature (Lond.)* **226**, 261-262

MCNUTT, N. S., CULP, L. A. & BLACK, P. H. (1971) Contact inhibited revertant cell lines isolated from SV-40 transformed cells II. Ultrastructural study. *J. Cell Biol.* **50**, 691-708

MIDDLETON, C. A. (1972) Contact inhibition of locomotion in cultures of pigmented retina epithelium. *Exp. Cell Res.* **70**, 91-96

PINTO DA SILVA, P. (1972) Translational mobility of the membrane intercalated particles of human erythrocyte ghosts. *J. Cell Biol.* **53**, 777-787

SATIR, B., SCHOOLEY, C. & SATIR, P. (1972) Membrane reorganization during secretion in *Tetrahymena. Nature (Lond.)* **235**, 53-54

SIDMAN, R. L. (1961) in *The Structure of the Eye* (Smelser, G. K., ed.), Academic Press, New York & London

SINGER, S. J. & NICOLSON, G. L. (1972) *Science (Wash. D.C.)* **175**, 720-731

WEISS, P. (1947) The problem of specificity in growth and development. *Yale J. Biol. Med.* **19**, 235-278

WOHLMAN, A. & ALLEN, R. D. (1968) Structural organization associated with pseudopod extension and contraction during cell locomotion in *Difflugia*. *J. Cell Sci.* **1**, 105-114

WOLPERT, L. & GINGELL, D. (1968) Cell surface membrane and amoeboid movement, in *Aspects of Cell Motility (XXII Symp. Soc. Exp. Biol.)*, pp. 169-198

General discussion II

DIRECTED MOVEMENTS

Harris: A few years ago, Carter (1965) developed a method for making substrates with various adhesivenesses to cultured cells by evaporating palladium in a vacuum onto cellulose acetate through stencils which determined the pattern of the deposited metal layer. When Carter cultured cells on such substrate, he found that these cells accumulated preferentially onto the metal-coated areas in preference to the uncoated cellulose acetate, as if the metal were more adhesive. Carter took these observations as evidence that cell spreading and movement is essentially passive and analogous to a liquid wetting a solid.

I have repeated Carter's work (Harris 1973*a*), using ordinary electron microscope grids as stencils and evaporating metal onto cellulose acetate, glass, ordinary non-wettable polystyrene, wettable 'Falconized' polystyrene (Falcon Plastics Company) and polystyrene sulphonated by contact with sulphuric acid. I have confirmed Carter's observations that fibroblasts preferentially accumulate on metallized areas and have also found that they 'prefer' metal to ordinary polystyrene. Also such cells accumulate on unmetallized areas of glass, wettable polystyrene or sulphonated polystyrene in preference to metallized areas. This relative 'preference' (in the order cellulose acetate or polystyrene, palladium metal, glass or wettable polystyrene or sulphonated polystyrene) is in itself evidence that the preferential accumulation is not due to differential toxicity but is actually a matter of relative strength of adhesion, as Carter concluded.

However, my observations do not support the idea of passive locomotion and suggest instead that cells move onto the more adhesive substrata as the result of a tug of war between several actively-pulling ruffled lamellae around their margins, each pulling outward in a different direction. For example, a cell which straddles a boundary between cellulose acetate and palladium, with one 'foot' on each side of the boundary, will tend to break its adhesion to the cellulose acetate more frequently than it will detach from the metal. Thus the cell will move preferentially onto the palladium. Cells placed on a grid pattern of metallized squares smaller than themselves will straddle two or more of these squares, with one foot (ruffled lamellae) in each.

Porter: Has anyone used carbon?

Harris: When I evaporated carbon onto glass in a grid pattern and plated cells onto this, I found no preferential adhesion, as though the glass and carbon were equally adhesive (Harris 1973*a*). I interpret the relative order of preference of cells for different substrata (given above) as an operational measure of the relative adhesiveness of these materials to the cell surface. Curiously, these results seem to contradict some of the accepted ideas about the chemistry of cell adhesion. For example, the highly negatively charged glass and sulphonated polystyrene seem to be *more* adhesive than the uncharged polystyrene, even though the cell surface itself is negatively charged and is thought to be repelled by negatively charged substances.

Allison: Do different cell types behave similarly?

Harris: All cell lines (of over a dozen tested) showed the same order of preference.

Curtis: What do you know about the surface charge density of the non-glass surfaces?

Harris: I have not measured these directly, but sulphonate groups will surely have a strong negative charge, and the oxidized surface of the palladium should have a slight negative charge.

Curtis: Even if charge is important, I do not believe you can base a theory on that alone.

Gingell: I agree, and also feel that we cannot learn anything about adhesive energies unless we are sure that the surfaces have similar roughness. If two surfaces had identical adhesive energy/unit area, but one (say metal) was rougher, one would expect the cells to end up on metal simply because they could get a better grip on the rougher material than on the smoother surface. If the surfaces you mention were equally smooth and if there were either an identical thin layer of adsorbed serum protein or no protein, from our calculations (Parsegian & Gingell 1972) I would have thought that the order of preference arising from intermolecular attractive forces would be: palladium > glass > plastic. But that is not what is found. Obviously we need to know about the electrostatic repulsion in each case before predicting adhesiveness. So although the system is appealing, I think it is too complex for conclusions to be drawn.

Curtis: How thick are these deposited layers?

Harris: Their thickness is about 100 nm or less. Incidentally, the deposited metal layers can be made rougher by increasing the angle of deposition; however since palladium is *less* adhesive than glass, contrary to Dr Gingell's theoretical calculations, this discrepancy cannot be explained by roughness. Also, isn't it true that the attractive van der Waals dispersion forces would exert an attraction only *perpendicular* to a smooth surface and not parallel to it?

Gingell: If the surfaces are absolutely smooth, van der Waals forces will have no net tangential component. Suppose the cell is adhering to a flat surface with a separation of about 5 nm between the substrate and the cell surface (i.e. the attractions balance the repulsions, so that the cell is suspended in fluid). If it then spreads, the cell will just slide around like somebody trying to get a grip on ice. But there will be a considerable energy barrier to overcome in order to get the cell off that surface. The actual force will vary according to the direction. The force required to slide the cell off the edge of the substrate will be several orders of magnitude smaller than the force required to remove it perpendicularly, providing there is no peeling. But I am sure cells usually peel off.

Harris: But actually the adhesion forces which allow a cell to spread must be directed parallel to the substratum surface.

Gingell: If there is no material between the substrate and the cell, I would imagine that there would be some resistance.

Abercrombie: I thought the interesting aspect for cell behaviour was that haptotaxis was not a strict taxis.

Harris: Yes, if this accumulation of cells on a certain substrate (haptotaxis) is due to a 'tug of war' rather than to a guiding or reorientation of cell movement, then, strictly speaking, the phenomenon is a kinesis rather than a taxis.

Abercrombie: You found this on the gradient, did you?

Harris: On a gradient of substratum adhesiveness, cells wander erratically with a gradual net accumulation of cells on the more adhesive areas.

Allison: Is it a persistent random walk?

Harris: It was a random walk with a bias. Cell adhesions on the upward (more adhesive) side of the gradient break less frequently than those on the lower (less adhesive) side, so gradual movement occurs up the gradient, but cells do not orient with respect to it.

Trinkaus: This is important. According to Carter (1965), the cells move directly up the gradient; but you find that this is not so.

Harris: Carter's illustration shows a rather densely populated culture with many confluent cells; possibly his cells retracted up the gradient in clumps rather than as single cells.

Gail: We are discussing a haptotactic field, but we could also consider chemotaxis, for example. The word taxis denotes directed movement with a velocity (or even an acceleration) toward a fixed point. Thus taxis and persistence are distinct concepts, the latter describing random walkers which move toward no particular fixed point but which change their direction of motion less frequently than 'pure' random walkers.

A tactic field can manifest itself in time lapse studies in several ways, and it is important to detect tactic influences since these can invalidate the random walk model and the use of D^* to measure motility (Gail & Boone 1970). Any accumulation of cells at a certain point in the time lapse field suggests taxis towards that point. Persistent random walkers should remain more or less homogeneously dispersed throughout the field. A second clue to the presence of taxis is the plot of mean square displacement against time as shown in Fig. 3 of my paper (p. 289). In the presence of taxis, this locus never becomes linear, but instead $\langle T^2 \rangle$ increases with the second power (or a higher power) of t for large t. This is best appreciated by regarding the total displacement, $\mathbf{T}$, as the vector sum of a random displacement, $\mathbf{S}_r$, and a directed tactic displacement, $\mathbf{S}_t$. In other words, the total displacement represents the superimposition of the tactic motion on the basic random walk motion. Then

$$\mathbf{T} = \mathbf{S}_t + \mathbf{S}_r$$

So

$$\mathbf{T} \cdot \mathbf{T} = \mathbf{S}_t \cdot \mathbf{S}_t + \mathbf{S}_r \cdot \mathbf{S}_r + 2\mathbf{S}_t \cdot \mathbf{S}_r$$

Taking expectations and recognizing that the direction of the random displacement is uncorrelated with that of the tactic axis, we find

$$\langle T^2 \rangle = \langle S_t^2 \rangle + \langle S_r^2 \rangle$$

Now $\langle S_r^2 \rangle$ is approximately $4D^*t$ for large t, and

$$\begin{aligned} \langle S_t^2 \rangle &= (Vt + \tfrac{1}{2}At^2)^2 \\ &= V^2t^2 + VAt^3 + \tfrac{1}{4}A^2t^4 \end{aligned}$$

where V is the magnitude of the initial tactic velocity and A is the magnitude of a possible tactic acceleration which may be zero. Here $\langle T^2 \rangle$ is evidently a second (at least) degree polynomial in time instead of varying linearly as in the case of the persistent random walk without taxis.

A third way to detect taxis and to measure the strength of the directionality is to measure the angle between the tactic axis and the displacement vector of the cell. If many such angles (corresponding to many cells) are measured, one can test for the presence of taxis by showing that these angles are not uniformly distributed on the circle (Greenwood & Durand 1955). Calling these angles θ_i for the ith cell and using the convention that a positive value corresponds to a clockwise deviation of the ith cell's displacement from the tactic axis and a negative value to counterclockwise deviation, we can obtain the quantity $U = \frac{1}{n} \sum_{i=1}^{n} \theta_i^2$ as a measure of the spread of these angles and U^{-1} as a measure of the concentration of the motion near the tactic axis (n is the number of cells observed). U^{-1} is large whenever the tactic motion predominates over the random motion. If σ^2 denotes the variance of the θ_i, nU/σ^2 is approximately distributed as a χ^2 variate with n degrees of freedom when the angles θ_i are small. To test whether a chemotactic substance 1 imposes greater directionality of motion than a chemotactic substance 2, one might measure U_1 on n_1 cells in the presence of substance 1 and U_2 on n_2 cells exposed to substance 2. Under the null hypothesis that the two chemotactic stimuli impose the same directionality on the motion, U_2/U_1 has an F distribution with n_2 and n_1 degrees of freedom. If the observed value U_2/U_1 exceeds the critical value of the corresponding F distribution (Snedecor & Cochran 1967) we have statistically significant evidence that the chemotactic substance 1 imposes greater directionality on the cells' movements than does the chemotactic substance 2. These comments assume that all cells are followed for the same fixed time interval since directionality really measures the relative effects of tactic and random movements and tactic effects predominate for long enough observation times. Thus measuring the angles between the cells' displacements and the tactic axis (over a fixed time interval) allows one to test for the presence of taxis and to compare two or more tactic fields to see which imposes greater directionality on the cells' movements.

SUBSTRATE FORM

Wolpert: As somebody who has never worked with tissue cultures but has always looked at living embryos, I can see how one might become preoccupied with artificial systems which provide splendid experimental material. Also the importance of cell adhesion and directed cell movements might have been overrated, and this seems particularly true of the intensively studied sorting-out phenomena. Sorting out might be a highly artificial situation and not provide any real clues about what is happening in the embryo. It is not clear to me to which phenomenon in the embryo sorting out is relevant. Haptotaxis and sorting out have placed a great deal of emphasis on explanations in terms of the differential adhesiveness. What I want to put to you is a rather different way of looking at cell contact and movement in development.

Some time ago, we studied the movement of the primary mesenchyme in the sea-urchin embryo (Gustafson & Wolpert 1963, 1967). These cells move about on the inner face of the embryo wall more or less randomly, but eventually take up a well-defined pattern. How do they take up this pattern? The cells have many pseudopods which continually make and break contact with this wall. Originally we put forward a haptotactic explanation; making and breaking contact, the cells reach a particular position because the pseudopods make more stable contacts there than elsewhere. In other words, we suggested a differential adhesiveness in the cell wall, and we were always being asked to try to define the chemical basis of this variation in adhesiveness.

There are two reasons for thinking of an alternative explanation. The first is the discovery of junctions that give functional coupling (Furshpan & Potter 1968). Within minutes of cells making contact with each other, a channel may be formed between them, so that the cells can communicate with each other's internal constituents. A most elegant and conclusive demonstration is that of Pitts (1972, and personal communication). He held two sets of fibroblasts on cover slips together for a few minutes and then separated them. No strong adhesive contact was made, but using appropriate metabolic markers he found that material had passed across the membranes. This means that when cells make contact we should not, as in the past, only consider the surface properties, but can make use of the internal properties of cells. The stability of a cell-to-cell contact might thus be quite unrelated to adhesiveness, but would depend on factors inside the cell.

The second reason for considering an alternative explanation is the possibility that cells make use of what I have called positional information (Wolpert 1971). This could be present as some sort of gradient, say, a concentration gradient. An alternative explanation to that of differential adhesiveness would be that where the cells come to make the most stable contact is at a particular level of the gradient. This would be a property of the moving cells: that is, the cells move least when they make contact by functional coupling, with the level of the gradient for which they are set, and tend to stay there (Wolpert & Gingell 1969).

Curtis: Here I agree with you. Our experiments (pp. 171-180) suggest that adhesion is not very important for behaviour, though it might be different if adhesive differences are enormous. The large differences I quoted caused some surprise, but there could be still larger differences in adhesiveness. Without stating the mechanism, a strong adhesion (for example an epoxy resin joint) might have an energy of adhesion up to possibly 10 erg/cm^2, in contrast to the energy of adhesion just sufficient to resist the force of Brownian motion, about 10^{-10} erg/cm^2; so there might be a 10^{11}-fold range of energies. Thus Harris (pp. 3-20) and Gail (pp. 287-302) may be looking at more extreme effects than in our own work.

Steinberg: There exists one case in which the nature of the positional information has been discovered. When a droplet of water and a droplet of oleic acid (or any immiscible liquids) touch, a reproducible effect ensues. The droplet of oleic acid spreads, to a characteristic extent, around the droplet of water. There is indisputably positional information in this system, since one phase consistently adopts an internal position and the other an external position. For over 100 years this has been known to be due to the specific interfacial free energies of the liquid interfaces. In a pure liquid, the surface free energy represents the reversible work required to expand the outer surface by a unit amount. The new external surface is created by bringing internal material to the outside. The outer surface area is increased by breaking the internal bonds, and consequently the specific interfacial free energy of a liquid body is the direct measure of the cohesive energy of its components. So here, where the nature of the positional information is known, we can say with assurance that is consists of the intensities with which the subunits adhere to one another.

Wolpert: This is not the sense in which I used 'positional information'!

Gustafson: Directed movements within an embryo are presumably not only determined by the varying adhesiveness of the substrate. Consider the possibility that different substrates have a varying capability to induce contraction of attaching pseudopods. If a cell has many pseudopods attaching at different substrates and only one of the substrates elicits strong contraction of the attaching pseudopod, the cell will preferentially move in one direction. This idea is supported by observations on the developing sea-urchin embryo. A number of cells attach with their pseudopods along the main ciliated band and line up there (e.g. the primary mesenchyme cells, the pigment cells and the neurons). That the impulse-generating ability of the ciliated band is particularly high fits with the observation that the specific cholinesterase activity is very strong in this zone. Pseudopods attaching at other sites may therefore not contract as efficiently as those attaching to the ciliated band. We have some experimental support for this view. In larvae treated with certain cholinolytics, the migration of the primary mesenchyme becomes aberrant, and many pseudopods appear to remain in an uncontracted state. However, our hypothesis certainly requires much more extensive investigations.

Trinkaus: I would like to return to the relation of spreading cells to their substratum during normal development. In *Fundulus* the outer epithelioid layer of the blastoderm, called the *enveloping layer*, spreads on the periblast during the first phase of epiboly. The periblast is, of course, an autonomously spreading syncytial layer, a living substratum. By ultrastructural studies, we (Betchaku & Trinkaus 1973) have shown that only the marginal cells of the enveloping layer form junctions with the underlying periblast. Although the detailed structure of the junctions between these cells and the periblast are most difficult to resolve because they are so convoluted, they appear to consist mainly of close junctions with 4–10 nm separating the apposed plasma membranes and an occasional focal tight junction. In all stages, both before and during epiboly, these junctions are confined to the most marginal region of these marginal cells. In this respect, these spreading cells *in vivo* are strikingly similar to fibroblasts (Harris 1973*b*) and epithelial cells (DiPasquale 1972) *in vitro*. The adhesions of fibroblasts and epithelial cells to their substrata, whether glass or plastic, are also strictly marginal. The width of this junction in *Fundulus* varies, depending on the marginal activity of the enveloping layer. Before epiboly, during the blastula stage before the enveloping layer has begun to spread, the junction is rather extensive, 1.5–1.8 μm wide. During the first phase of epiboly, in contrast, when the enveloping layer is actively spreading over the wide marginal periblast and shows undulating activity on its upper surface extending to the advancing margin (ruffling?) (Trinkaus *et al.* 1973), the marginal junction narrows considerably to form a band only 0.4–1.0 μm wide. It should be emphasized that this junction is much simpler than the junctional complex between enveloping-layer cells behind the blastoderm margin, where, in addition to tight and close junctions, one finds desmosomes, gap junctions, and interdigitations of apposed plasma membranes separated by a gap of 15–20 nm (Lentz & Trinkaus 1971). It seems likely that the comparative simplicity of the enveloping layer–periblast junction is related to the fact that marginal enveloping layer cells are actively spreading over the periblast, whereas enveloping layer cells do not spread over each other.

During the second phase of epiboly, when the marginal region of the enveloping layer has stopped undulating and the blastoderm is being hauled passively over the yolk by the actively expanding periblast (Trinkaus 1971), the entire marginal region of the marginal enveloping-layer cells becomes embedded in the periblast and the extent of the marginal junction increases markedly to a width of 2.0–2.3 μm. It seems likely that the remarkable extent of this junction serves to resist the tension exerted by the spreading periblast and keep the margin of the enveloping layer in constant contact with the periblast.

Inasmuch as the periblast serves as the substratum for the actively spreading enveloping layer during the first phase of epiboly, the fine structure of its surface is a matter of some interest. Accordingly, Betchaku & I (1973) recently studied the periblast surface with the scanning electron microscope. In contrast to the smooth surface of glass or plastic over which cells spread so well in culture, the periblast surface at the beginning of epiboly is highly convoluted and covered with innumerable microvilli about 5–7 μm long. Since both deep cells and marginal cells of the enveloping layer use the periblast surface as a substratum for locomotion, the role of these abundant microvilli in this process will be of considerable interest.

Although the periblast surface in late epiboly remains convoluted, it is much less so than in the earlier phases, and the long microvilli are no longer present. This is consistent with the hypothesis that expansion of the periblast surface during epiboly is due in part to unfolding of its highly convoluted surface (Trinkaus 1971).

Wessells: This same suggestion has been made by Szollosi (1970), who found that microvilli originally present on the surface of blastomeres from Cephalopod embryo disappear as the cleavage furrow forms. Szollosi proposed that microvilli serve as a storage site for membrane needed to form the cleavage furrow.

Goldman: Did you see distinct tight junctions in your electron micrographs?

Trinkaus: Yes. Between cells of the enveloping layer, behind the margin of the blastoderm, we always see tight junctions apically. Proximal to this, the plasma membranes diverge somewhat and then come together again to form another tight junction or so. Subjacent electron-dense material is invariably associated with these tight junctions (Lentz & Trinkaus 1971).

BASIC MECHANISMS

Curtis: One system in which adhesion is probably functional in morphogenesis is the sponge *Ephydatia*, where van de Vyver (1970) discovered clones. The feature of these clones is that they will not fuse; two sponges start to fuse on contact but then separate, leaving a zone with non-adherent cells between them. The control of this system lies in the sponge's production of factors which specifically decrease adhesiveness of heterologous cell types. It is reasonable to assume that these factors are diffusible, that concentration gradients are set up and that in this way adhesions of heterologous cells are controlled, possibly leading to morphogenesis. I do believe that adhesion is important here and is controlled by diffusible factors; I intend to investigate this (see also Curtis & van de Vyver 1971).

Steinberg: With regard to the possible role of passive movement or haptotaxis in cell movements, we believe that adhesive differentials are important in *guiding* certain cell rearrangements and cell sorting. However, they might actually *drive* such rearrangements. To investigate this, we made mixed monolayers, thinking then that contact inhibition would prevent *active* cell movements, and that we would be able to see whether the cells sorted (Steinberg & D. Garrod, unpublished findings, 1970). As it turned out, contact inhibition did not prevent active movements; the cells did sort and show active movements.

Addition of cytochalasin, which we also thought would inhibit active cell movements, did inhibit cell sorting in ordinary mixed aggregates (Wiseman & Steinberg 1971; Steinberg & Wiseman 1972), as found by Sanger & Holtzer (1972) and Maslow & Mayhew (1972). Independently, Armstrong & Parenti (1972), using three different cell combinations, found that in two cases, cytochalasin inhibited cell sorting but, in the third combination, sorting went on in cytochalasin—but with an important difference. The cell concentrations were such that in the absence of cytochalasin, one of the tissues would be sorted into a single island totally enveloped by the other tissue type. In the presence of cytochalasin, the same sorting took place but, instead of a single internally segregating island, a group of internal clusters were formed. This suggests that the cells in these internal clusters were unable to send out extensions to touch cells in other clusters for cluster fusion. I suggest that active cell motility is an important contribution to the cell movements during sorting, but that surface forces can in some cases contribute sufficiently to actually propel cell rearrangements to a certain extent. The facts point to both active motility and interfacial forces combining to cause cell translocation in sorting. There is active motility, but differences in strengths of cell adhesions not only guide the rearrangements but also to some extent even help pull the cells along.

Bray: The idea that actin might be anchored into membranes (pp. 349, 350) could have important consequences for structures that show complex movements of membranes, particularly if the actin is attached to the surface at an angle. This would give the local region of membrane a preferred direction of movement and, if adjacent regions were aligned, then it might help to explain how coordinated movements can occur.

For example, when two membranes in which the actins were of opposite polarity came close enough, short lengths of soluble myosin could pull the two membrane faces in opposite directions (see Fig. 1). If one of these faces were on a vesicle and the other in the external membrane then the vesicles would be propelled in a constant direction. In a growth cone, the distribution of these actin 'arrows' could be such as to direct vesicles constantly towards the front edge of the cone from sites of micropinocytosis at the base of the cone to sites of incorporation at its leading edge. Such a flow of membrane might generate the kinds of 'ruffling' motion observed in both fibroblasts and growth cones [see Harris (pp. 3-20) and myself (pp. 195-210)].

The idea of membrane polarity could also be used to explain the formation of morphological features such as the filopodia of growth cones. These long thin structures have a fairly regular diameter, close to 0.2 μm, and often contain vesicles, particularly of the tubular variety (pp. 195-208). If the arrows are distributed in the membrane as in Fig. 2, then any vesicle within a filopodium

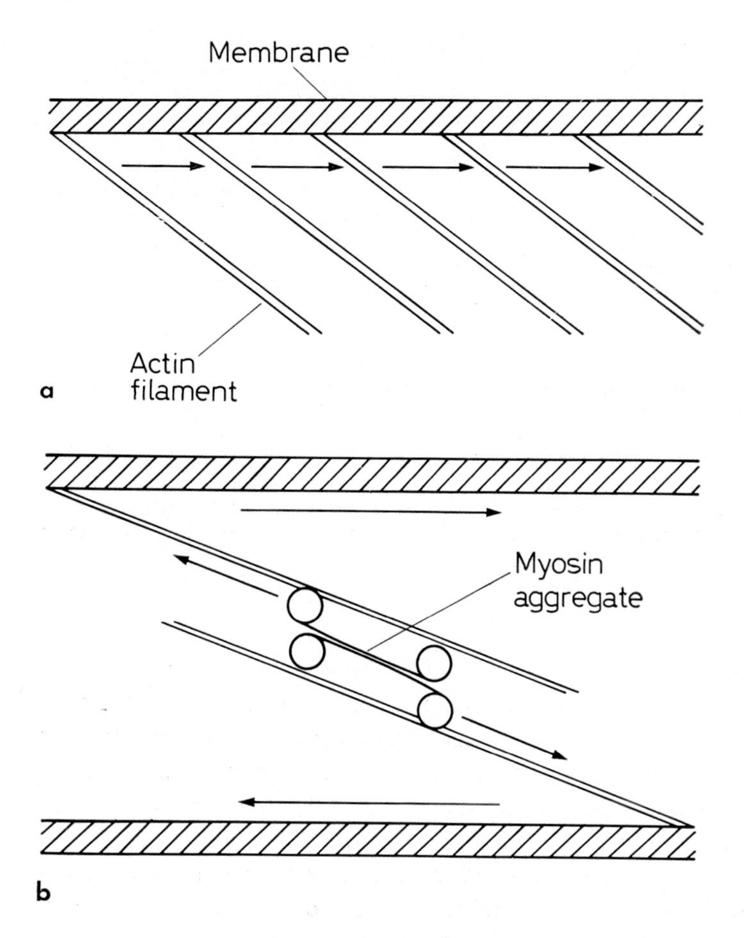

FIG. 1 (Bray). (a) A region of membrane to which filamentous actin molecules are attached is shown schematically. The actin molecules lie at an angle to the membrane and so give it a preferred direction of movement (indicated by the arrows.) (b) Two membrane surfaces of opposite polarity can be moved against each other by small aggregates of 'myosin'.

will be propelled to its tip. There they could fuse with the surface membranes: note that this eversion will change the direction of the arrows. Automatically, the arrows on the vesicle would become aligned with those on the filopodium and so the structure would be regenerated, thereby constituting a system which would take up vesicles at its base and assemble them into long filopodia of well-defined diameter.

Although the details remain to be worked out, this kind of model could be extended to account for much of the behaviour and morphology of a growth

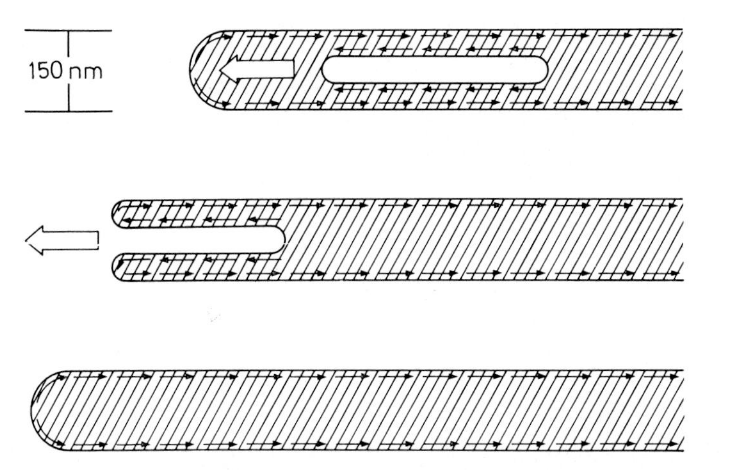

FIG. 2 (Bray). The tip of a filopodium is shown schematically in cross-section with the polarity of its membranes indicated by arrows. The figure shows the transport of a vesicle to the tip of the filopodium and its fusion there.

cone. For example, if vesicles could move against microtubules as well as against other membrane surfaces, the vesicles could thus be transported down the axon and also contribute to the membrane cycling in the growth cone.

Porter: Doesn't this happen as well in the differentiated synapse? There membrane is fed into the presynaptic membrane all the time, and from there it flows back to be incorporated again at the base of the bouton.

Bray: Yes. On the basis of horseradish peroxidase studies, Heuser & Reese (1972) have suggested that the membrane added to the surface by synaptic vesicles is taken up again at the periphery of the bouton.

Trinkaus: Does the presynaptic membrane have a unit membrane structure?

Porter: It has a unit membrane which is about 7 nm thick.

Harris: Are any of these vesicles large enough to be seen with the light microscope? In fibroblasts we see the pinocytic vesicles moving rearward from the margin but no vesicles moving forward toward the margin.

Bray: The vesicles are much smaller than pinocytic vesicles—perhaps 50 nm in diameter.

Wessells: This model seems to account for almost everything we know about growth cones, as well as much of what goes on in ruffled membranes. A major problem is to explain how the cell surface can move over cylindrical microspikes or flat plate-like ruffles. In your two-dimensional representation of filopodia, Dr Bray, the membrane moves up one side and down the other. To do this, the three-dimensional geometry of a cylinder requires the presence of discontinuities where membrane moving up one side slips past that moving down the other side. Such an arrangement seems intuitively unlikely to me. That is the importance of the concept of the fluid-mosaic model for the cell membrane. It is much easier to imagine proteins or filament insertion points moving up and down microspikes or ruffles with the lipid remaining in relatively fixed position. Clearly we must learn how to mark the cell surface and trace movement so as to test the elegant Bray proposal.

References

ARMSTRONG, P. B. & PARENTI, D. (1972) Cell sorting in the presence of cytochalasin B. *J. Cell Biol.* **55**, 542-553

BETCHAKU, T. & TRINKAUS, J. P. (1973) Marginal contacts of the *Fundulus* enveloping layer with the periblast before and during epiboly, in press

CARTER, S. B. (1965) Principles of cell motility: the direction of cell movement and cancer invasion. *Nature (Lond.)* **208**, 1183-1187

CURTIS, A. S. G. & VAN DE VYVER, G. (1971) The control of cell adhesion in a morphogenetic system. *J. Embryol. Exp. Morphol.* **26**, 295-312

DIPASQUALE, A. (1972) *An Analysis of Contact Relations and Locomotion of Epithelial Cells*, Ph. D. Dissertation, Yale University, New Haven, Connecticut, USA

FURSHPAN, E. J. & POTTER, P. D. (1968) Low resistance junctions between cells in embryos and tissue culture. *Curr. Top. Dev. Biol.* **3**, 95

GAIL, M. H. & BOONE, C. W. (1970) The locomotion of mouse fibroblasts in tissue culture. *Biophys. J.* **10**, 980-993

GREENWOOD, A. J. & DURAND, D. (1955) The distribution of length and components of the sum of *n* random unit vectors. *Ann. Math. Stat.* **26**, 233-246

GUSTAFSON, T. & WOLPERT, L. (1963) The cellular basis of morphogenesis and sea urchin development. *Int. Rev. Cytol.* **15**, 139

GUSTAFSON, T. & WOLPERT, L. (1967) Cellular movement and contact in sea urchin morphogenesis. *Biol. Rev. (Camb.)* **42**, 442-498

HARRIS, A. K. (1973*a*) Behaviour of cultured cells on substrata of variable adhesiveness. *Exp. Cell Res.* **77**, 285-297

HARRIS, A. K. (1973*b*) Location of cellular adhesions to solid substrata. *Dev. Biol.*, in press

HEUSER, J. & REESE, T. S. (1972) Stimulation induced uptake and release of peroxidase from synaptic vesicles in frog neuromuscular junctions. *Anat. Rev.* **172**, 329-333

LENTZ, T. L. & TRINKAUS, J. P. (1971) Differentiation of the junctional complex of surface cells in the developing *Fundulus* blastoderm. *J. Cell Biol.* **48**, 455-472

MASLOW, D. E. & MAYHEW, E. (1972) Cytochalasin B prevents specific sorting of reaggregating embryonic cells. *Science (Wash. D.C.)* **177**, 281-282

PARSEGIAN, V. A. & GINGELL, D. (1972) *J. Adhes.*, in press

PITTS, J. D. (1972) Direct interaction between animal cells in *Cell Interactions* (Silvestri, L. G., ed.), pp. 277-285, North Holland, Amsterdam

SANGER, J. W. & HOLTZER, H. (1972) Cytochalasin B: effects on cell morphology, cell adhesion and mucopolysaccharide synthesis. *Proc. Natl. Acad. Sci. U.S.A.* **69**, 253-257

SNEDECOR, G. W. & COCHRAN, W. G. (1967) *Statistical Methods*, pp. 560-563, Iowa State University Press, Ames, Iowa, USA

STEINBERG, M. S. & WISEMAN, L. L. (1972) Do morphogenetic tissue rearrangements require active cell movements? The reversible inhibition of cell sorting and tissue spreading by cytochalasin B. *J. Cell Biol.* **55**, 616-634

SZOLLOSI, D. (1970) Cortical cytoplasmic filaments in cleaving eggs. *J. Cell Biol.* **44**, 192-210
TRINKAUS, J. P. (1971) Role of the periblast in *Fundulus* epiboly. *Ontogenesis* **2**, 401-405
TRINKAUS, J. P., RAMSEY, W. S. & BETCHAKU, T. (1973) Surface activity and spreading of surface cells of the *Fundulus* blastoderm during epiboly, in press
VAN DE VYVER, G. (1970) La non-confluence intraspécifique chez les spongiaires et la notion d'individu. *Ann. Embryol. Morphog.* **3**, 251-252
WISEMAN, L. L. & STEINBERG, M. S. (1971) The reversible inhibition of cell sorting and tissue spreading by cytochalasin B *(11th Ann. Meet. Am. Soc. Cell Biol.)*, New Orleans, Louisiana, p. 328, Abstracts
WOLPERT, L. (1971) Cell movement and cell contact. *Sci. Basis Med. Annu. Rev.* 81-98
WOLPERT, L. & GINGELL, D. (1969) The cell membrane and contact control in *Homeostatic Regulators (Ciba Found. Symp.)*, pp. 241-259, Churchill, London

Conclusion

M. ABERCROMBIE

Strangeways Research Laboratory, Cambridge

An attempt to summarize the wide-ranging discussion would not be sensible, but a few remarks on the state of our subject might suitably conclude the Symposium.

First of all, one area of general agreement should be singled out: that the actomyosin of metazoan cells is a key component in their locomotion. This is important because reasonable and testable molecular hypotheses are thereby encouraged. But while the arrangement of actin fibrils in the cell is beginning to emerge, it is not yet at all clear where and how the shearing process between actin and myosin is applied to produce locomotion.

There were also a few areas of lively discord: about how the cell surface behaves during movement, and about how contact inhibition is to be interpreted in terms of the crudest features of the locomotory mechanism. Though agreement is lacking, at least we can say that there is enough substance in these areas to build allegiances with.

In other areas we did not know enough to disagree with much conviction. A good deal of the discussion of the locomotory mechanism was concerned with the transfer of cell material forwards in relation to adhesions with the substratum. Somewhere in this transfer actomyosin has its role. Material is protruded in front of the anteriorly situated region of adhesions and (though not in a nerve fibre) drawn forward from the hind end of the cell towards the region of adhesions. The discussion suggested various ways, by no means incompatible, by which the transfer of material might be made. The cytoplasm might be driven along by the shearing process between actin and myosin. It might be carried by the transport activity of the microtubules. The forward protrusion might be due to a squeeze exerted by a contractile cortex, or to the deformation of a contractile lattice that might fill the cytoplasm at the front end. The drawing forward of the hind end of the cell might be a contractile process of the conventional muscular kind. The relative importance of these various processes is unknown, and nobody feels as yet strongly committed to a view.

Finally, there was an area of notable silence. In order to shift itself in relation to the substratum in such a way as to achieve continuing locomotion, a cell has to form attachments to the substratum and to release them in an organized way. On this the symposium had rather little to say. The need to account for this part of the mechanism emphasizes what the undoubted role of microtubules also shows us, that there is more to cell movement than an arrangement of the classical muscle proteins.

I think we all feel that progress up the scale of knowledge from silence to agreement has been much encouraged by this Symposium.

Index of contributors

Entries in **bold** type indicate papers; other entries are contributions to discussions

Indexes compiled by William Hill

Subject index